Autodata

Car Repair Manual

ORT

Compiled and Written by

TONY STUART-JONES &
ALAN GROVE

Ford Escort from 1980

1.1 Van
1.1 Base Saloon
1.1 E Saloon
1.1 Base Estate
1.1 LE Saloon
1.1 L Saloon
1.1 L Estate

1.3 Van
1.3 Saloon
1.3 L Saloon
1.3 L Estate
1.3 GL Saloon
1.3 GL Estate
1.3 Ghia Saloon

1.6 Van
1.6 L Saloon
1.6 L Estate
1.6 GL Saloon
1.6 GL Estate
1.6 Ghia Saloon
1.6 XR3 Saloon

The Author would like to acknowledge the help from the following companies in the preparation of this repair manual:

Ford Motor Co. Ltd
Champion Spark Plug Co. Ltd
KL Automotive Products Ltd
Tanfield Towing Bracket Co Ltd
Sykes Pickavant Ltd (Special tools)
Speedograph Ltd (Test equipment)
Lucas Electrical Ltd
Hella Ltd
Sound Service (Oxford) Ltd

Autodata Car Repair Manual for Ford Escort

Compiled and written by Tony Stuart-Jones & Alan Grove
Illustrations: Eddie Kent

Layout and paste-up: Mandy Way/Anne Morgan/Carolyn Banfield
Composing: Simer Stuart-Jones/Asmita Patel

Published by Autodata Limited
St. Peter's Road, Maidenhead, Berkshire SL6 7QU. England

ACRM 291

ISBN 0-85666-308-5

Printed in England by Spottiswood Ballantyne Ltd

Introduction

The Autodata Car Repair Manual is designed to guide you through all the stages of repair or service jobs on your car - from a simple engine oil and filter change right through to the removal and overhaul of the engine.

MANUAL LAYOUT

Easy reference of the appropriate section dealing with the part of your car to be checked or repaired is provided by the Contents pages and the individual chapter headings. Each chapter contains easy-to-follow repair sequences together with clear line drawings, cross referenced with the text, showing what fits where.

A comprehensive Index page at the end of the manual gives quick reference to components and assemblies.

MOT TEST

A special section is devoted to passing the MoT test, with cross reference to the chapters concerned. Check over all the items detailed in this section before submitting your car for its MoT test to avoid any needless failure.

ROUTINE MAINTENANCE

The regular maintenance operations are contained in a complete chapter and are forwarded by a Service Schedule identifying all the maintenance items required and showing the appropriate service intervals.

To ensure that your car is set up to give maximum performance and economy a comprehensive Tune-Up chapter follows the Routine Maintenance. From the information contained in both chapters you will be able to carry out all the regular maintenance and adjustment operations required to keep your car running as efficiently, economically and safely as possible.

TROUBLE SHOOTERS

To assist you in making a correct fault diagnosis a special Trouble Shooter is included at the end of each appropriate chapter. These Trouble Shooters provide details of symptoms and possible causes and will help in tracking down problems as and when they arise.

TECHNICAL DATA

Technical information required for specific operations is contained in the text throughout the manual, to make each section as complete and easy-to-follow as possible.

At the end of the manual a Technical Data section is provided to give a comprehensive listing of the technical specifications likely to be needed by the DIY motorist.

SPECIAL TOOLS

Certain repair jobs covered in the manual require the use of special tools not normally found in a DIY toolkit. When such tools are required we tell you in the introduction to each repair operation. If the special tool is likely to be available from your local tool hire shop then we tell you. Equally, if the job can only be done with a tool which is unique to your make of car then we advise you.

SPECIALIST SERVICES

In some cases the non-availability of spare parts and the need for special tools means that the best solution is to fit an exchange or specialist overhauled component, we then give the procedure for removal and replacement of the unit.

The need for specialised equipment to carry out some operations will require you to take your car to your local garage or service centre. Wheel alignment, as an example, can be checked using DIY equipment, but a full front suspension geometry check can only be undertaken by a garage or tyre specialist having the necessary equipment. In such cases we advise you accordingly.

Contents

History & Identification

SEPTEMBER 1980

New Escort range introduced, with new body styling in 3 - or - 5-door hatchback versions, front wheel drive and 4 - speed gearbox. Available in Basic, L, GL, and Ghia levels of trim and equipment. 1.1. litre models have 1117 cc OHV Fiesta engine, 1.3 and 1.6 models have new overhead cam CVH engine with aluminium cylinder head and hydraulic tappets. All models have laminated type windscreen, outboard face level vents with warm air facility, illuminated heater controls and rocker switches, concealed inertia type seats belts, two-speed front wipers and five warning lights on instrument panel as standard. GL and Ghia models have additional warning sensors to monitor engine oil, coolant level, front brake pad wear, low fuel level and screen washer level, with wiring circuit check micro-processor.
Escort 35 and 55 van versions announced.

NOVEMBER 1980

Escort Estate version introduced, available in 1.1 Basic, 1.1 L, 1.3 L, and GL, 1.6 L and GL trim level and equipment.
XR3 hatchback high performance version also introduced, powered by uprated, twin-choke carburettor 1.6 CVH engine, uprated suspension with gas-filled shock absorbers, alloy wheels with low-profile tyres, front air dam and rear tailgate spoiler.

SEPTEMBER 1981

L and GL models now have A - frame headrests on front seats as standard. GL also has nearside door mirror as standard. All models now available with 'Econolight' fuel economy display system as option.

MARCH 1982

All 1.6 litre models now have a 5-speed gearbox as standard, optional extra on 1.3 models only.

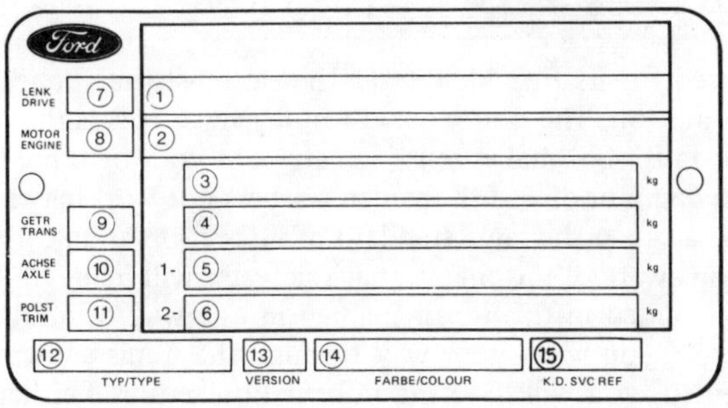

1. *National Type Approval number*
2. *Vehicle Identification Number*
3. *Gross vehicle weight*
4. *Gross vehicle and trailer weight*
5. *Permitted front axle loading*
6. *Permitted rear axle loading*
7. *Steering*
8. *Engine*
9. *Transmission*
10. *Axle (final drive ratio)*
11. *Interior trim*
12. *Model type and source*
13. *Version for special territories*
14. *Body colour*
15. *KD reference*

Fig. 1 Details of the VIN plate

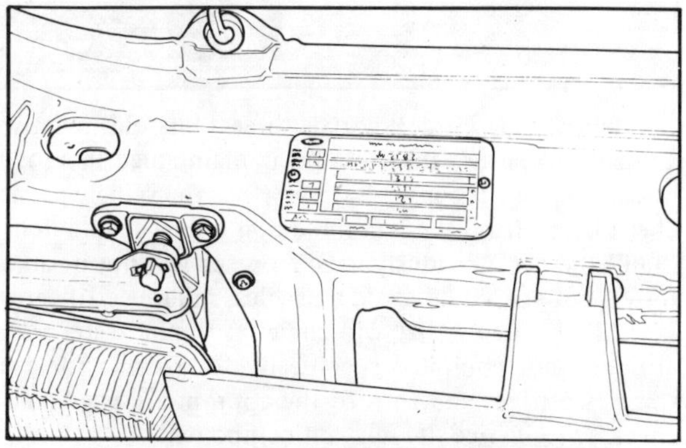

Fig. 2 VIN plate location on front panel

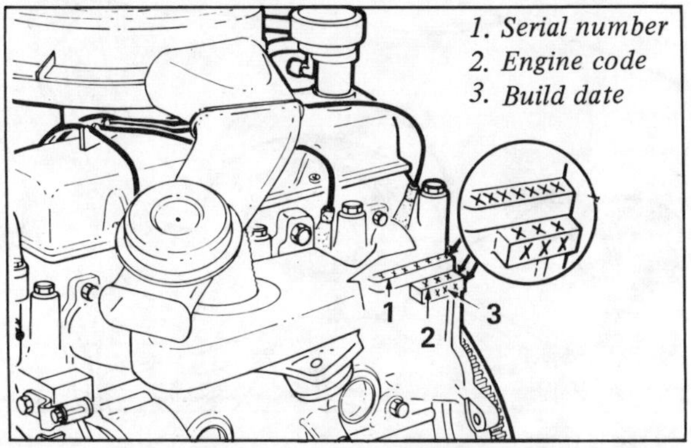

Fig. 3 Location of OHV engine numbers

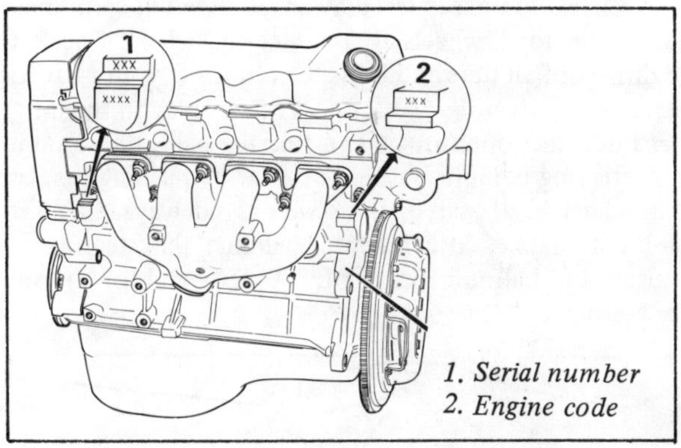

Fig. 4 Location of CVH engine numbers

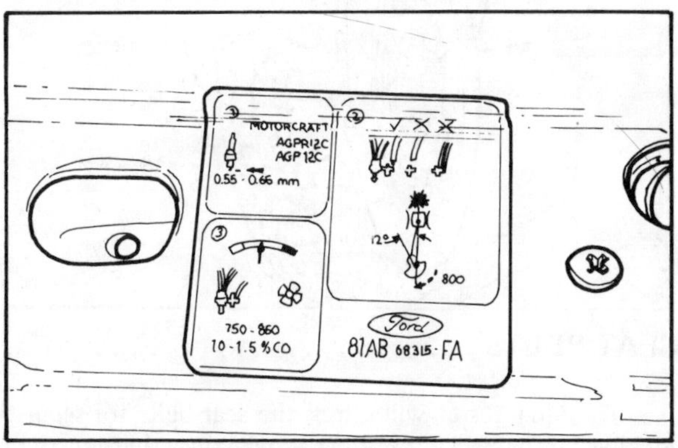

Fig. 5 Location of tuning decal

VIN PLATE (Figs. 1 & 2)

The Vehicle Identification Number plate is mounted on the front panel above the radiator. This plate carries the EEC type approval number, the VIN, alpha-numerical code, various component identification numbers, the maximum weight and loadings as well as the paint and trim codes.

ENGINE NUMBER (Figs. 3 & 4)

The engine number, code and build date is stamped on the side of the engine block at the clutch housing end on OHV engines (Fig. 3).

On CVH engines, the engine numbers are stamped on the side of the engine block, the serial number to the left of the exhaust manifold, and the code number on the right of the manifold (Fig. 4).

TUNING DECAL (Fig. 5)

Basic tuning information is detailed on a decal stuck to the front panel above the radiator. This information relates to that particular engine model and will vary according to the model.

Pass the MoT

When your Ford Escort is three years old, it must be submitted for its first MoT test. The car must then be tested annually to ensure that it is being maintained in a road worthy condition. The test fee paid to the garage covers the cost of carrying out the inspection whether the vehicle passes the test or fails, so it makes sense to carry out your own pre-test check beforehand. Bear in mind that even a simple item like a brake lamp or one of the screen washers not working could "fail" the car. Obviously a DIY owner will not be able to examine a car to the same standard of inspection as an official tester, but you could avoid a needless failure certificate just by being aware of the checks that the tester will make.

All the items that will come under the tester's scrutiny are included in this repair manual in one way or another, although not compiled specifically for passing the test. However, if you work your way through the items shown on these two pages and turn to the appropriate page number referred to, you will have the information required either to check or to service the relevant components.

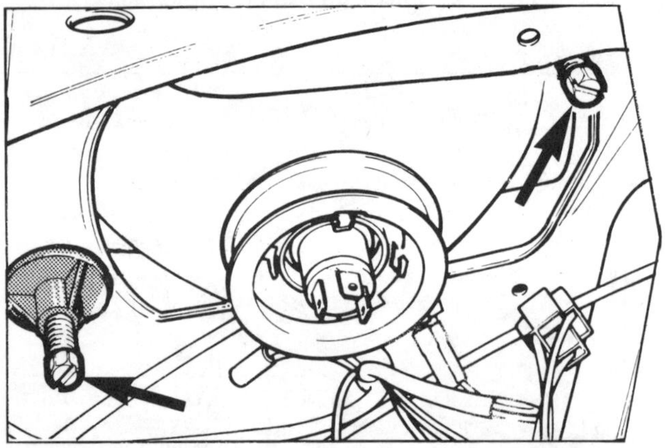

STEERING .... Pages 20, 119

Check for excessive play in all steering components from the road wheels to the steering wheel. Check the tightness of all nuts and bolts. Check for any unusual stiffness in the steering operation. Feel for play in the universal joint, rack mountings and steering ball joint. Examine the steering column bushes for wear by pushing the steering wheel back and forth as well as sideways. Check the column bracket attachment. Examine the steering rack gaiters and ball joint gaiters for security and for any splits or tearing.

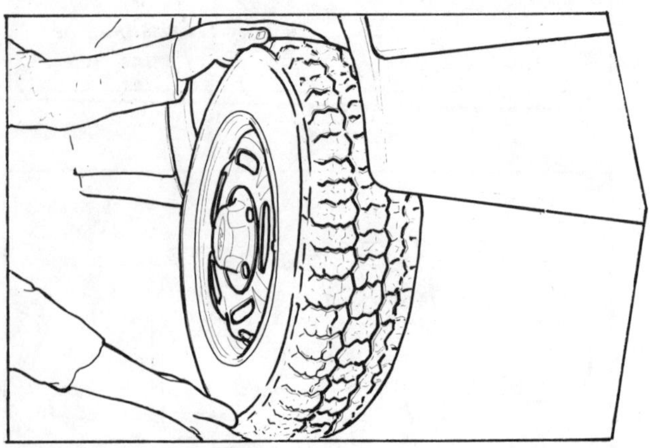

LIGHTING EQUIPMENT .... Pages 23, 154, 180

All external lights must be in working order including the headlamp main and dipped beam - and visible from a reasonable distance. Light lenses and reflectors must not be damaged or missing. The indicators must flash at the correct rate - between one and two flashers per second - and the panel warning lights must also be functioning. Headlamps must be correctly aligned so as not to dazzle other drivers. The stop lamps must illuminate whenever the foot brake is used. It is always worthwhile carrying a set of bulbs in the car. Check the bulb ratings - a list will be found in the TECHNICAL DATA.

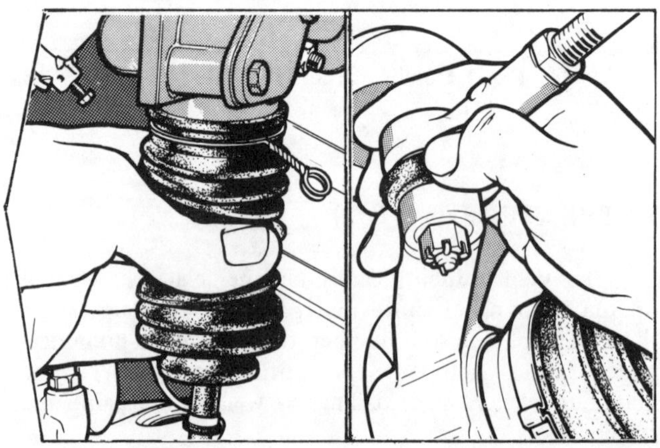

WHEEL BEARINGS .... Pages 20, 130, 136

Raise and support each wheel in turn and check for bearing slackness or roughness by rotating the wheel. A worn bearing will either be heard or felt at the tyre as the wheel turns. Grasp the wheel at the top and bottom then rock it to check for excessive or insufficient bearing clearance. If the bearings appear to be damaged or worn, or if there are any signs of grease leakage from the oil seals, the assembly should be overhauled. Although a check of the wheel bearings should be made at every annual service, the condition should be checked whenever a wheel is off the ground, for whatever reason.

SEAT BELTS .... Pages 23, 178

The MoT tester will check the seat belts for security and the fabric for chafing and any obvious damage. Check belt locking mechanism by pulling slowly then sharply, the belt should lock. The belt should retract smoothly once released. Check the belt and stalk mounting points for security. There should be no structural rust near the seat belt mounting points.

SUSPENSION . . . . Pages 20, 127, 136

The vehicle will have to be raised and supported to check the suspension. Using a suitable long lever or screwdriver to give leverage, check for excessive play in all the suspension joints. Check the condition of all shock absorber units - looking for fluid leakage and the security of the upper and lower mountings. Examine the drive shaft bellows. The drive shaft joints can be checked by trying to move the joint shafts in opposite directions, looking for movement between the two shafts or joints. Under road test conditions a worn joint will be heard as a "knock" under initial acceleration or on overrun.

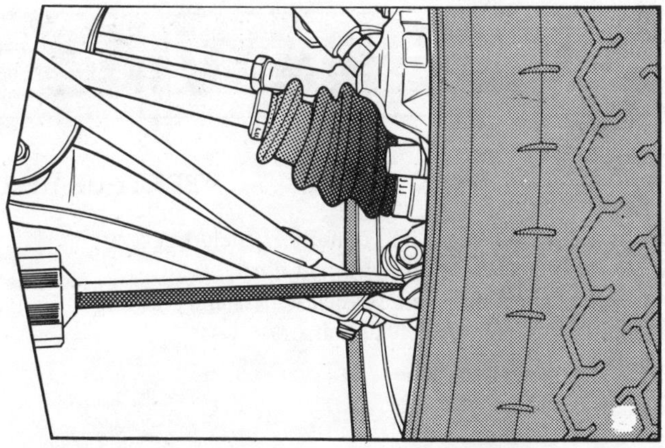

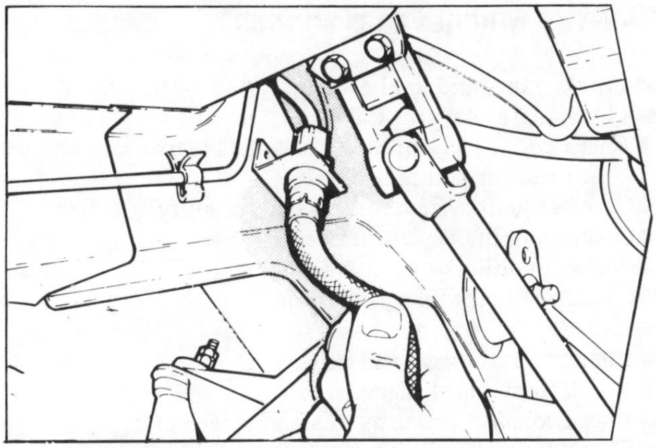

BRAKES. . . . . Pages 14, 16, 141

Check the operation of the brakes and handbrake. The pads and linings should be free of contamination and of adequate thickness. Check for brakes pulling to one side and ascertain the cause Also check the condition of the flexible brake hoses (illustrated), looking for signs of cracking and for corrosion on the rigid metal pipes. Check that the brake servo is working correctly (if fitted). The testing station will use a roller brake tester to check the efficiency of each wheel, including the handbrake mechanism. The brakes should not stick on after the handbrake is released. Check around the master cylinder for brake fluid leakage.

WHEELS AND TYRES. . . . . Page 18

Check the condition of all tyres, including the spare. Check the tread depth around the circumference of each tyre and inspect the side walls - looking for cuts, bumps and bulges. See that all tyres are inflated to the correct pressures (this could affect the brake test).
Check the tyre type - steel and fabric belted radial tyres should not be mixed on the same car. Examine all the road wheels, checking for damage or distortion. Make certain that wheel bolts are correctly located and tightened adequately.

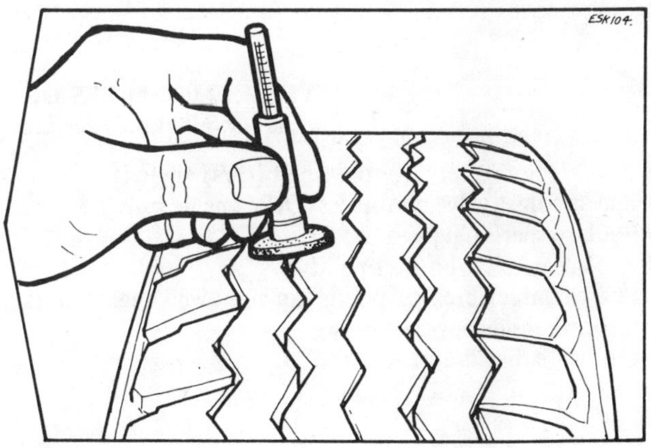

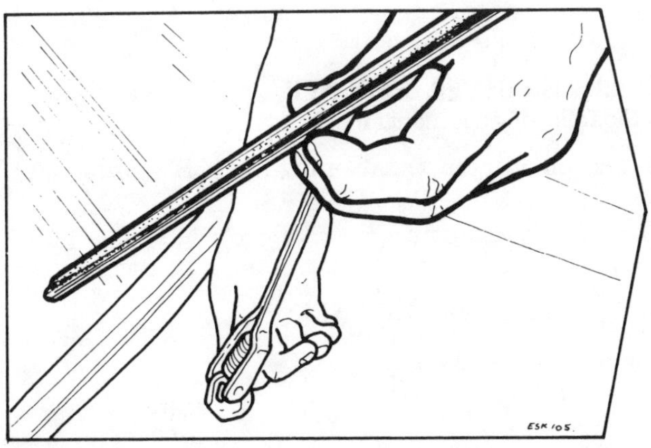

GENERAL . . . . Pages 23, 48, 68, 160, 175

The windscreen wipers and washers should be working efficiently. Bear in mind that wiper blades do deteriorate and it is a sensible rule to replace them or the rubbers at each once a year. The horn should also operate clearly. Check that exhaust system does not leak or make an excessive amount of noise. Check the body for any damage or corrosion likely to render the car unsafe, especially all the important load bearing areas - steering, suspension and transmission supports.

NOTE: The items mentioned on these pages are a guide so that the keen DIY owner can check his or her Escort before submitting it for the MoT test. Although it is based on the official MoT check list at the time of publication, it is only a guide and should be treated as such.

Service Schedule

WEEKLY OR WHEN REFUELLING

- Check tyre pressures and condition, including spare
- Check operation of all lights and horn
- Check operation of windscreen wipers and washers
- Check condition of driving mirrors
- Check windscreen, headlamp and tailgate washer levels
- Check coolant level
- Check engine oil level
- Check brake fluid levels

EVERY 6,000 MILES OR EVERY 6 MONTHS, WHICHEVER IS SOONER

- Clean battery terminals, smear with petroleum jelly
- Check/top up battery electrolyte
- Drain and refill engine oil
- Replace engine oil filter
- Clean oil filler cap
- Check condition and adjustment of alternator drive belt (OHV engine only)
- Clean and reset spark plugs (contact breaker ignition cars only)
- Check/adjust contact breaker points dwell angle (contact breaker ignition cars only)
- Check/adjust ignition timing (contact breaker ignition cars only)
- Check/adjust carburettor idle settings
- Check brake fluid level warning system operation
- Check fuel system for leaks
- Check cooling system hoses and brake servo vacuum hose for wear or perishing
- Check exhaust system for leaks and security
- Check front brake pads for wear
- Check condition of front brake discs
- Check rear brake linings for wear
- Examine brake system hoses and pipes for damage
- Check tyres for wear and damage
- Check front wheel alignment
- Check condition/security of steering rack unit
- Check steering rack unit and bellows for leaks
- Check condition of suspension steering joints

EVERY 12,000 MILES OR YEARLY, WHICHEVER IS SOONER
As for 6,000 mile service, plus the following additional items

- Check/adjust alternator drive belt (CVH engine)
- Check/adjust valve clearances (OHV engine only)
- Replace spark plugs
- Replace contact breaker points
- Reset contact breaker points gap and dwell angle (contact breaker ignition cars only)
- Lubricate distributor
- Clean distributor cap and leads
- Check rotor arm and HT leads for condition
- Check/top-up manual transmission oil level
- Check drive shaft gaiters for security and damage
- Lubricate all door locks and hinges (not steering lock)
- Check headlamp alignment
- Check operation of seatbelt mechanism

EVERY 24,000 MILES OR 2 YEARS, WHICHEVER IS SOONER
As for 12,000 mile service, plus the following additional items

- Replace crankcase emission valve (CVH engine only)
- Replace air filter element
- Clean oil filler cap, ensure orifice free (OHV engine only)

EVERY 36,000 MILES OR 2 YEARS, WHICHEVER IS SOONER

- Drain cooling system, flush and refill

EVERY 36,000 MILES OR 3 YEARS, WHICHEVER IS SOONER

- Change brake fluid
- Inspect brake flexible hosing, replacing if necessary

Routine Maintenance

INTRODUCTION . . . . [1]

The more knowledge you have about your car, the better equipped you will be to put things right if and when they go wrong. Tackling the servicing or routine maintenance yourself will give you the opportunity to get to know your car, at the same time saving you money.

The aim of this section is to guide you through the service items to be found in the interim, yearly and two-yearly services. The Service Schedule on the opposite page lists all the service checks and adjustments with the intervals at which they should be carried out. Cars which cover a low annual mileage should be serviced on a time basis instead of mileage. The sequence in which the items are given has been arranged to give a reasonably logical order of working around and under the car. It is therefore recommended that this be followed where possible. Note however that certain safety related items should be checked more regularly than specified.

Apart from the necessary materials - oils, filters, plugs, etc. - the most important requirement is time. If time is limited, the service can be split up into two or more sections and the remaining items carried out at a later date. The advantage of this is that the service can be spread over several weekends rather than doing it all in one go. In this way there will be no need either to hurry or skimp over any items as each is important in its own way, even if it is only a check or inspection.

Few tools, apart from normal hand tools (spanners, screwdrivers, oil can etc.) will be required. However, some tools will be specifically required for servicing. Given below is a list of tools that it may be worthwhile buying if servicing is to be carried out on a regular basis.

Oil filter strap wrench (manual cars)
13 mm socket or ring spanner
Battery hydrometer
Anti-freeze hydrometer
Tyre pressure gauge
Torque wrench

Items listed in the Service Schedule are covered in detail either in this, or the next chapter. For convenience, all items directly concerned with engine performance and economy - e.g. plugs, points, valve clearances - are covered in the following chapter, TUNE-UP. This enables a complete engine tune to be carried out as a separate operation if required. The remainder of the routine maintenance checks and operations are contained in this chapter, although major overhaul operations are described in the relevant chapters.

Before starting work, read through both of these chapters carefully so that you are aware of the work entailed and the tools and parts required. All relevant data, such as capacities, clearances, etc., where not included in the text, can be found in TECHNICAL DATA at the end of the manual.

JACKING PROCEDURE . . . . [2]

The Escort is fitted with four jacking points on the underbody side sills. These can be located by indentations in the sills, or plastic wheel arch ends on XR3 models. These jacking points are designed only for use with the jack supplied with car (Fig. A:1), when changing a wheel only and should not be used on its own when working underneath the car. When wheel changing, one wheel should be kept beneath the car to act as a cushion should the jack slip. A heavy duty 'scissors' type jack is supplied with van versions only.

If a trolley jack is available, the complete front or rear of the car can be lifted, giving easier access to the underbody. Support pads, for example, sections cut from old tyres should be used between the top of the jack and the support members. At the front a wooden block, shaped as illustrated (Fig. A:2) will be required to prevent damage to the front panel. Position the jack at one of the points shown on the illustration (Figs. A:3 & 5). Van versions (unladen) can be lifted at the rear by jacking in the centre of the beam axle, clear of the brake valve.

Before raising any part of the car, ensure that the wheels are securely chocked, the handbrake is on and the car is in gear. Once raised, axle stands should be positioned beneath the points shown in Figs. A:4 & 6. Support pads should be used, as described previously. This is especially important when using the sill points. Do not allow the car's weight to rest upon the rear trailing arms or leafsprings.

ENGINE OIL & FILTER................[3]

Oil Level Check

The auxiliary warning system automatically checks the oil level when starting the engine (CVH engine cars only). The system will only work if the ignition is turned on after a stop of not less than 3 minutes. The oil level can be checked manually, but about 3 minutes should be allowed after running the engine for the oil to drain back into the sump.

The dipstick, (Fig. A:7) is located behind the engine on the left hand side of the engine compartment (viewed from the front of the car). Do not disconnect the lead to the dipstick (CVH engines only) withdraw from the mounting tube. Wipe the measuring end of the dipstick with a piece of clean lint-free rag or tissue and re-insert it, then withdraw it immediately and examine the level.

If the level is below the upper mark on the dipstick, top up with the recommended grade of oil, given in the TECHNICAL DATA. The approximate quantity of oil required to raise the level from the lower to upper mark is 1.75 pints (1,0 litre). Do not overfill the engine with oil, as this may result in oil leaks, increased oil consumption and damage to the hydraulic tappets. Fill through the filler cap in the rocker cover (Fig. A:8).

Changing Engine Oil & Filter

The engine oil and filter should be changed at the recommended intervals or more frequently under severe operating conditions. The most severe type of operation, and that which gives rise to a sludge formation inside the engine, is light engine loading, slow engine speeds and short journeys where the engine never reaches normal operating temperature. High speeds over long distances are generally kinder to the engine. Modern multigrade engine oils contain additives which go a long way towards preventing sludge formation, but even these have certain limitations.

The oil should be changed when the engine is warm, after a run. Unless you have access to a pit, it will be necessary to raise the front of the car, either by jacking up as described previously, or by driving the front wheels up car ramps. Place a tray under the sump drain plug. This is located at the rear side of the sump pan (Fig. A:10). An old 5 litre oil can with the side cut out is ideal for this (Fig. A:11). Undo the drain plug using a 13 mm socket or ring spanner, while the oil is draining, clean the plug and fit a new washer. When the oil has drained, refit the plug and the tighten to a torque of 15 to 20 lb ft (2,1 to 2,8 kg m). Do not overtighten as difficulty will be met when draining oil the next time.

The oil filter is of the throw-away cartridge type, located on the rear side of the engine. On the CVH engine, the filter points rearwards, on the OHV engine the filter is mounted sideways on the oil pump.

Although it is possible to remove the cartridge from above, there is better access from below. Place a container below the cartridge to catch any oil spilled. A strap wrench may be needed to start the filter moving (Fig. A: 12). If a strap wrench is not available, drive a screwdriver right through the filter casing to provide leverage. The latter is a somewhat messy way, so have a drip tray handy to catch the oil remaining in the filter.

Thoroughly clean the filter sealing flange on the engine to remove all traces of oil and dirt. Make sure the rubber sealing ring on the new filter is correctly located and then apply a smear of clean engine oil to the sealing ring.

Offer up the new filter through the access hole and screw the filter onto the spigot until the sealing ring makes contact with the engine, then tighten it a further ¾ of a turn. Do not overtighten the filter as this may distort the sealing ring and give rise to oil leakage.

The engine oil should be added in two stages. Pour in to the engine filler cap the amount of oil corresponding to the sump capacity given in the TECHNICAL DATA and run the engine for a few seconds. Stop the engine and top up to the maximum mark on the dipstick. The new filter cartridge will have absorbed approximately 0.6 pints (0,3 litres).

NOTE: It is illegal to dispose of old oil by pouring it down the drain or burying it in the garden. Most local councils have a facility for oil disposal and use should be made of this. Alternatively it may be possible to get rid of it through your local garage.

Oil Leaks

Check all joints on the engine which can be seen from above for leakage. Also check the underside of the engine for leaks at the following places: oil drain plug, sump joint, engine to gearbox flange joint, oil filter mounting flanges, timing to chain or belt casing, drive shaft oil seals and gear selector mechanism.

If any evidence of oil leakage is found, the area should be wiped clean then the engine run to confirm the source. If the leak is serious, remedial action should be taken as soon as possible.

GEARBOX OIL ....[4]

The gearbox and final drive units share a common oil supply. The oil level should be checked regularly by undoing the combined filler/level plug at the front of the gearbox housing by the side of the mounting bracket (Fig. A:13). When checking the oil level, the car should be on level ground. Remove the plug and check that the oil is level with the bottom of the plug hole, check with a finger through the hole. If the oil level is below the hole, top-up the level with the specified grade of oil-see TECHNICAL DATA. Gearbox oil is usually supplied in flexible plastic containers, allowing the oil to be squeezed in from below. Replace the filler level plug and tighten it to a torque of 17 to 22 lb ft (2,3 to 3,0 kg m).

NOTE: Lack of gearbox oil can only result from leakage and if the level is low, check for the source of the leak and rectify.

BATTERY ....[5]

Electrolyte Level

The level of the battery electrolyte should be checked

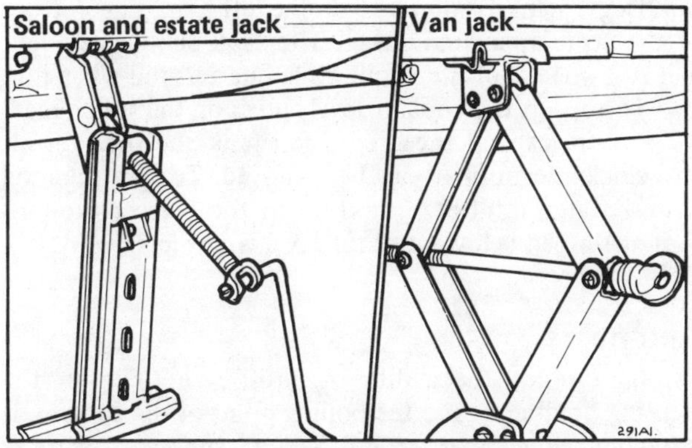

Fig. A:1 Lifting car with jack supplied

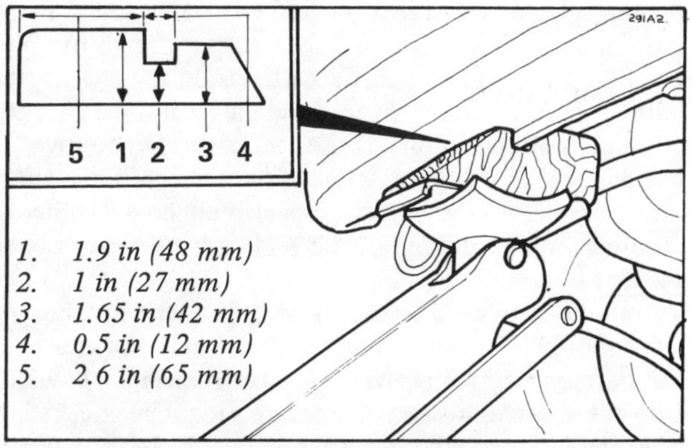

Fig. A:2 Lift the front end using a block

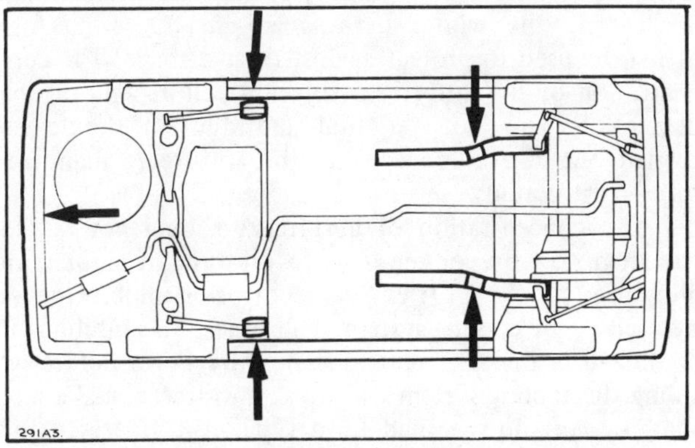

Fig. A:3 Trolley jack lifting points, saloon and estate

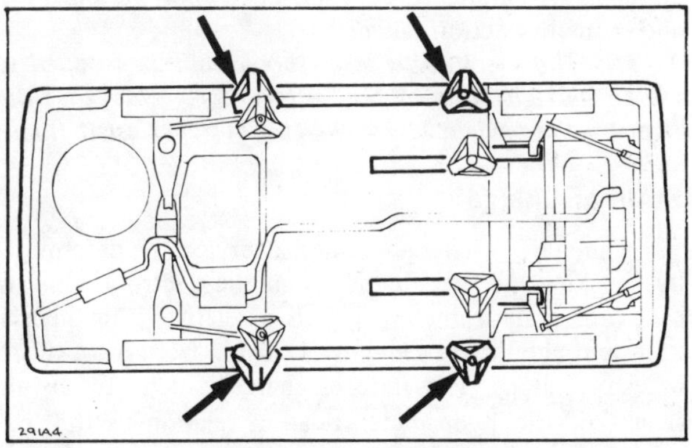

Fig. A:4 Axle stand location points, saloon and estate

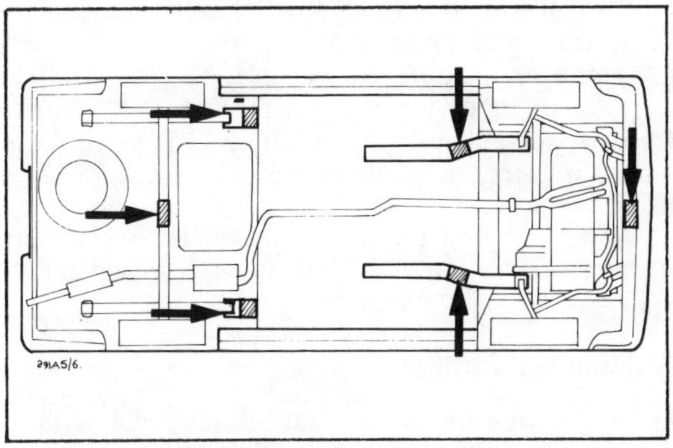

Fig. A:5 Trolley jack lifting points, van

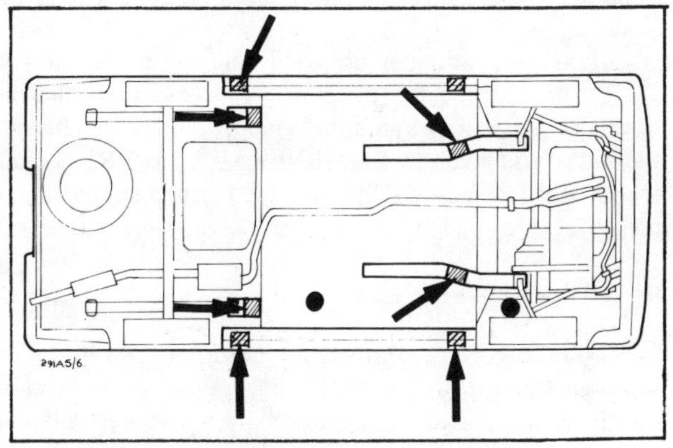

Fig. A:6 Axle stand location points, van

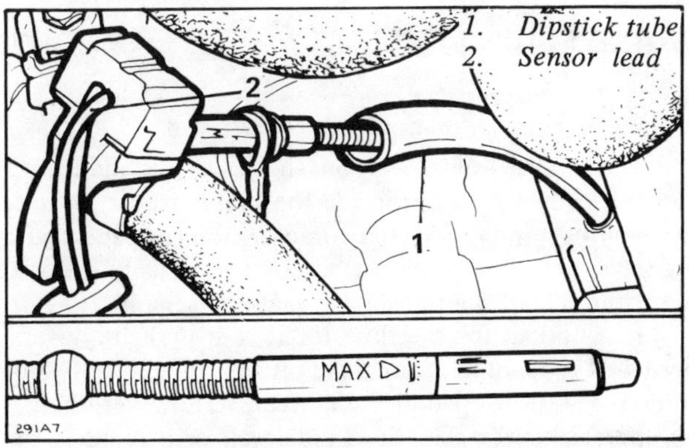

Fig. A:7 Dipstick location at rear of engine

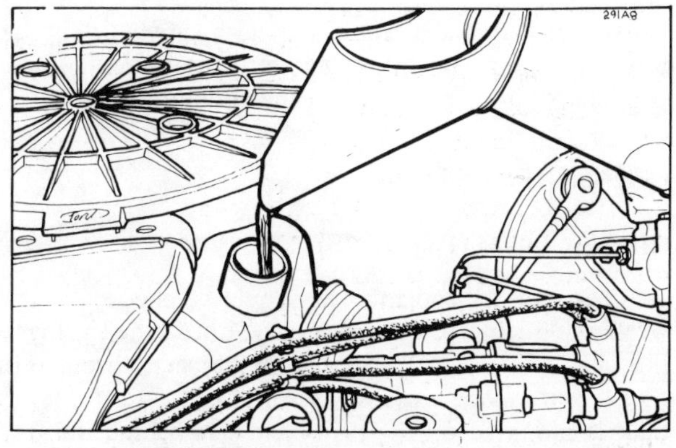

Fig. A:8 Top up and fill with engine oil through rocker cover cap

periodically, and distilled (de-ionised) water added if the level in any cell is below the separators, or the bottom of the filling tube on trough fill batteries. In some cases, the battery case is translucent to allow the level to be checked without the need for lifting the vent cover. Do not overfill the battery. It is good practice to run the car immediately after topping up the battery, especially in cold weather to ensure thorough mixing of the acid and the water and so prevent freezing.

If the battery is found to need frequent topping up, steps should be taken to find out the reason. For example, the battery may be receiving an excessive charge, in which case the charging system charge rate should be checked. If one cell in particular needs topping up more than the others, check the condition of the battery case. If there are signs of an electrolyte leak, the source should traced and corrective action taken.

NOTE: The electrolyte level should not be topped up within half an hour of the battery being charged, other than by the vehicle's own charging system, lest it floods.

State of Charge

The state of charge of the battery can be determined by checking the specific gravity of the electrolyte in each cell with a hydrometer. This is a relatively inexpensive item and should be generally available from most motor accessory shops. The state of charge is indicated by the height of the hydrometer float in relation to the scale (Fig. A:14).

A specific gravity reading of at least 1.275 should be obtained if the battery is fully charged, or 1.120 if discharged.

If a low reading is obtained, the battery should be charged up and the specific gravity rechecked after leaving the battery for about an hour. The procedure for charging the battery is given in the GENERAL ELECTRICS chapter. If the reading is still low, the battery should be replaced.

Battery Connections;

To ensure good contact, the battery cables should be tight on the battery posts. If the battery posts or cable terminals are corroded, indicated by white deposits, the cables should be disconnected and the terminals and posts cleaned with a soda solution and wire brush. When reconnecting the cables to the battery posts, a thin coating of petroleum jelly (not grease) should be applied. The battery earth strap and the engine earth strap should also be checked for proper connection and condition.

COOLING SYSTEM ....[6]

Coolant Level (Fig. A:15)

The radiator coolant level should be checked at least weekly and topped up as necessary. Check the level when the engine is cool. The level can be seen through the translucent plastic expansion tank. On 1.1 litre cars, the tank is mounted on the right-hand side behind the front panel (seen from the front of the car), on 1.3 and 1.6 cars the 'Degas' tank is mounted on the left hand side in front of the strut upper mounting. The tank should be about half full when cool and about three-quarters full when hot.

To top up the coolant level, press on the spring-loaded pressure cap on the expansion tank and then turn it anti-clockwise until it can be removed. Top up the level with a water/anti-freeze mixture in the correct proportions, described below, until the level is correct (Fig. A:15). Refit the pressure cap.

Anti-freeze

Because of the ability of anti-freeze to lower the freezing point and raise the boiling point of the coolant, it is recommended that an 'All Season' type anti-freeze is used permanently in the cooling system to afford maximum protection against both freezing and overheating.

During the winter months an anti-freeze mixture MUST be used to protect against frost damage. The concentration of the anti-freeze solution will depend on the degree of protection required and dilution should be carried out in accordance with the anti-freeze manufacturer's instructions.

The concentration of anti-freeze should not fall to less than 45% by volume. Due to the type of aluminium alloy used on the CVH engine, a corrosion inhibitor must be used in the cooling system at all times. This inhibitor is contained in the anti-freeze specified by Ford. Before refilling the cooling system with fresh anti-freeze, make sure that it is of the specified Ford type. Anti-freeze to the Ford specification is coloured Pink, see TECHNICAL DATA.

The concentration of anti-freeze can be measured using an anti-freeze hydrometer. The specific gravity of a 45% anti-freeze mixture, for example, is 1.069. Note that the mixture should be changed regularly - see Service Schedule, for although the anti-freeze will continue to work, the corrosion inhibitors wear out and must be replaced.

Note that when topping up the cooling system, the correct proportion of anti-freeze to water should always be added.

Draining and Refilling

Full details of the coolant draining and refilling procedure are given under the appropriate heading in the COOLING SYSTEM chapter later in this manual.

BRAKE FLUID LEVEL ....[7]

Checking

The brake fluid level on the Escort is monitored automatically by a float inside the brake master cylinder reservoir cap and connected to an instrument panel warning light. The fluid level can be checked visually by looking at the side of the translucent plastic reservoir.

It is normal for the fluid level to drop slightly over a period of time, due to the brake pistons moving outwards to compensate for pad or lining wear. If, however, there is an appreciable drop in the fluid level, or it requires frequent topping up, this indicates that there is a leak in the

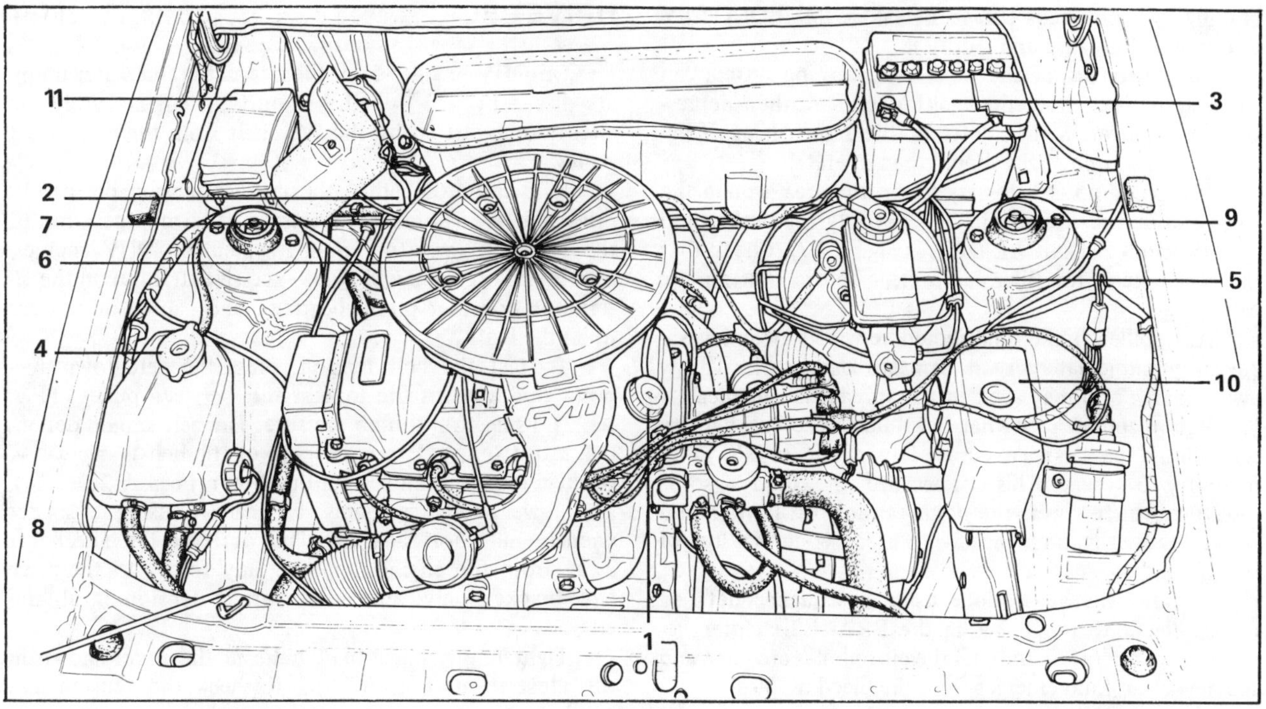

1. *Oil filler cap*
2. *Engine dipstick*
3. *Battery*
4. *Degas tank filler cap*
5. *Brake fluid reservoir*
6. *Air cleaner*
7. *Crankcase emission filter*
8. *Alternator drive belt*
9. *Strut top mounting*
10. *Windscreen wash reservoir*
11. *Fuse box*

Fig. A:9 Escort engine bay, CVH engine type illustrated

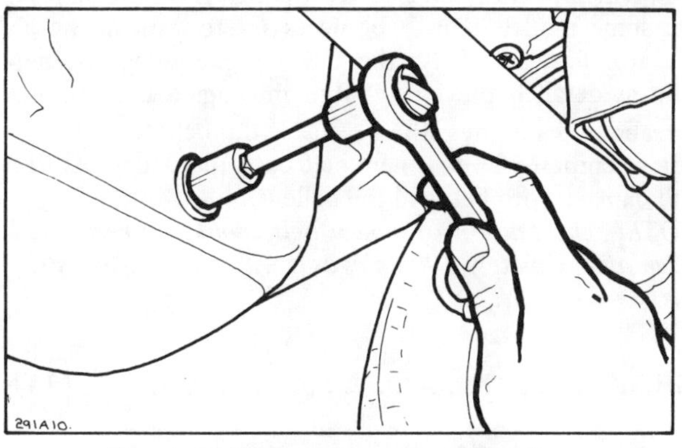

Fig. A:10 Engine sump plug removal

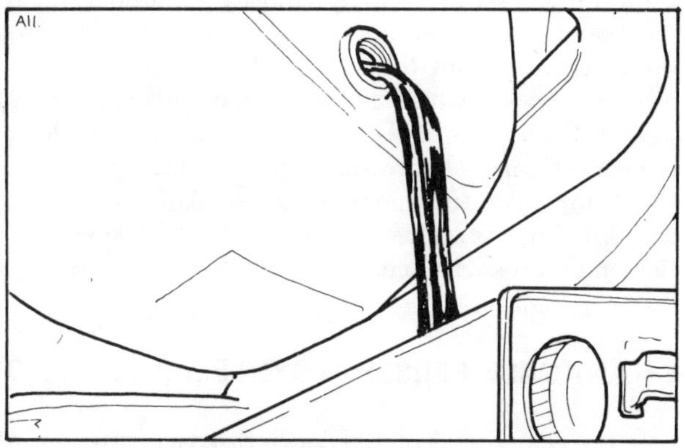

Fig. A:11 Drain engine oil into an old oil can

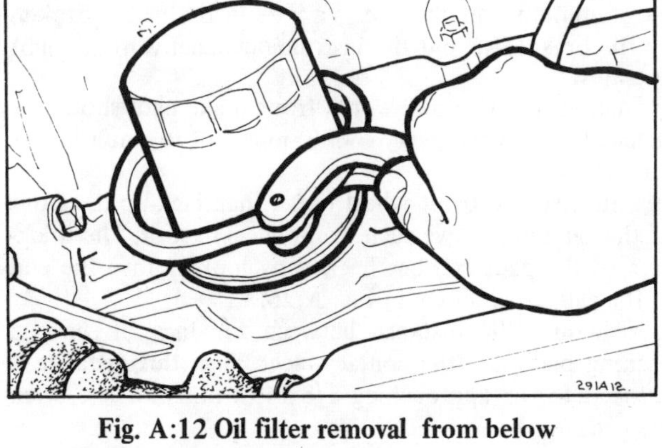

Fig. A:12 Oil filter removal from below

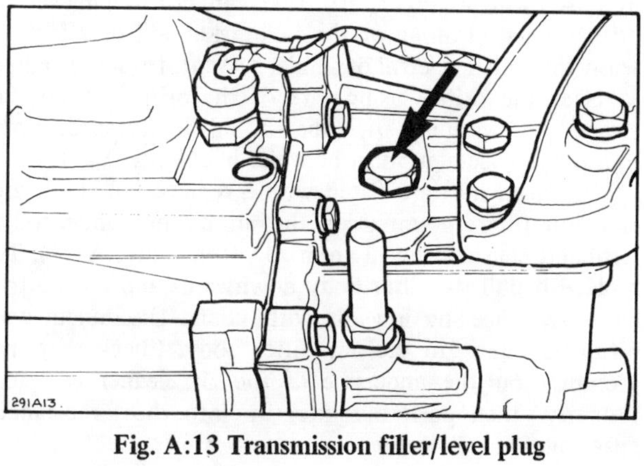

Fig. A:13 Transmission filler/level plug

braking system. Steps should be taken IMMEDIATELY to establish the cause and rectify it.

The operation of the fluid level warning system can checked by switching the ignition on and then depressing the plunger in the centre of the reservoir cap (Fig. A:16). The warning light should then operate.

If topping up is necessary, clean the area around the filler cap before unscrewing it. Use only the brake fluid specified in TECHNICAL DATA (Fig. A:17). Check that the vent hole in the filler cap is clear before refitting the cap.

Any spilled brake fluid should be wiped up and washed away immediately as it damages paintwork. Throw away the rag or it may accidentally spread the chemical.

It is recommended that the fluid in the braking system should be changed completely every 36,000 miles or every 3 years. This is because the fluid absorbs moisture which lowers the boiling point of the fluid and could result in vapour locks with a consequential loss of braking.

A fluid change is done using a sequence detailed under Bleeding the Brakes in the BRAKES chapter, repeating the operation until all the air bubbles are gone and the new clean fluid emerges from the bleed nipples.

AIR CLEANER........................[8]

The air filter fitted inside the air cleaner housing should be replaced every 24,000 miles or two years. If the car has been used extensively in dusty conditions, the filter should be changed more often. Remove the four crosshead screws holding the lid (Fig. A:18). The filter cover on XR3 models is held by two 14 mm nuts (Fig. A:19). Pull off the lid, levering off with a screwdriver from the tag if necessary and remove the element. Wipe the inside of the casing clean, to remove any accumulation of oil, dust and dirt. Fit a new element and refit the lid. Do not overtighten the crosshead screws.

CRANKCASE EMISSION SYSTEM........[9]

The OHV engine oil filler cap is fitted with a removable calibrated orifice which should be regularly cleaned. Pull the centre pipe from the filler cap and remove the orifice by unscrewing it (Fig. A:20). Take care not to damage the orifice when cleaning or the calibration will be affected. Wash the unit in petrol or similar solvent. Do not attempt to clear the orifice using wire. If the orifice cannot be cleaned by washing, the tube should be replaced. Note that the replacement must be white in colour.

CVH engines are fitted with a replaceable crankcase emission filter mounted on the air cleaner underbody, left-hand side as viewed from the front (Fig. A:21). To replace it pull the filter body downwards and release the clips by squeezing inwards with pliers. Disconnect and refit the pipes to the new filter body. Check that the grommet on the underside of the air cleaner is fitted properly, then push the filter up into the air cleaner, ensuring that it is secure.

DRIVE BELT........................[10]

On OHV engines both the alternator and water pump are driven by a 'V' - belt, but on CVH engines, only the alternator is driven by the 'V' - belt from the crankshaft pulley.

It is important that the correct tension of the 'V' drive belt is maintained to ensure efficient operation of the electrical, and the cooling system on OHV engines. Too great a tension will place excessive strain upon the alternator or water pump bearings, and cause undue wear upon the belt.

To test the belt tension, press the belt down at a point mid-way on the longest run between pulleys (Fig. A:22), using firm thumb pressure. The belt should deflect 0.4 in (10 mm). If retensioning of the belt is necessary, slacken the alternator mounting bolts (Figs. A:23 & 24). The lower bolts can only be reached using a socket wrench and short adaptor. Pull or lever the alternator away from the engine, applying any force to the drive end bracket only, until the correct tension is obtained.

Tighten the adjustment bolts in the order shown in the illustration (Fig. A:25). Recheck the tension and readjust as necessary.

Check the condition of the drive belt periodically and replace it if it is excessively worn, stretched or showing signs of fraying. The belt should be replaced as a matter of course every 2 years or 24,000 miles.

To replace the drive belt, proceed as detailed for adjustment previously and detach the belt from the pulleys. On some models it may be necessary to remove the alternator heat shield. Fit the new belt and adjust the tension as detailed previously. Run the engine at a fast-idle for about 5 minutes, then recheck the belt tension. Do not attempt to lever a new belt onto the pulleys as this will damage both belt and the pulleys.

NOTE: The tension of a new belt should be rechecked after approximately 100 miles as it will stretch after initial use.

BRAKES...........................[11]

Front Brake Pads

Escort GL, Ghia and XR3 models are fitted with a brake pad wear warning light. As the pad wears down to the minimum pad thickness, a wire in the pad is broken by the brake disc and the instrument panel warning light operates.

On all Escort models, the front brake pads should be checked for wear every 6000 miles or 6 months as follows:

Slacken the front wheel bolts, then jack-up the front of the car and remove each front wheel in turn. The thickness of the pads can be checked by looking into the end of the caliper as shown in Fig. A:26.

Measure the distance between the face of the pad backing plate and the contact face. When this distance is reduced to, or approaching 1/8 in (3 mm), or less, then the pads must be replaced.

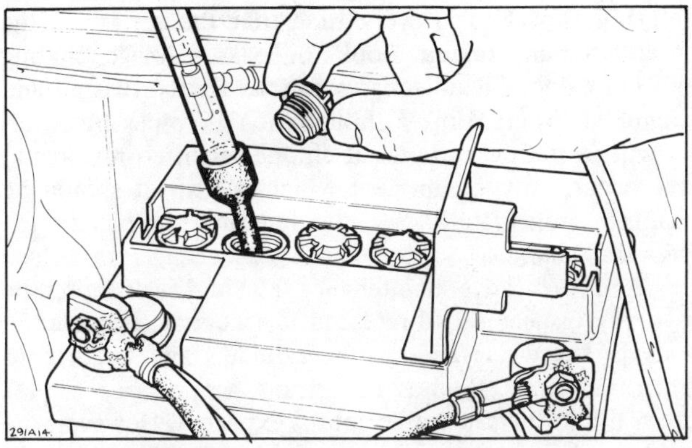

Fig. A:14 Check battery charge and level

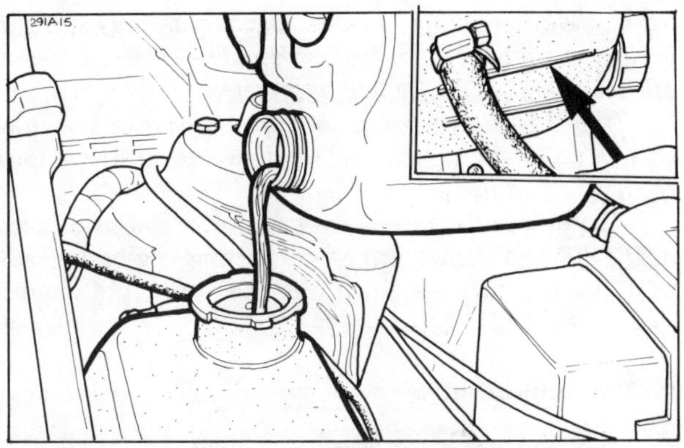

Fig. A:15 Top up degas/expansion tank

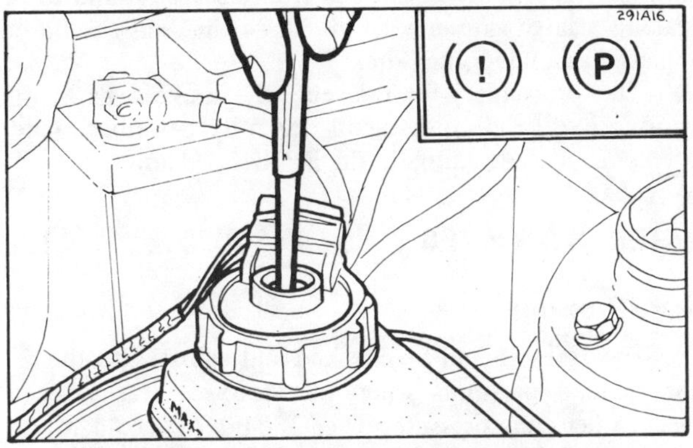

Fig. A:16 Test brake fluid warning float

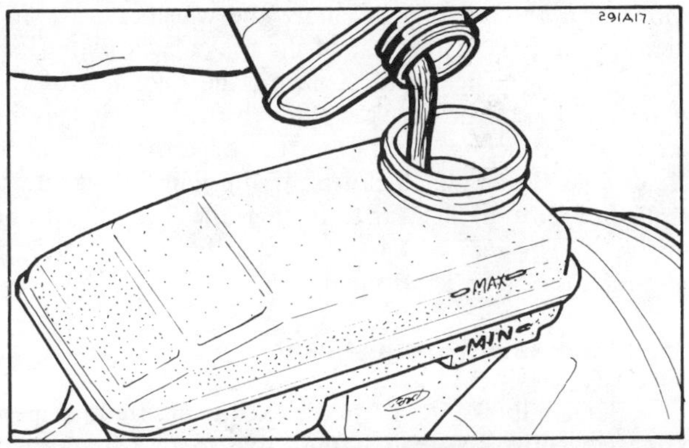

Fig. A:17 Brake master cylinder reservoir top up of level

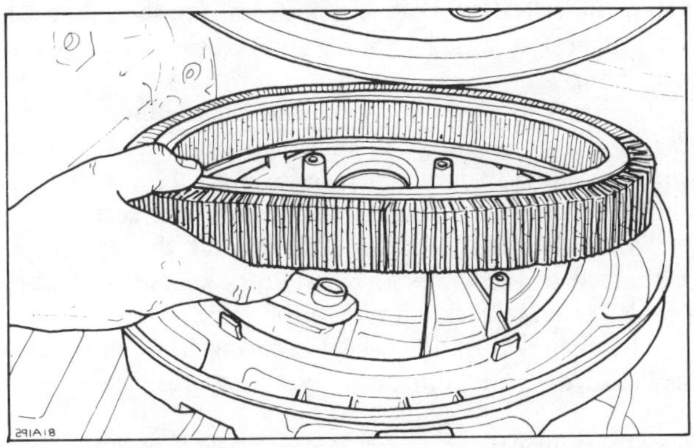

Fig. A:18 Air filter element replacement, CVH engine

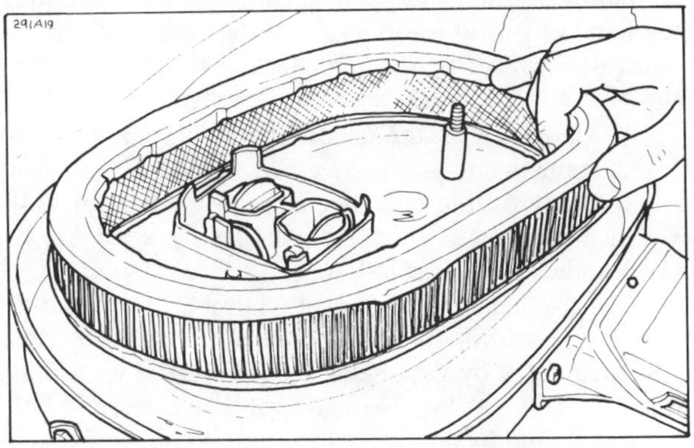

Fig. A:19 Air filter element replacement, XR3

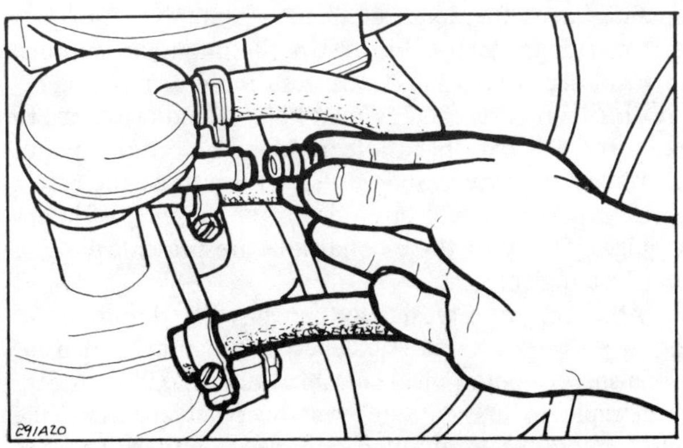

Fig. A:20 OHV engine oil filler cap calibrated orifice

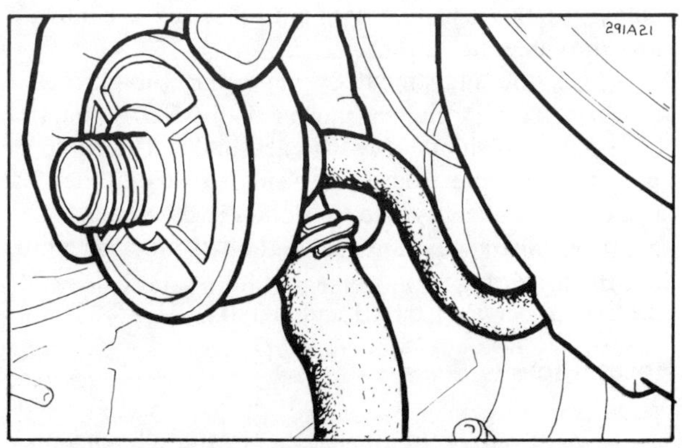

Fig. A:21 CVH engine crankcase emission valve

NOTE: Whenever one or more pads require replacement, both pads on both front brakes should be replaced as a set otherwise brake balance will be affected.

If the pads are not to be replaced, make sure that enough pad material remains to allow the car to run until the next scheduled service check.

The procedure for replacing the brake pads is detailed in the BRAKES chapter. After checking the brake pads, refit the road wheels and lower the car to the ground. Tighten the wheel bolts.

Rear Brake Linings

The thickness of the rear brake linings can be checked without removing the drums. Slacken the rear wheel bolts then raise and support the rear of the car as detailed previously in this chapter. Remove the wheel. Locate and remove the plug at the rear of the brake back plate (Fig. A:27). Using a light and a mirror, the thickness of the shoe linings can be checked through the hole. It may be necessary to scrape paint from the edge of the shoe to determine the lining thickness. If the lining thickness is 1/16 in (1,5 mm) or near to it, then the shoes should be replaced. The rear brake shoes are self-adjusting, and do not require regular adjustment.

Handbrake Cable Adjustment

The handbrake should be checked to ensure it is operating correctly. Check that the cable follows its correct run and is correctly located in its guides.

Lubricate the cable sparingly with grease at the equaliser bracket and where the cable enters the outer sleeves. Lubricate all pivot points and check that they are operating freely.

As stated above, the handbrake should not normally require adjustment as any free-play should be taken up automatically by the rear shoe adjuster mechanism. However, if movement at the handbrake lever is excessive the cable can be adjusted as follows:

With the handbrake lever fully released check that the relay levers within the brake assemblies are in contact with their stops by depressing the plungers at each brake backplate (Fig. A:28). No perceptible movement of the plungers should be felt.

Adjustment is effected at the left-hand abutment bracket on the transmission tunnel (Fig. A:29). Adjust the cable using the adjusting nuts until all slack is eliminated from the cable.

Check the adjustment by depressing the plungers at each backplate. A plunger movment of 0.020 to 0.040 in (0,5 - 1,0 mm) should exist at each brake. The cable can be moved by hand from side to side to even up the clearance if biased one way due to friction in the system.

If the clearance is unequal at the two plungers, ensure that the total plunger movement at both sides when added together is between 0.040 and 0.080 in (1,0 - 2,0 mm).

Hoses, Pipes & Seals

A visual check of flexible hoses and all rigid brake pipes should be carried out at every 6,000 mile or 6 monthly service, also every time that the car is off the ground for any reason. Look for leaks, damage, chafing and corrosion. Flexible hoses can be gently twisted and examined for cracking. Although the rigid pipes are covered with a plastic material, if chipped then corrosion can still occur. Any component which is suspect should be changed immediately. See the relevant sections in the BRAKES chapter.

As preventative maintenance, all fluid seals should be regularly inspected and replaced if necessary. Because the seals in the wheel and master cylinders are designed to work with brake fluid as a lubricant, a small seepage past the seal into the protective boot is expected. However, the amount of fluid at this point should be minimal. A drain hole is fitted to the master cylinder boot from which a minimal seepage is acceptable. If any seal is found to be leaking a large amount of fluid all of the seals should be replaced as a safety measure.

The procedures for replacing the brake seals in the various hydraulic components are fully described under the appropriate headings in the BRAKES chapter.

WHEELS & TYRES ....[12]

Tyre Pressures

The tyres should be checked and adjusted to the recommended pressures where necessary at least once a week. Check the pressures when the tyres are cold as tyre pressures may increase by as much as 6 lb/in^2 (0,4 kg/cm^2) when hot. The recommended inflation pressures are given in TECHNICAL DATA. The spare tyre should be set at the highest specified pressure and adjusted as necessary after fitting.

Incorrect inflation pressures will cause abnormal tyre wear and may result in premature failure. There is an average loss of 13% tread mileage for every 10% reduction in inflation pressure below the recommended figure.

When checking pressures ensure that the dust caps are refitted to the valves as, apart from keeping out dirt, they also provide a second seal to the valve.

The tightness of the wheel securing bolts should be checked at the same time as the tyre pressures.

Tyre Inspection

The tyres should be checked periodically for wear or damage. Check the depth of tread, preferably with a proper tread depth gauge. In the UK the minimum permissible tread depth is 1 mm, but tyres should be replaced before this level is reached as road holding and resistance to punctures will have been affected long before this point.

Check the tyre casing visually for cuts in the casing fabric, exposure of ply or cords, or the presence of lumps or bulges. If any of these conditions are present, the tyre should be replaced.

Abnormal tyre wear may be caused by improper inflation pressures, wheel imbalance, misalignment of front or rear suspension, or mechanical irregularities (Fig. A:30). When rapid or uneven tyre wear becomes apparent, the cause should be established and dealt with as soon as possible.

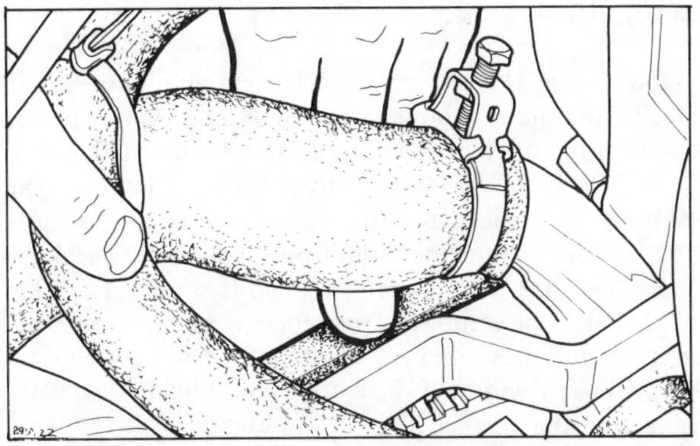

Fig. A:22 Test alternator drive belt

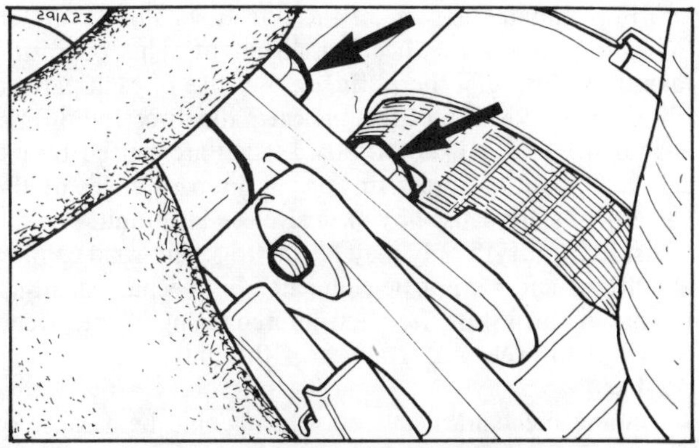

Fig. A:23 Alternator upper mounting bolts

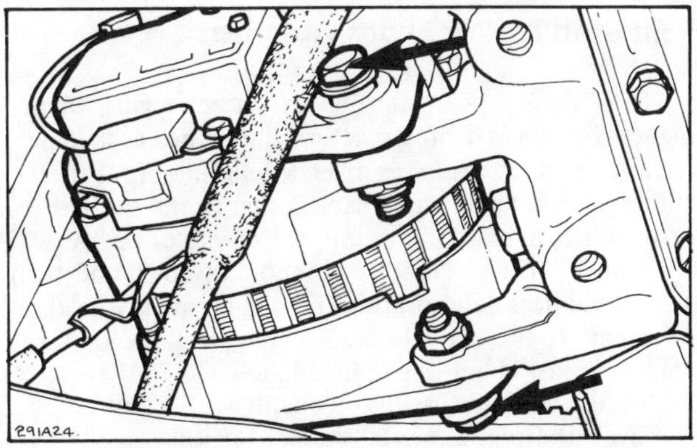

Fig. A:24 Alternator lower mounting bolts

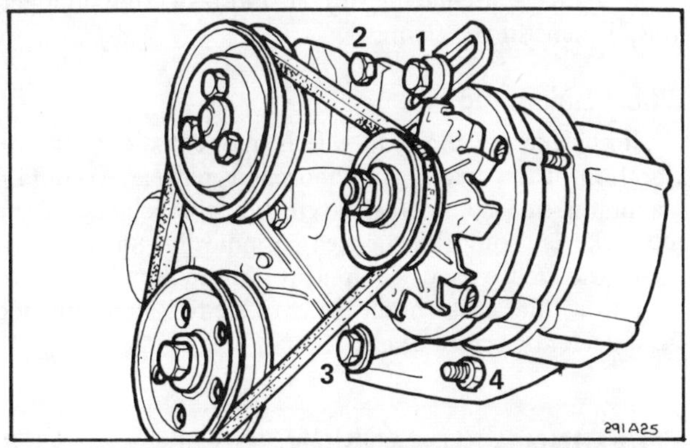

Fig. A:25 Alternator mounting bolt tightening order

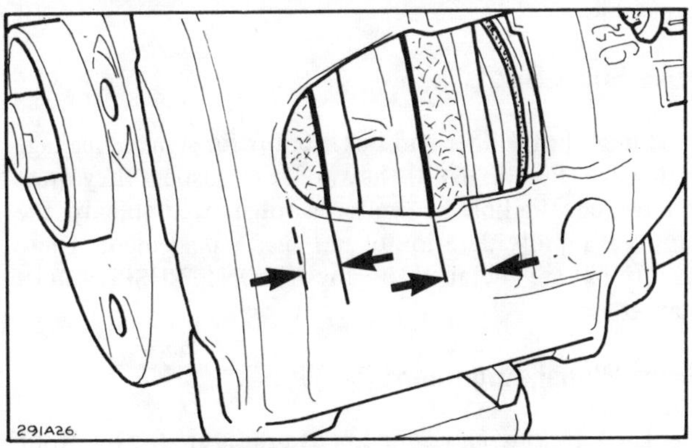

Fig. A:26 Front brake pad thickness

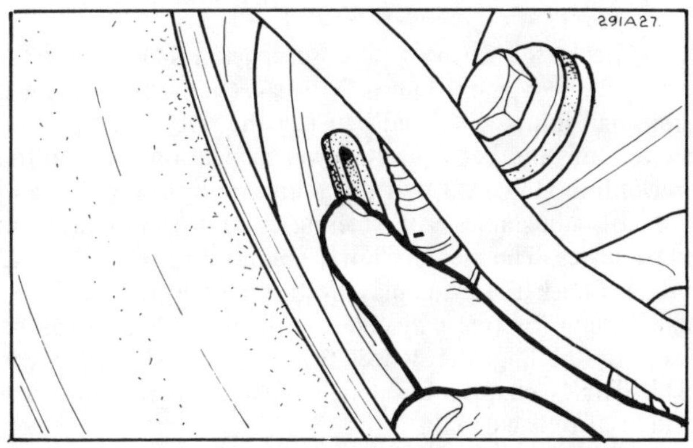

Fig. A:27 Rear brake backplate showing blanking plug

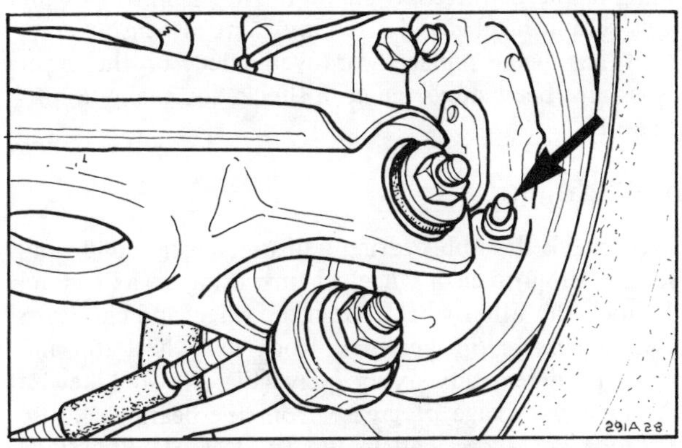

Fig. A:28 Check brake plunger for free play

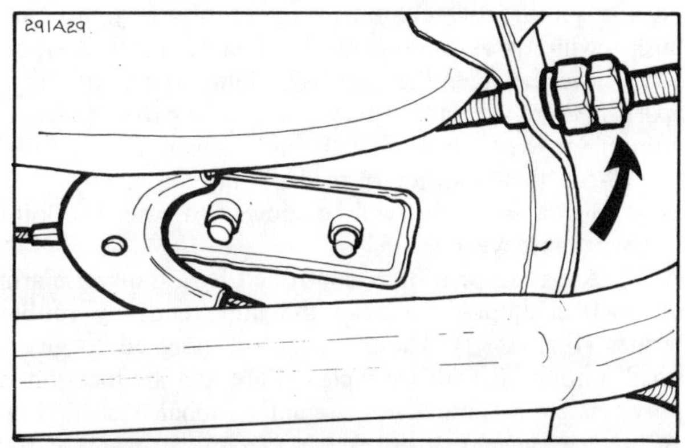

Fig. A:29 Handbrake cable adjuster from below

Fins and feathering on the tyre tread surface are an indication of severe wheel misalignment. This condition takes the form of a sharp 'fin' on the edge of each pattern rib, and the position of this indicates the direction of misalignment. Fins on the outboard edges are caused by excessive toe-out, whereas fins on the inboard edges of the pattern ribs are caused by excessive toe-in. Finning on the nearside front tyre only may be due to severe road camber conditions and cannot be eliminated by mechanical adjustment. In this event, frequent interchanging of the affected wheel to even out tyre wear is the only solution (Fig. A:31).

Some mechanical defects which could be a cause of abnormal tyre wear are: loose wheel bearings, uneven brake adjustment, distorted brake discs, excessive loosenless or damage in the suspension, loose steering connections, or bent steering arms.

Wheel Balancing

Imbalance of the road wheels may cause axle tramp, vibration in the steering or abnormal tyre wear. To obtain maximum ride comfort and tyre life, the balance of the road wheels should be checked periodically. Since specialised knowledge and equipment are required for this operation, the work should be entrusted to an authorised Ford dealer or tyre specialist.

STEERING & SUSPENSION .... [13]

Ball Joints & Bushes

Check the steering and suspension ball joints for excessive movement indicating wear. On the front suspension, ball joints are fitted between the track rod ends and swivel hub, and between the lower suspension arm and the swivel hub (Fig. A:33). These joints are sealed for life and are not adjustable. The steering column is mounted on two bushes. The steering lower column has two universal joints. Check these for play by turning the steering wheel and feeling for free play. Wear in the upper bushes can be seen by pulling the wheel from side to side. See the STEERING chapter for details of steering column bush and joint replacement.

The suspension lower arm is mounted on the body by a rubber bush, as is the anti-roll bar. With the car's wheels on the ground, watch the movment of the joints and bushes while an assistant turns the steering wheel. Position a tyre lever between the suspension joints and gently lever apart. If there are signs of play or lift, new parts will be required. See the FRONT SUSPENSION chapter for details.

Check the condition of the ball joint dust covers, as if these are damaged dirt will be allowed to enter the joint and accelerate wear.

The rear suspension is controlled by a transverse arm on each side, positioned on the inner ends by rubber bushes (Fig. A:32). These bushes are designed to give a small amount of sideways play. Fore and aft location is provided by a trailing rod, similarly rubber bushed (Fig. A:34). The weight of the car is taken by a pair of coil springs and damped by shock absorbers.

Steering Rack

Check that the steering rack is securely mounted and that all the nuts and bolts are present and are not loose. Examine the convoluted bellows at each end of the rack (Fig. A:35). If the bellows are torn or otherwise damaged they must be replaced and the unit refilled with lubricant. If the bellows are in good condition but have been leaking around the clips, it is possible to top them up. The unit should ideally be removed from the car to carry out this operation, but it should be possible to inject grease into the unit in-situ with a syringe after loosening the bellows. Ensure that the correct lubricant is used. This is specified in the TECHNICAL DATA. Ensure that the clips are retightened after refilling.

Front Strut & Rear Shock Absorbers

The shock absorbers do not need regular maintenance, but their damping action should be checked periodically as it deteriorates over a long period. Test the action by pressing down on each corner of the car in turn. If the unit is in good condition, the corner of the car will spring up again and stop. If it rebounds down and then stops, the unit is deteriorating and will eventually need replacement. If the car rebounds again, then the unit has failed and must be replaced immediately.

See the relevant suspension chapter for details.

Check the rear shock absorbers for leakage. If there are any signs of fluid seepage, the unit should be replaced. A minimal seepage of fluid from the front strut units is acceptable.

Drive Shaft Boots

Check both inner and outer boots for signs of leakage or damage (Fig. A:36). If they are at all suspect they must be renewed. If a hole in a gaiter is not spotted quickly, the joint will rapidly deteriorate and need replacement. These operations are detailed in the FRONT SUSPENSION chapter.

Front Wheel Alignment

The checking and adjustment procedure for the front wheel alignment is fully described under the appropriate heading in the STEERING chapter. The alignment of the wheels should be checked after any front suspension component has been replaced or overhauled, or the front wheels have been subjected to a shock, for example hitting a kerb.

Wheel Bearings

To check the hub bearing adjustment and condition, raise and support each wheel in turn using an axle stand and chock the other wheels. Spin the wheel and check for any roughness in the bearings. Rock the wheel top and bottom to detect any excess play (Fig. A:37) Likewise look for any leakage of grease from the bearings. These operations are described in the FRONT SUSPENSION and REAR SUSPENSION chapters.

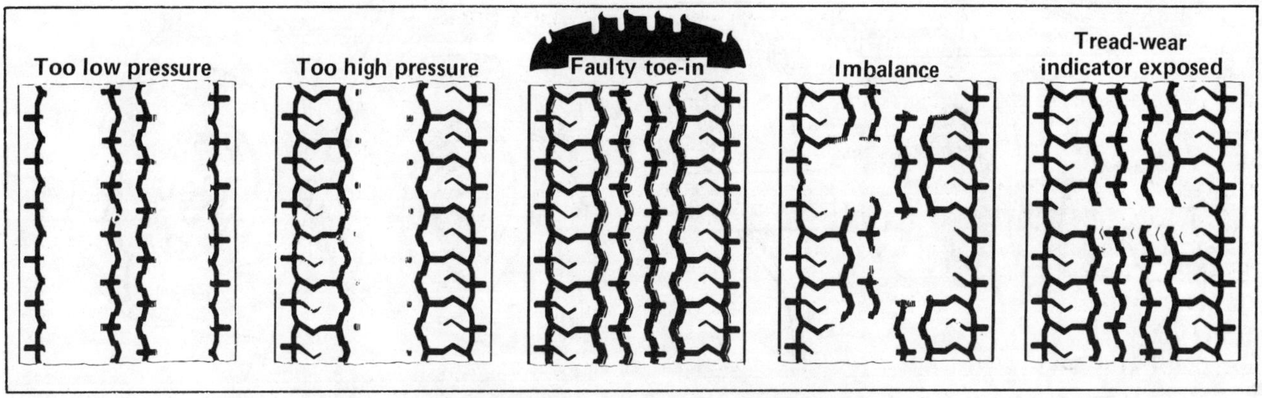

Fig. A:30 Examples of tyre tread wear

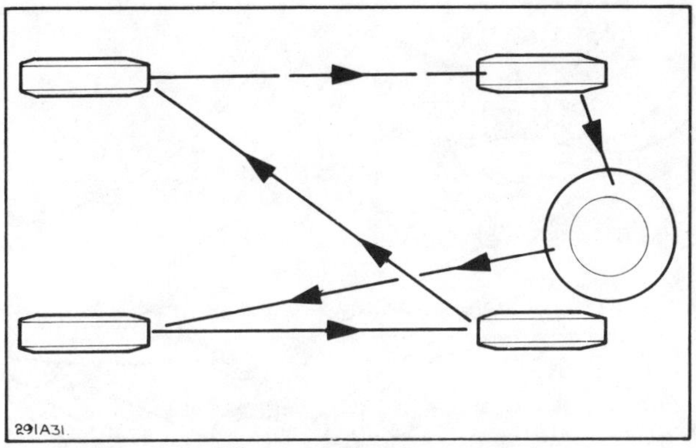

Fig. A:31 Wheel rotation to even out wear

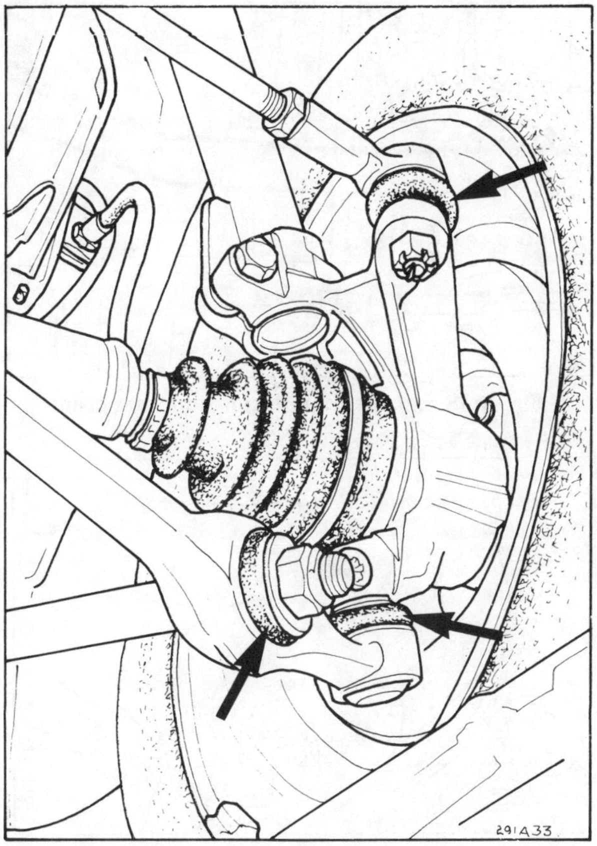

Fig. A:33 Front suspension assembly ball joints and bushes

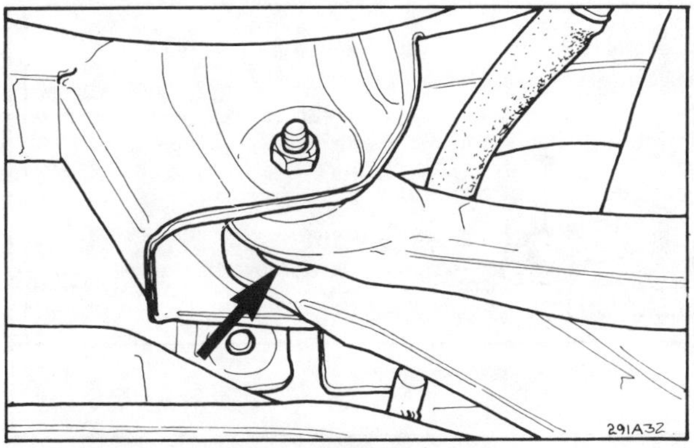

Fig. A:32 Rear suspension arm inner bush

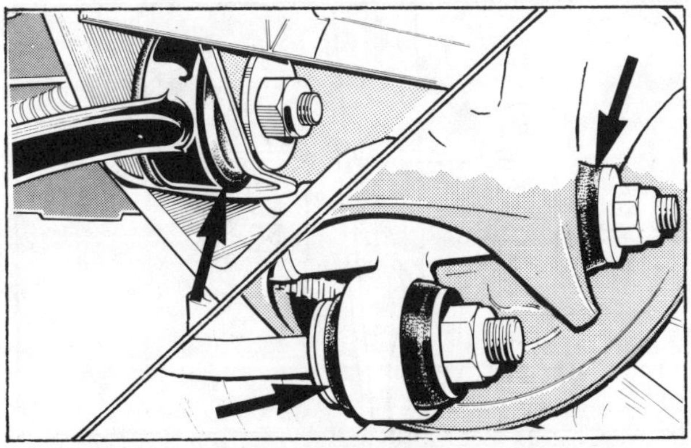

Fig. A:34 Rear suspension arm outer bushes

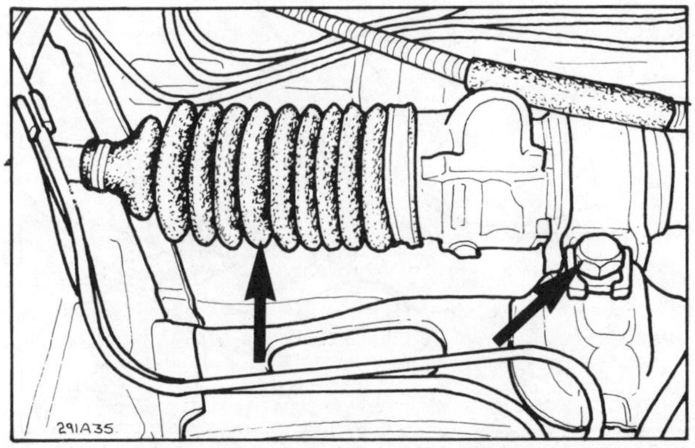

Fig. A:35 Steering rack driver's side mounting and gaiter

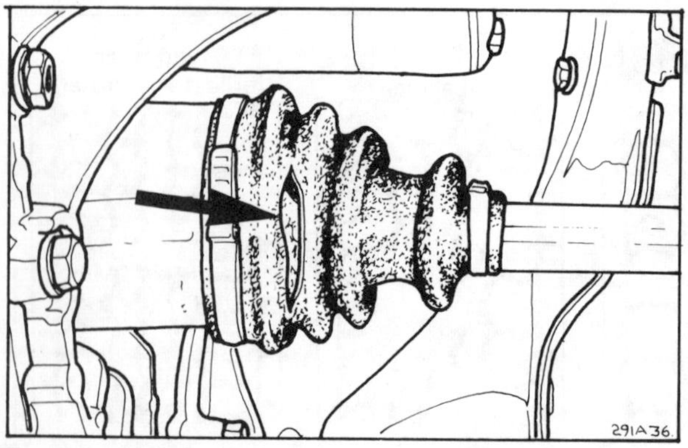

Fig. A:36 Split driveshaft boot

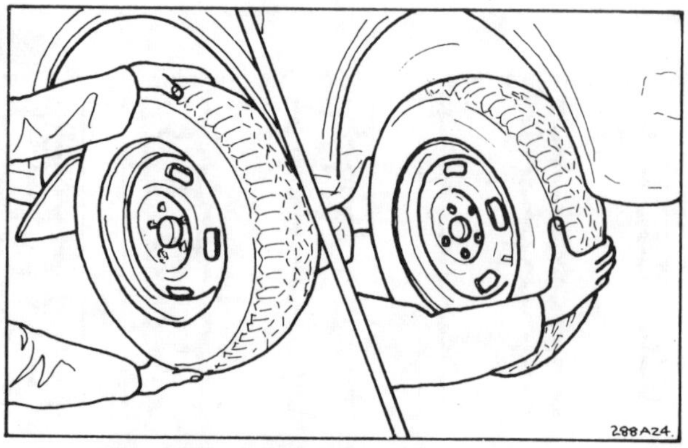

Fig. A:37 Rock wheel to detect play

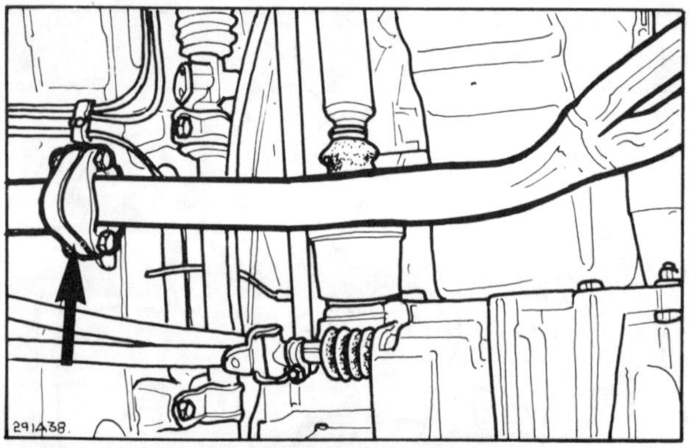

Fig. A:38 Exhaust front section and front clamp

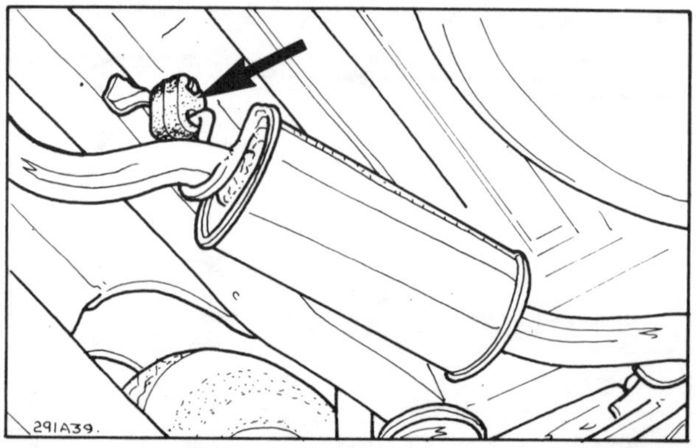

Fig. A:39 Exhaust rear section and rubber mountings

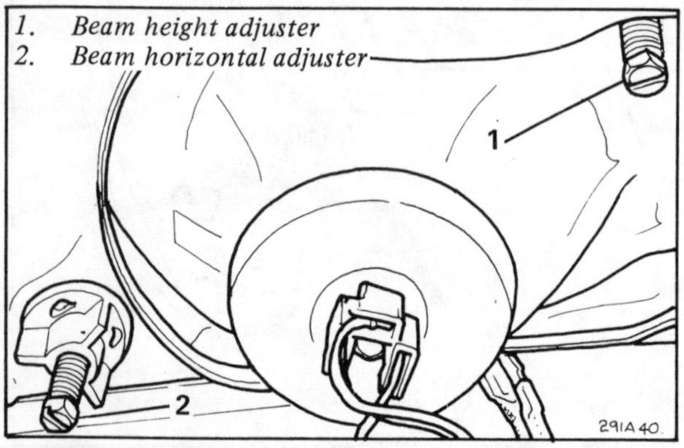

Fig. A:40 Headlamp adjusters, near side shown

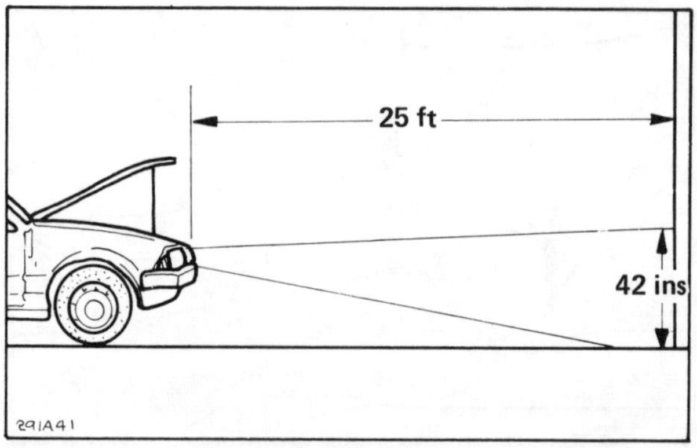

Fig. A:41 Headlamp alignment height maximum setting

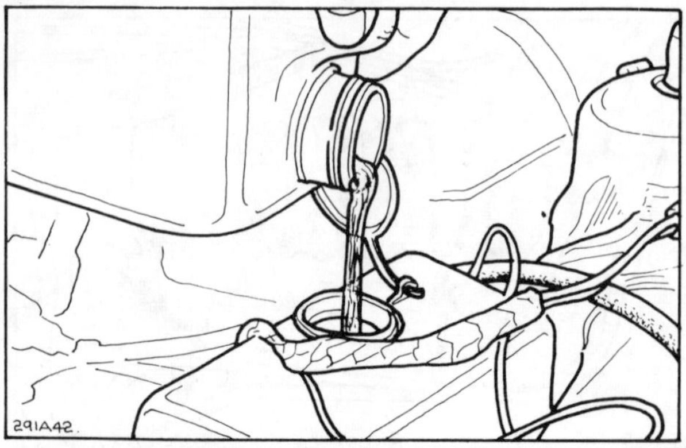

Fig. A:42 Topping up windscreen washer reservoir

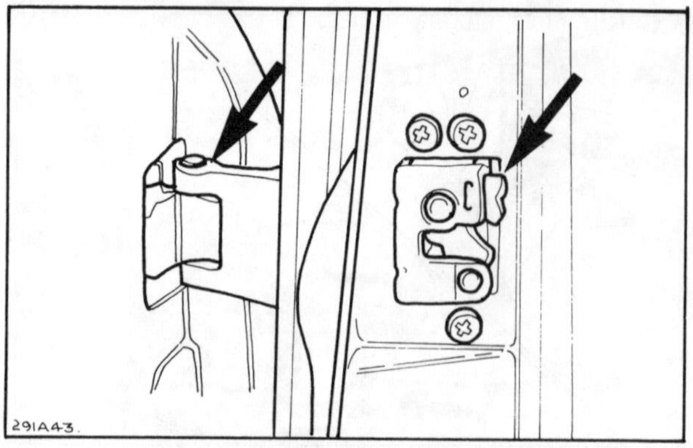

Fig. A:43 Lubricate hinges and locks

EXHAUST SYSTEM . . . . [14]

The exhaust system should be checked periodically for leaks and security. It is a good idea to spend a few minutes examining the system whenever work is being carried out under the car, always ensuring that the car is supported by axle stands or on ramps and the wheels chocked.

Check the alignment of the system to ensure that none of the mounting points are under strain (Fig. A:38). Inspect the exhaust pipes and silencer boxes for damage, corrosion or signs of exhaust gases blowing. Any leakage can be detected by running the engine at fast idle and placing a gloved hand over the tailpipe to pressurise the system. Any leaks can then be heard. Badly rusted components can be detected by tapping the pipes and boxes with a screwdriver handle or similar light tool. Take care not to strike the system with a heavy tool as this may damage it.

Check the rubber mountings for splits and ensure that they are adequately supporting the system (Fig. A:39). Exhaust system replacement is dealt with in the ENGINE chapter.

LIGHTS & INSTRUMENTS . . . . [15]

The lamps, horn and indicators should be checked periodically to ensure that they are functioning correctly. The function of the instruments is best checked under road test conditions. Bulb replacement procedure is described in the GENERAL ELECTRICS chapter.

Headlamp Alignment

Headlamp alignment is usually carried out in a garage using optical beam setting equipment, to which the DIY owner is unlikely to have access. However, it is possible to achieve a satisfactory setting if the following procedure is followed.

The headlamps should be so aligned that they will not dazzle a person standing more than 25 feet (7,5 metres) from the front of the car, and with an eye level of 42 inches (106 cm) or more above the ground. This can be checked after dark by positioning the car at the specified distance from a wall, standing on level ground. Covering each headlamp in turn, set the beam height and direction by turning the adjuster screws with a screwdriver or spanner. The lower adjuster controls the horizontal movement, the upper adjuster controls beam height (Fig. A:40).

WINDSCREEN WIPERS & WASHERS . . . . . [16]

Wiper Blades

The wiping speeds, intermittent wipe action and park position of the wiper blades as well as the condition of the blades and rubbers should be checked periodically.

Blades which are contaminated with insect or oil deposits should be removed and cleaned with a hard brush and detergent solution.

Worn blades will cause streaks and unsatisfactory cleaning of the glass. The wiping edge of the blades must not be perished or torn. Wear of the blades will increase under conditions of dust, air pollution and when used on a frozen or dry screen.

The wiper blades are retained on the arms by plastic clips. To release a clip, push the plastic tongue down and push the wiper blade down the arm. Fit a new blade in the reverse order. Check the new blade operation.

Washer Reservoir Top-up

The windscreen washer reservoir should be checked and topped up weekly or more often if in regular use. The reservoir is mounted on the right hand side of the engine bay (seen from the front of the car) at the front (Fig. A: 42). If a washer system is fitted to the back of saloon cars, the reservoir is behind the trim panel on the left of the boot compartment. On estate cars it will be necessary to fold back the spare wheel cover panel to reveal the reservoir.

The reservoirs should be filled with a mixture of water and washer fluid. This will improve the washing action as well as stopping the fluid freezing in cold weather. Do not use any other fluid such as anti-freeze.

SEAT BELTS . . . . [17]

Although seat belts are fitted to the car, it does not necessarily mean that they are working efficiently or, indeed, that they are capable of doing their job when they are needed. This is why the condition of the seat belt is now included in the MoT test when due. There are some simple checks to be made on a regular basis to ensure that they are in working order. They are as follows:

1. Pull each seat belt against its anchorage to see that it is properly secured to the vehicle structure.
2. Examine carefully the condition of the webbing, looking for cuts or obvious signs of deterioration.
3. Fasten each seat belt locking mechanism and then try to pull the locked sections apart. Operate the mechanism, whilst pulling on the belt to determine that the mechanism releases when required.
4. Check the condition of the attachment fittings and adjusting fitting on each belt for distortion or fracture.
5. As far as practicable check the condition of the vehicle structure around the seat belt anchorages - this will be best carried out from below the vehicle.
6. Pull a section of the webbing from the reel unit and then release it to see that the webbing automatically winds back.

GENERAL LUBRICATION . . . . [18]

Locks, Hinges & Linkages

All hinges and locks should be regularly lubricated to prevent wear and possible seizure. Use a pump type oil can containing engine oil. Remove the plastic cap then put a few drops onto each hinge and latch (Fig. A:43). Wipe up any excess oil to prevent staining and dripping.

The lock barrels in the doors and tailgate should be lubricated by dipping the key in lock-lube oil or graphite powder and inserting it into the lock. Do not attempt to lubricate the steering column lock.

Tune-Up

INTRODUCTION .... [1]

Difficult starting, poor performance and excessive fuel consumption are some of the problems associated with an engine which is badly worn or out of tune. This is why at every major service the various relevant components of the engine should be checked and adjusted as specified in the Service Schedule.

Engine tune-up has been deliberately presented as a separate chapter independent of the ROUTINE MAINTENANCE so that if any problems occur between services the engine can be attended to on its own.

The following checks and adjustments have been arranged in logical sequence and it is advised that they be followed in the order given when carrying out a complete engine tune-up.

Often it is the condition or adjustment of only one component which is at fault, for example the ignition timing or idle adjustment, and consequently it will not be necessary to carry out a complete engine tune-up. However, it is usually only by a process of elimination that the fault can be traced and rectified. To assist in pinpointing the source of any trouble a Non-start Trouble Shooter is included at the end of this chapter.

The Escort requires a minimum of special tools and equipment for tuning the engine. The new Ford CVH engine requires little servicing to keep the engine in tune. The valve clearances are set automatically by hydraulic tappets. Electronic ignition is fitted as standard (except 1.1 CVH engine) and the ignition timing need not normally be disturbed. The valve clearances on the OHV engine should be checked following the Service Schedule, as should the contact breaker points gap. Lubrication of the distributor, carburettor adjustments and the other tasks outlined in this chapter apply to both types of engine.

A selection of spanners and screwdrivers will be necessary for most work. The nuts and bolts are metric, so metric spanners or sockets should be used. A plug spanner or socket should be used for spark plug removal. Because of the design of the CVH cylinder head, a 14 mm deep reach socket or box spanner will be necessary. A normal 14 mm plug socket or box spanner should be used for the OHV engine plugs. A socket is preferable for setting the tightening torque with a torque wrench. If the cylinder compression pressures are to be checked, a compression test gauge should be obtained. A set of feeler gauges for setting the spark plug gaps and valve clearances are essential. Static setting of the ignition timing requires a test lamp, but dynamic testing requires a stroboscopic timing light.

Useful aids for tuning the engine are a tachometer or rev counter and a dwell meter for accurately setting the contact breaker points gap on the OHV engine. There are a variety of tune-up analysers available, often combining these functions with a points condition indicator and the usual electrical ranges of amps, volts and ohms. It is advisable to use an exhaust gas analyser to set the carburettor idle speed and CO levels, although this tool is not essential.

SPARK PLUGS .... [2]

Inspection

The spark plugs should be removed and examined periodically as recommended in the Service Schedule. *NOTE: CVH engines only. Do not remove the plugs when the engine is hot, or the threads will be damaged.*

When disconnecting the HT leads from the plugs, grasp the moulded cap and pull it off the plug. Do not pull on the plug lead itself, otherwise the core inside the lead may be damaged. The spark plug leads have numbers on the casings. However, if an unmarked set of replacement leads has been fitted, label each lead with its respective cylinder number to ensure it is refitted in its correct position.

Blow or brush any dirt away from around the plug base before removing the plug from the cylinder head.

Use a proper spark plug socket or box spanner to remove the spark plugs. CVH engine cars are fitted with recessed 14 mm spark plugs, requiring a special long reach plug spanner. Take great care to avoid damaging the white porcelain insulator on the plug during removal (Fig. B:1).

Inspect the condition of the centre electrode and insulator nose and the outer earth electrode of each plug, as this can give a good indication as to the general state of the engine. Typical examples of spark plug conditions are shown in Fig. B:3.

Electrode Gap (Fig. B:2)

Spark plugs which are in good condition and with low mileage can be cleaned, preferably with a proper sandblast cleaner, but a stiff wire brush will also do. Hold the spark

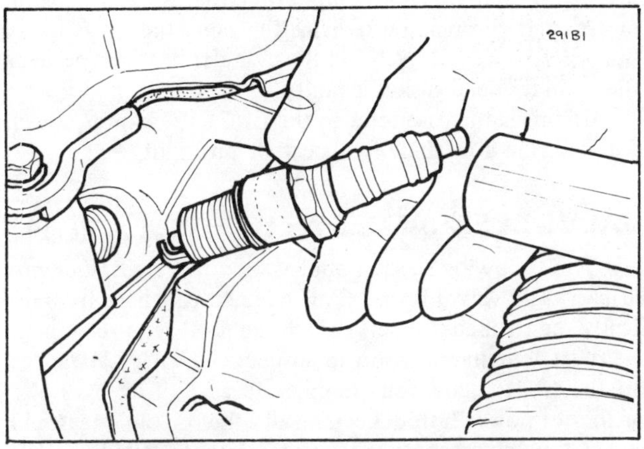

Fig. B:1 Spark plug removal, CVH engine

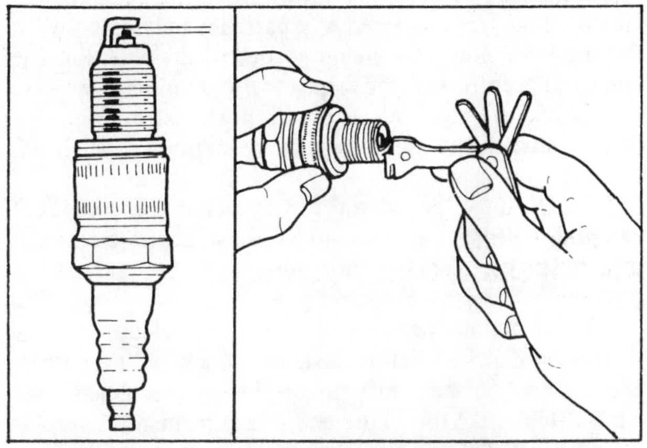

Fig. B:2 Check and adjuster spark plug electrode gap

1. NORMAL – If all plugs are lightly coated with tan or grey deposits and electrode wear is normal, good engine and ignition condition is indicated. Plugs can be cleaned, re-gapped and re-fitted, replace at mileage shown in Service Schedule.

2. HEAVY DEPOSITS – White or yellow deposits on the electrodes and plug shell, which are easily flaked off, are usually the result of fuel or oil additives. These change the chemical nature of the deposit to minimise misfire. Re-use plug alter cleaning.

3. CARBON FOULING – Deposit has dry black appearance. May be corrected by using next hotter grade of plug. First check for sticking valves or damaged ignition leads (if only one or two plugs are affected): clogged air filter or a faulty choke.

4. OIL FOULING – Black wet deposits resulting from plug short circuiting due to excessive oil entering combustion chamber. In high mileage engines the piston rings, valve guides or cylinder bores may be excessively worn. Hotter plugs can temporarily cure problem but severe cases require engine overhaul.

5. OVERHEATING – White or blistered insulator nose and badly eroded electrodes indicates hotter than normal operating temperatures. Engine overheating or incorrectly set ignition timing may be cause. If individual plugs overheat check for coolant blockage, worn distributor parts or manifold leaks. Replace affected plugs.

6. PRE-IGNITION – Electrodes melted, probably caused by excessive temperatures resulting in pre-ignition. This could result in engine damage. Check ignition timing, intake manifold for air leaks, distributor for worn parts, and possible causes of overheating. Make sure that plug is of correct heat range. Plug must be renewed.

Fig. B.3 Typical spark plug conditions

plug with the electrode end downwards while brushing the electrode surfaces to avoid any particles lodging between the insulator nose and the outer shell of the plug. File the end of the centre electrode square with a points file.

Check the electrode gap with a gap setting gauge or feeler gauges. The gap is specified in the TECHNICAL DATA.

If necessary, adjust the gap by bending the OUTER electrode. NEVER attempt to bend the centre electrode, otherwise the ceramic insulator may be cracked or broken.

When fitting new spark plugs, the electrode gaps should be checked before installing them in the engine. Ensure the replacements are of the correct grade - see TECHNICAL DATA at the end of the manual for specifications.

Apply a few drops of clean engine oil to the plug threads before fitting. Do NOT overtighten the spark plugs when installing them - this is most important.

The spark plugs fitted to the OHV engine are of the taper seat type, so the tightening torque must not be exceeded when refitting. A special pre-set torque wrench for fitting these spark plugs is available. The plugs fitted to the CVH engine are sealed by washers. Screw them in finger-tight followed by a maximum of quarter turn with the plug spanner.

COMPRESSION CHECK.................[3]

Valuable time can be wasted trying to tune an engine which is badly worn. This is particularly applicable in the case of an engine which has covered considerable mileage.

The engine compression pressures should be within the specifications given in the TECHNICAL DATA. Note that when testing CVH engine cars, a tester with a long extension between nozzle and gauge will be needed.

1. The engine should be warm before attempting to measure the compression pressure. The OHV engine should be at its normal operating temperature. The CVH engine cylinder head should be warm to the touch but not hot. Remove the spark plug leads and plugs, as described in the Spark Plugs section.
2. Disconnect the ignition coil low tension lead, at the positive (+) terminal.
3. Screw or push the compression tester into the number 1 plug hole and, with the throttle wide open, turn the engine over with the starter motor (Fig. B:4). If the compression tester has to be held in position by hand then hold it firmly to ensure that there is no leakage of compression. Make a note of the reading. An assistant will be needed inside the car to hold the throttle open and to operate the ignition switch.
4. Carry out the same test on each cylinder in turn and compare the readings with each other and with the specifications at the TECHNICAL DATA.

If one or two cylinders are lower than the others, it is likely that a cylinder head valve or piston and bore is worn. Either can be confirmed by carrying out another compression test on the suspect cylinder, only this time adding a small amount of oil to the cylinder beforehand. This is called a 'wet' test as opposed to the previous 'dry' test. If the wet test causes the compression reading to rise, it is likely that there is a fault with either the bore, piston or rings. If the reading remains the same then it is likely that there is a faulty valve in the cylinder head, or perhaps the cylinder head gasket is faulty.

If this is the case refer to the ENGINE chapter, which will describe how these faults can be put right.

VALVE CLEARANCES.................[4]

The following section applies only to the OHV engined cars. The valves in the CVH engined cars have hydraulically controlled tappets which do not require adjustment. If you have reason to suspect the valve clearances on the CVH engine for example; because of excess noise or loss of power (after checking all other possible reasons) check the oil pressure as described in the OHV ENGINE chapter, under 'Oil Pump'.

Check the OHV engine valve clearances as follows:

1. The valve clearances should be checked when the engine is completely cold, and has not run for several hours. Remove the air cleaner assembly; held by the larger two screws on the cleaner lid and by a bolt on the rocker cover. Lift the cleaner body away and disconnect the emission tubing from the filler cap. Disconnect the HT leads to the spark plugs, then remove the plugs as described previously. Disconnect the vacuum hose from the inlet manifold.
2. Remove the four bolts holding the rocker cover. Remove the cover and gasket. If the cover will not move, jar it off with the heel of the palm. Do not use a hammer or attempt to lever the cover off, or the casing may be distorted.
3. Rotate the engine as by the method described in the Ignition Timing section. During adjustment of the valve clearances, rotate the engine in the normal direction only (clockwise, as seen from the crankshaft pulley end of the engine).
4. By turning the crankshaft to and fro a little and, at the same time, observing the rocker assembly, it can be determined whether the valves for No. 1 and No. 4 cylinder are rocking. The two rocker arms will be moving in opposite directions, the exhaust just closing and the inlet just opening if the valves of No. 4 cylinder are rocking, the valve clearances at No. 1 cylinder should be checked.
5. The setting is correct when the appropriate thickness of feeler gauge is a neat sliding fit between the end of the valve stem and the pad on the rocker arm.

If adjustment is necessary, rotate the adjusting nut on the push rod end of the rocker arm until the correct gap is obtained (Fig. B:5). The adjuster bolts are self-locking and only a suitable size of ring spanner or socket in good condition should be used on them.

When adjustment at No. 1 (or No. 4) cylinder is complete, turn the crankshaft through half a turn and check the clearances at No. 2 or No. 3 cylinder, which ever is appropriate, in the same manner. Continue turning the crankshaft half a turn at a time and checking the appropriate valve clearances until all four cylinders have been done.

The valve arrangement, from the front, is: Ex - In - Ex - In - Ex - In - Ex - In.

The adjustment sequence is as follows:

Cylinder No. 4 rocking - Adjust cylinder No. 1

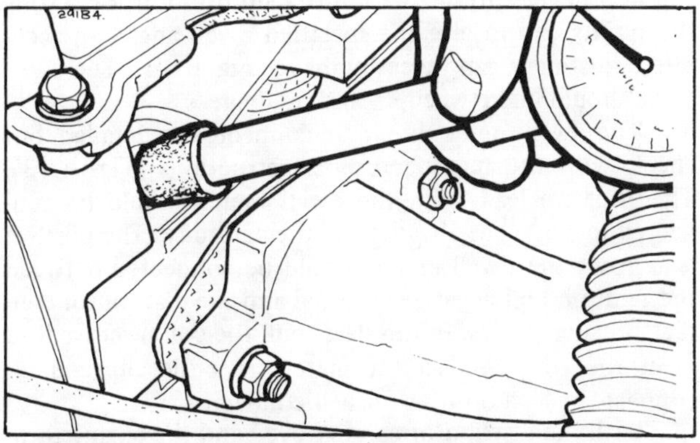

Fig. B:4 Check cylinder compression with a test gauge

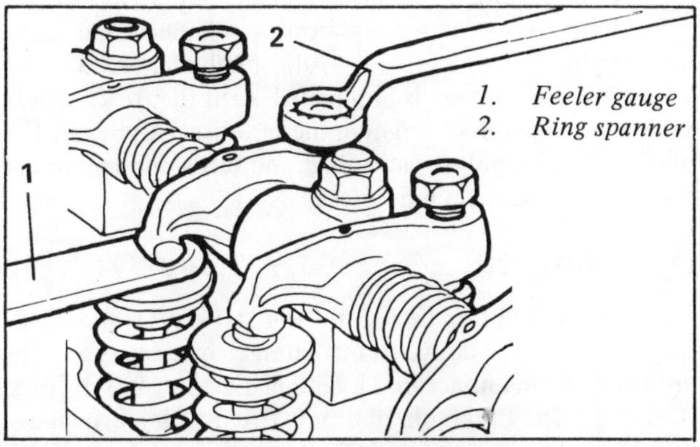

Fig. B:5 Adjust valve clearance

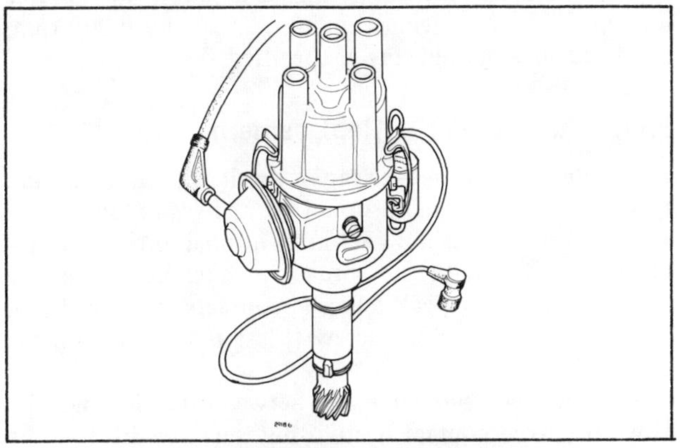

Fig. B:6 Distributor, OHV engine

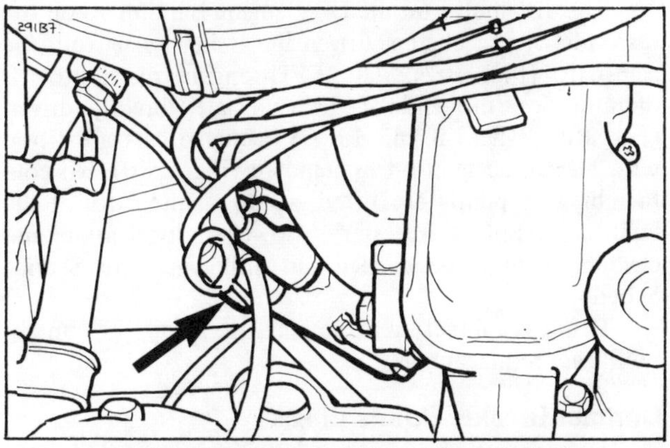

Fig. B:7 Distributor location, OHV engine

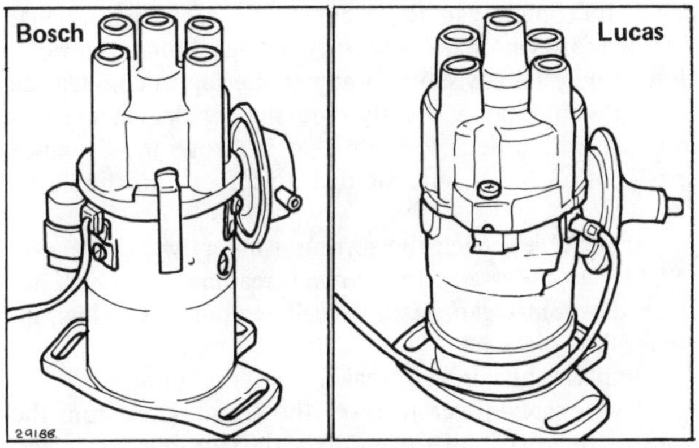

Fig. B.8 Distributor type, CVH engine, contact breaker ignition

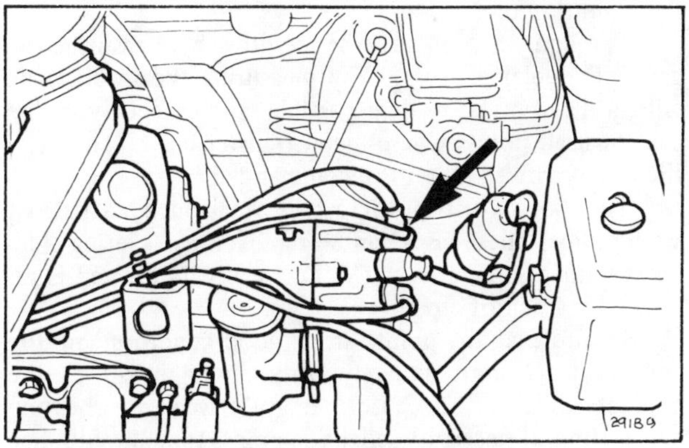

Fig. B:9 Distributor location, CVH engine

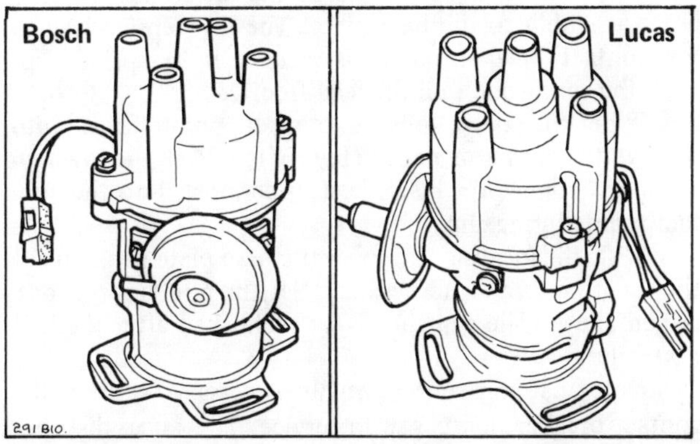

Fig. B:10 Distributor type, CVH engine, electronic ignition

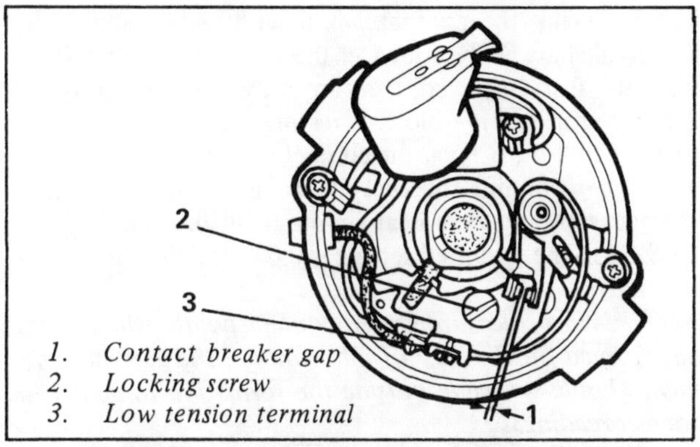

Fig. B:11 Lucas distributor points gap

Cylinder No. 3 rocking - Adjust cylinder No. 2
Cylinder No. 1 rocking - Adjust cylinder No. 4
Cylinder No. 2 rocking - Adjust cylinder No. 3

When adjustment is completed, refit the rocker cover with a new gasket and tighten the screws evenly. Refit the air cleaner assembly, spark plugs and leads and reconnect the breather and vacuum hoses.

DISTRIBUTOR........................[5]

The distributor fitted to the 1.1 CVH and OHV engines is of the conventional contact breaker type. The unit may be manufactured by either Lucas or Bosch (Figs. B:6 & 8). The Bosch distributor is fitted with a brown cap (CVH engines) or red (OHV engines), the Lucas unit is fitted with a Black cap.

The distributor on the OHV engine is mounted on the rear side of the engine, driven by a skew gear from the camshaft (Fig. B:7). The CVH engine distributor is mounted on the end of the camshaft, directly driven (Fig. B:9). The 1.3 and 1.6 CVH engine cars are fitted with electronic ignition as standard (Fig. B:10). No contact breaker points are fitted, so no maintenance of the ignition circuit is necessary. However, lubrication and cleaning should be carried out following the Service Schedule.

If Lucas contact breaker points are fitted, the contact points are self cleaning.

Contact Breaker Points Check

As the contact set is a relatively inexpensive item, it is recommended that it be replaced if any doubt exists as to the condition of the contacts.

The contact points can be set either by measuring the gap with feeler gauges, or by measuring the dwell angle with a dwell meter. Where possible, and especially in the case of used points where metal transfer has taken place, the use of a dwell meter is recommended.

The Bosch distributor cap is retained by a pair of spring clips. These clips can be removed by levering sideways with a screwdriver. The Lucas distributor cap is retained by a pair of crosshead screws.

Ensure that the points are in good condition and the contact faces abut squarely. Any misalignment of the contact surfaces will cause premature wear, overheating and pitting. If necessary, align the points to make full face contact by bending the fixed contact breaker. Do NOT bend the breaker arm.

To check the contact points with feeler gauges turn the crankshaft as described in the Ignition Timing section until the distributor contacts are fully open, i.e. until the breaker arm rubbing block is resting on the highest point of the cam lobe. Check the contact gap by inserting a feeler gauge of suitable thickness between the contact points - the gauge should be a neat sliding fit in the gap (Figs. B: 11 & 12). The specified setting is 0.016 to 0.020 in (0,4 to 0,5 mm).

NOTE: When measuring used contact points where a pip has formed on the face of one contact, the gap measurement should be made outside the formation to achieve a correct reading.

A dwell meter measures the contact breaker points gap as a degree of the distributor cam rotation. By setting the points gap to give the specified dwell angle, an accurate adjustment can be easily made (Fig. B:14). The dwell angle should be between 48 and 52 degrees.

The dwell meter should be connected as detailed, but check with the manufacturer's instructions (Fig. B:13). The negative lead from the dwell meter should be connected to a good earth point on the car body. The positive lead from the dwell meter should be connected between the ignition coil negative terminal and the distributor contact breaker points. In this case, pull the coil negative lead from the coil, connect the meter crocodile clip and reconnect the lead onto its pushfit connector.

With the distributor cap removed and the engine turning on the starter motor, the dwell angle should be as specified. Adjust the angle setting as described next in this section. With the engine running at idle and 2,000 rpm, the dwell angle should stay as specified.

Points Adjustment & Replacement

If the contact breaker gap or dwell setting varies from specification, adjustment should be made as follows:

1. Turn the crankshaft over in its normal direction of rotation by hand until the contact breaker heel is bearing directly on a cam peak and the contacts are fully open.
2. Slacken the locking screw (Figs. B:11 & 12) on the contact breaker.
3. Insert a screwdriver blade between the slot and pip. Move the fixed contact by twisting the screwdriver in the required direction.
4. When the gap is correct, tighten the lock screw and check the points gap for accuracy on each cam lobe. If one or more cam peak is worn, causing a points gap variation, it may be advisable to adjust the gap to compensate for the worn lobe by slightly reducing the clearance. However, where a lobe is so worn that it throws the clearance specification badly out, the distributor or cam should be renewed.
5. If possible, check the setting using a dwell meter and re-adjust as necessary. If the dwell reading is too high, reduce the points gap. If the dwell reading is too low, increase the points gap.

Replace the contact breaker points as follows:

1. Use a screwdriver to lever the spring clips from the sides of the distributor cap. Move the cap to one side and pull the rotor arm free.
2. Remove the single screw securing the contact breaker point assembly to the base plate. The low tension lead to the contact breaker points is secured by a spring clip. Push the spring and pull the lead free.
3. Wipe the new contact breaker point faces with methylated spirits or petrol (Fig. B:15). Push the low tension lead connector, black lead uppermost, into the contact points spring clip.
4. With the locating fork over the base plate peg, replace the securing screw and washers. Set the points gap as described above. The gap should be rechecked after the first 500 miles.

Note that it is not normally necessary to clean the contact breaker points gap in service. The Lucas distributor has a linkage on the points assembly allowing the

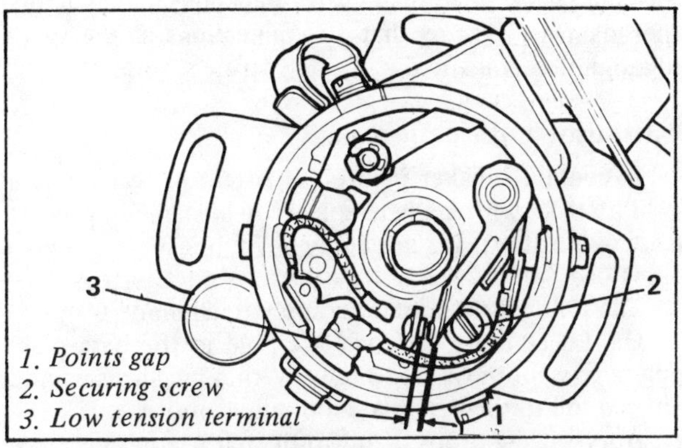

Fig. B:12 Bosch distributor points gap

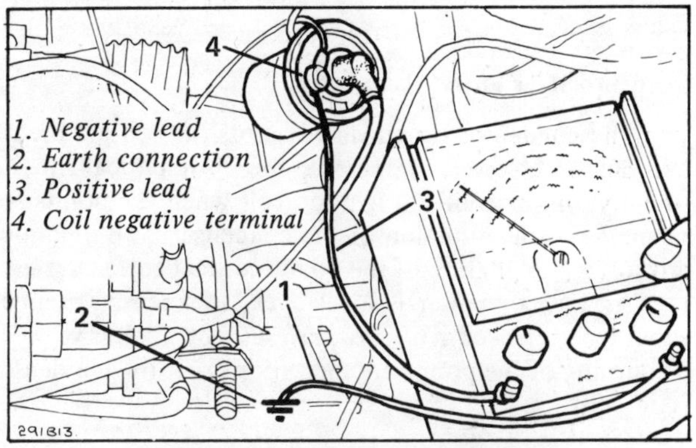

Fig. B:13 Dwell meter connections

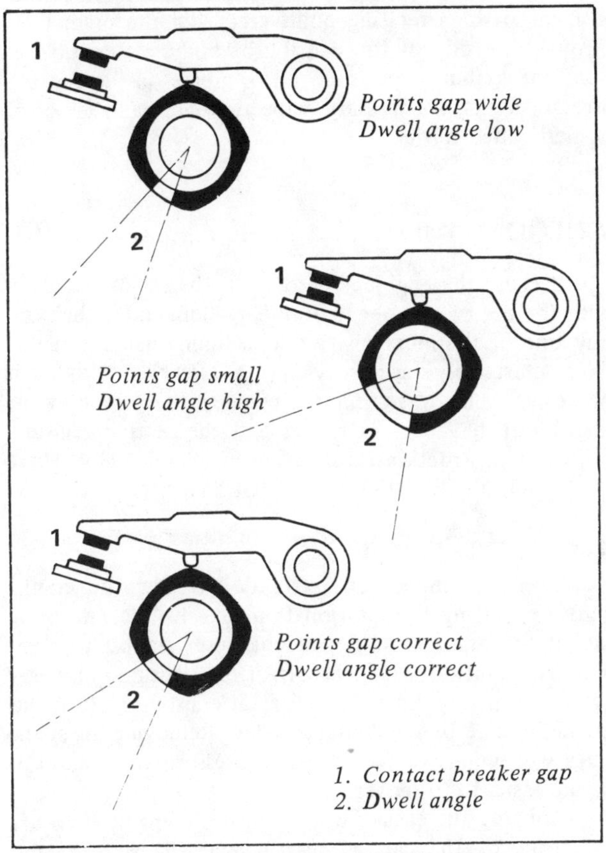

Fig. B:14 Dwell angle in relation to points gap

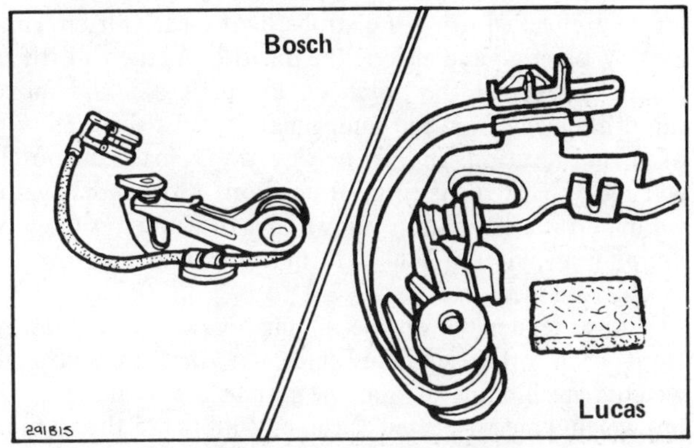

Fig. B:15 Distributor contact breaker points

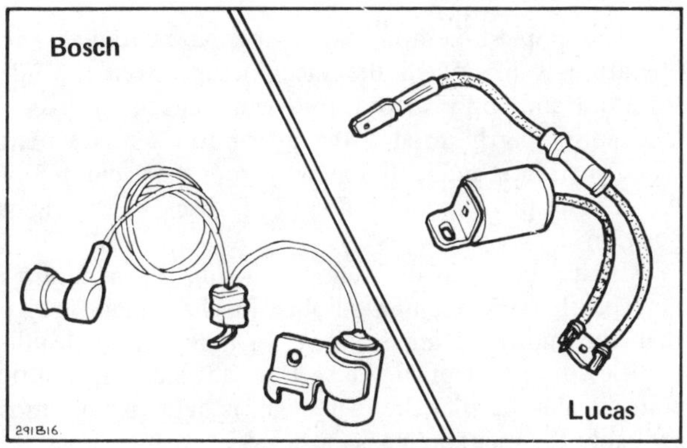

Fig. B:16 Distributor condenser set

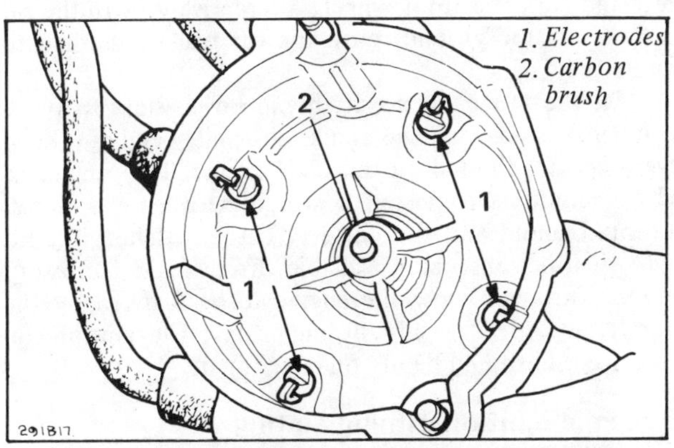

Fig. B:17 Check distributor cap and electrodes

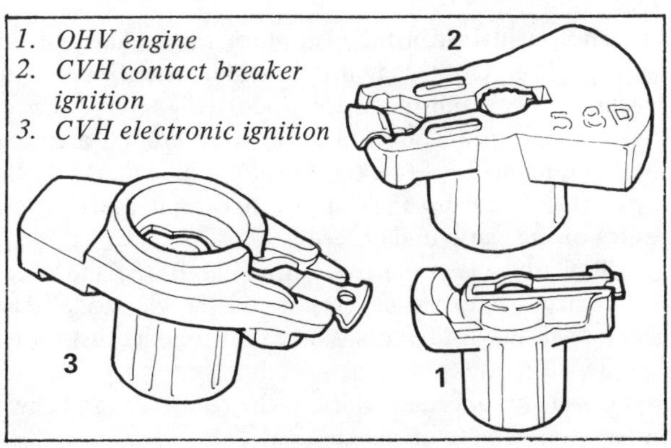

Fig. B:18 Rotor arm

contact faces to rub over each other, polishing the surfaces.

Condenser (Fig. B:16)

The condenser is connected across the contact breaker points. It serves to increase the current peak in the primary windings of the ignition coil when the points are open. A further function of the condenser is to minimise erosion and oxidation of the points contact surfaces.

As the condenser is a relatively inexpensive item it should be replaced if in any way suspect. Excessive wear or burning of the points are indications that the condenser is defective.

Distributor Advance Mechanism Check

The operation of the distributor advance mechanism may be checked using the strobe light (see Ignition Timing) by opening and closing the throttle. As the throttle is gradually opened the notch on the pulley should move smoothly away from the timing mark against the direction of engine rotation, and as the throttle is closed the notch will move back to its original position. Any tendency towards erratic operation shown by the notch suddenly jumping away from the mark indicates that the advance mechanism is faulty.

Any tendency for the timing marks to leap away from each other suddenly indicates that the advance weights are binding or that the springs are weak. If binding weights are suspected, clean and lubricate them, then check the advance mechanism again.

Vacuum Advance

The simplest method of checking the vacuum advance operation is to detach the vacuum pipe from the inlet manifold and apply suction to it while checking that the mechanism partly rotates the distributor contact plate. When suction is suddenly removed from the vacuum pipe, the contact plate should snap smartly back to its original position.

If the contact plate seems at all sluggish on return it may be that the vacuum advance mechanism is dirty or that the return spring is worn. In most cases, cleaning rectifies the problem. If the vacuum advance will not operate, it may be that the diaphragm is defective but most probably the vacuum line is damaged.

Distributor Cap, Rotor Arm and HT Leads

Thoroughly clean the distributor cap, inside and out with a clean cloth, paying particular attention to the spaces between the metal electrodes (Fig. B:17). Check that the electrodes are not excessively eroded and that there are no signs of tracking (shiny hairline marks on the cap surface). Ensure that the small carbon brush in the centre of the cap is undamaged.

Similarly, clean the rotor arm (Fig. B:18) and check for damage or excessive erosion of the electrode. Also check that the rotor arm is a neat fit on the distributor spindle without excessive play.

Clean the outside surface of the coil tower and check for signs of damage or tracking.

Wipe all grease, oil and dirt from the high tension leads and check them for signs of cracking, chafing or any other damage. Ensure that all connections at the spark plugs, ignition coil and distributor cap are secure.

Distributor Lubrication

If contact breaker ignition is fitted, the cam should be lubricated with a light smear of high melting point grease (Fig. B:19). Do not apply any lubricant to the cam wiping pad.

The following instructions apply to all models.

On Lucas units, through the hole in the base plate apply a few drops of oil to the governor mechanism. Add a drop of oil to the base plate mounting groove.

Apply a few drops of oil to the felt pad in the centre of the cam. Lightly smear the points pivot post with grease (Fig. B:20). Bosch distributors should be lubricated with a smear of high melting point grease on the pivot post through the side of the distributor body. After lubrication, check that there is no oil or grease on the contact breaker points. If necessary, wipe away with a little petrol or methylated spirits.

IGNITION TIMING . . . . [6]

Because the engine is designed to be set dynamically with the use of a strobe light, it is recommended that you buy one of the inexpensive strobe timing lights available from most good accessory shops. However, should it prove necessary to re-time the engine, for example if the distributor has been removed and the setting random, follow the instructions to time the engine manually. These are given in the ENGINE ELECTRICS chapter.

Turning the Crankshaft

Several methods can be used to rotate the crankshaft for setting the ignition timing, valve clearances or any of the other tasks dependent upon crankshaft position. The ignition should be off or the ignition coil negative (-) lead disconnected before attempting to turn the engine, in case of accidental starting. Removing the spark plugs will reduce the compression pressure and make the engine easier to turn over.

Probably the easiest way to turn the engine is to lift the front of the car, as described at the start of the ROUTINE MAINTENANCE chapter, put the car in 4th gear and turn the front wheel. Alternatively, with the car in gear on the ground, push the car back and forth to rotate the engine.

On the OHV engine, it should be possible to put a spanner or socket on the crankshaft pulley end nut, however access is limited on the CVH engine. The other possible method is to install a switch between the starter motor solenoid and the battery. These switches, in a kit with instructions, are available from some accessory shops. This will allow the engine to be cranked by the starter motor from inside the engine bay. The ignition coil negative lead should be disconnected first.

Dynamic Ignition Timing Setting

The contact breaker points gap or dwell angle must

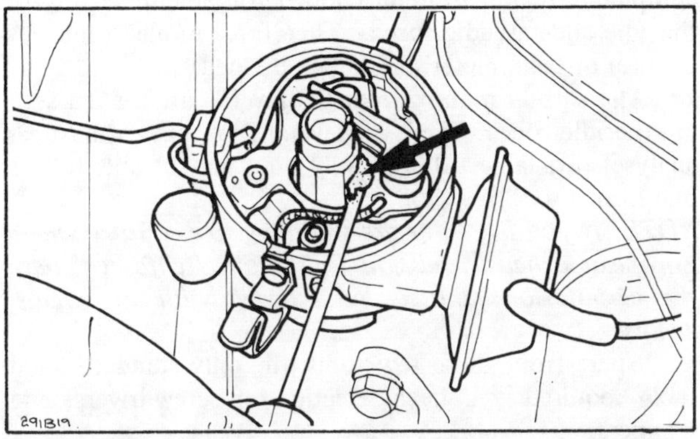

Fig. B:19 Lubricate contact breaker cam with grease

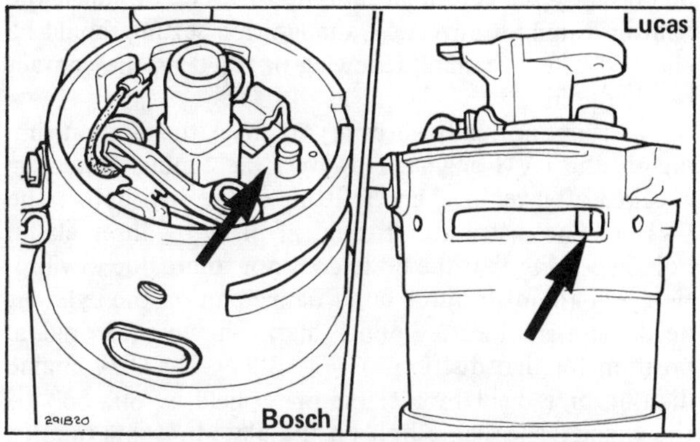

Fig. B:20 Lubricate distributor governer mechanism

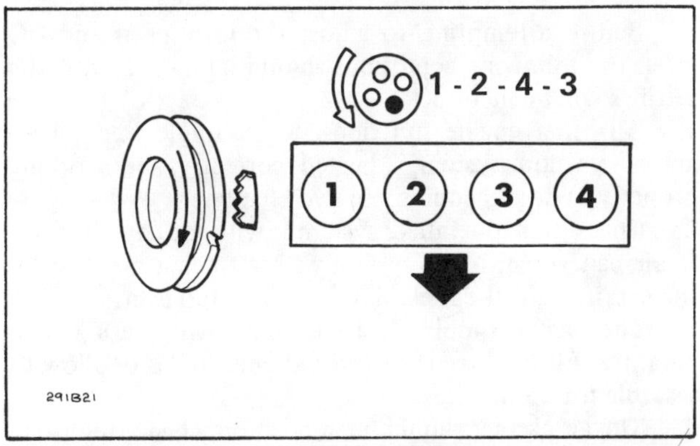

Fig. B:21 OHV engine ignition and cylinder numbering

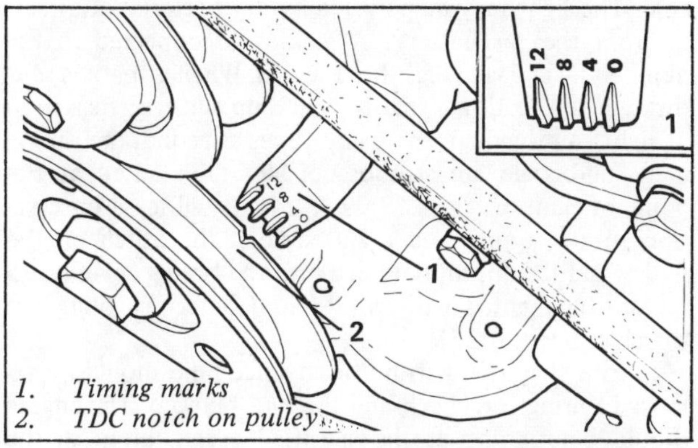

Fig. B:22 OHV engine ignition timing marks

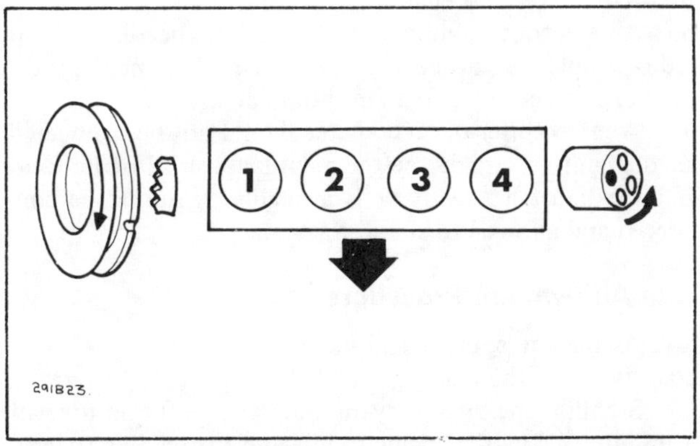

Fig. B:23 CVH engine ignition and cylinder numbering

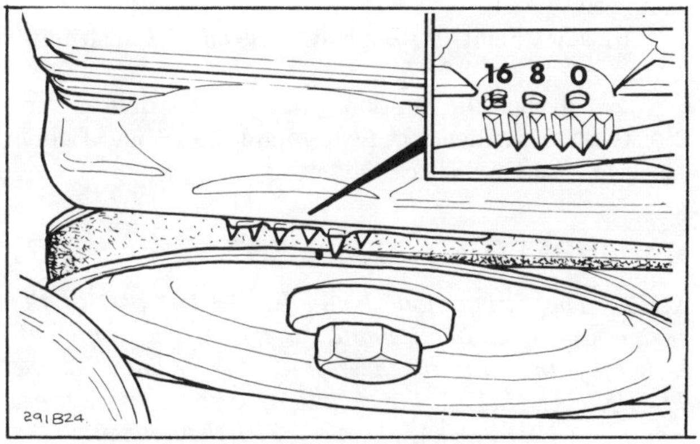

Fig. B.24 CVH engine ignition timing marks

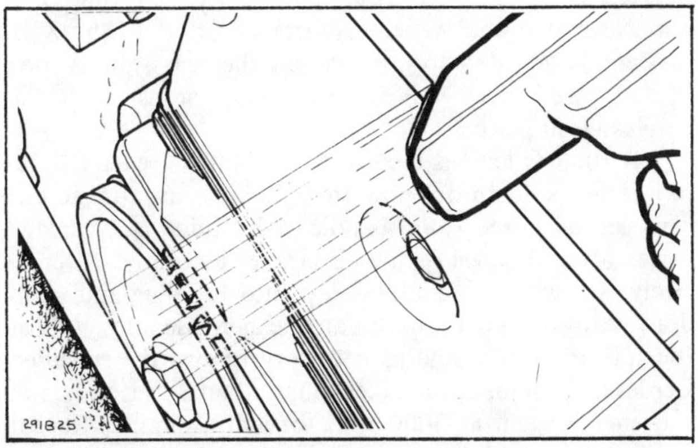

Fig. B.25 Use a stroboscopic timing light to check timing

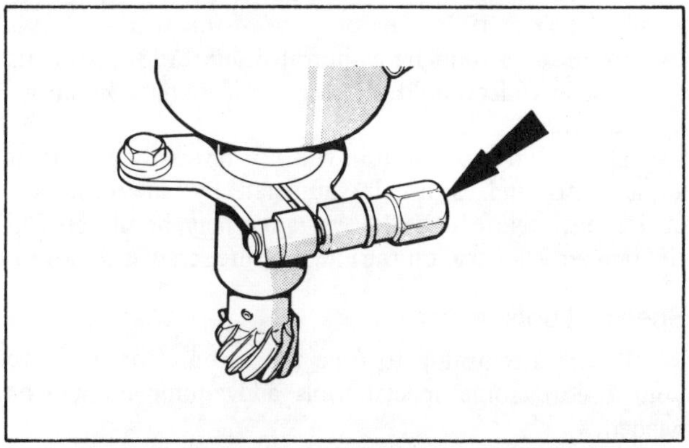

Fig. B:26 Distributor clamp bolt, OHV engine

be correctly set before attempting to check or adjust the ignition timing. Conversely, the ignition setting should be checked after cleaning, renewing or resetting the contact breaker points.

It is not normally necessary to check the ignition timing of the CVH engine. However, the timing should be checked after removal and refitting of the distributor. The CVH engine distributor is held in place by three clamp bolts, ensuring that the body does not rotate due to vibration. On the distributor body flange and on the cylinder head there is a locating punch mark, showing the original position for that distributor (Fig. B:27). The OHV engine distributor is held by a clamp plate, held by one bolt to the engine. The clamp bolt on the plate prevents the distributor body from rotating (Fig. B:26).

There are several types of dynamic timing light available. The light may be powered by dry batteries or directly from the car battery. A pickup is connected to the number 1 HT lead (Figs. B:21 & 23). When a spark is sent through this lead, a signal is picked up and triggers a flash of light. As this happens many times a second, the timing mark and scale are stroboscopically illuminated at the point of ignition (Figs. B:22 & 24). It will ease observation of the timing mark and scale if the notch on the pulley and the appropriate mark on the timing cover housing are highlighted with white paint before carrying out the check (Fig. B:25).

Note that the distributor advance pipe should be removed during the check and the end plugged. The engine should be running at the specified speed, given in the TECHNICAL DATA. A tachometer or rev counter may be needed to determine the engine speed, if the car is not so equipped.

If adjustment is required, slacken the distributor clamp bolt (Figs. B:26 & 27).

To advance the ignition, rotate the distributor body clockwise and anti-clockwise to retard the setting. Tighten the clamp bolt and check the setting.

CARBURETTOR ADJUSTMENT. . . . . [7]

NOTE: The carburation should be the last part of the engine tune-up to be attempted as the settings will be affected by the tune of the rest of the engine. See the TECHNICAL DATA for details of carburettor settings and the FUEL SYSTEM chapter for overhaul details.

As a requirement of the ECE and EEC vehicle 'Build' regulations, "tamperproof" carburettors have been fitted to all cars since 1976. The purpose of this requirement is to prevent indiscriminate carburettor mixture adjustments to be made which could increase exhaust emissions above a certain level.

The carburettor mixture will not normally go out of adjustment, and is usually adjusted by an authorised dealer only. Note that the warranty may be affected if the tamperproof seal on the idle mixture screw is removed.

Special Tools

Before attempting to tune the carburettor fitted to your Escort, some special tools and equipment will be needed.

The rev counter fitted to some models as standard equipment is not accurate enough for precise setting of the idle and fast idle speeds. Therefore, an electronic rev counter or tachometer should be obtained.

The carbon monoxide or CO level is used when setting the idle speed. There are several DIY type exhaust gas analysers available. Alternatively, you may be able to hire one from a tool hire shop.

NOTE: If you live in or are travelling to a country where European exhaust emission legislation is enforced, have the carburettor adjusted by a dealer with an exhaust analyser.

Apart from these few tools, the only standard hand tools required will be a selection of screwdrivers, and spanners.

Idle Adjustment, Preparatory Conditions

Before attempting to adjust the idle speed and CO level, the following conditions should be met. If not, the settings will be incorrect.

1. All other engine functions, for example valve clearances, ignition system, adjusted correctly and working properly as described in these two chapters.
2. Check that the intake system is without any air leaks. If air can get into the system without being metered by the carburettor, the fuel mixture will be too lean.
3. The engine should be at the normal operating temperature. Either drive the car for about 5 miles or allow to fast idle for 15 minutes.
4. The air cleaner should be in position when attempting to adjust the idle speed.
5. The choke valve should be fully open, if suspect see the FUEL SYSTEM chapter.
6. All electrical components such as the headlamps and radio should be turned off during the adjustment procedure. The electric cooling fan should be on.
7. A rev counter or tachometer should not be connected to the ignition circuit, see the manufacturer's instructions. If an exhaust gas analyser is available, it should be connected and allowed to stabilise.

Idle Adjustment Procedure

As the rev counter and exhaust gas analyser readings stabilise, note the idle speed and CO levels.

Stabilise the engine by running at 3,000 rpm for half a minute. Allow the engine to idle and the reading to even out. Note the idle speed and CO level.

Adjust the idle speed screw using a 10 mm ring spanner. Note that the Weber carburettor fitted to the XR3 versions is adjusted by a screw on the rear side. A long bladed screwdriver will be needed to adjust with the air cleaner in place.

If the CO level is incorrect, it will be necessary to remove the tamperproof plug from the idle mixture screw. This can be prised from its hole with a thin bladed screwdriver after removal of the air cleaner assembly. Alternatively, screw a self tapping screw into the plug and withdraw using pliers. Refit the air cleaner assembly. Adjust the idle mixture screw as necessary to give the specified CO level. If adjustment takes longer than 30 seconds, run the engine again at 3000 rpm for half a minute. Repeat until adjustment is complete.

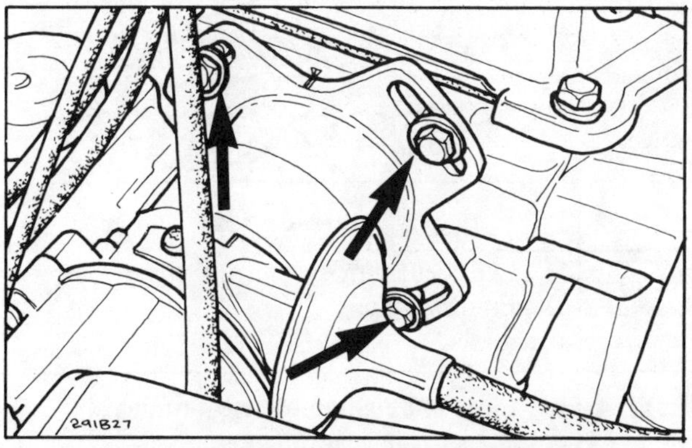

Fig. B:27 Distributor clamp bolts, CVH engine

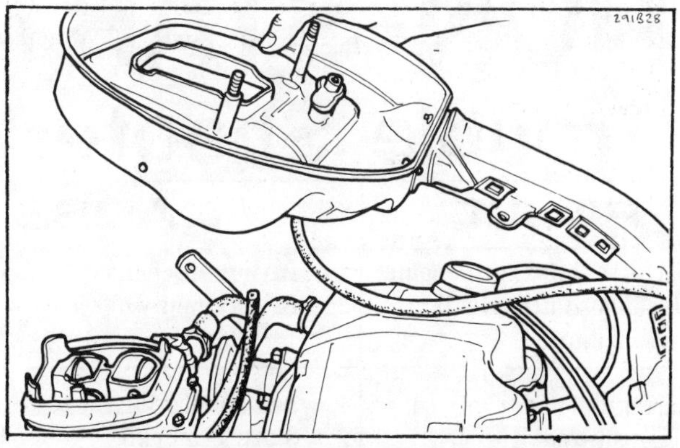

Fig. B.28 Air filter casing removal, XR3 shown

Fig. B.29 OHV engine idle speed adjustment

Fig. B:30 OHV engine idle mixture adjustment

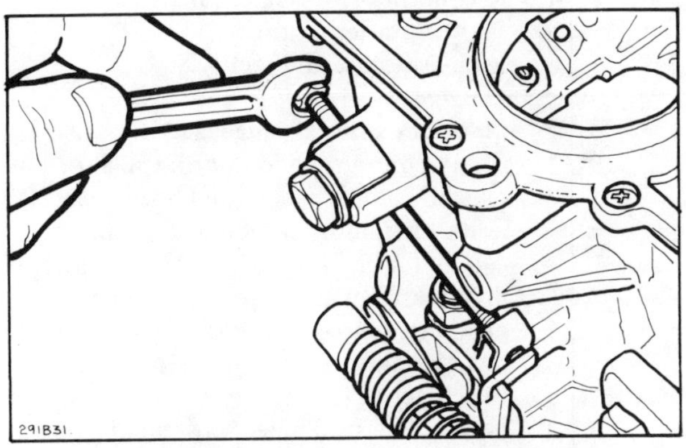

Fig. B:31 CVH engine idle speed adjustment

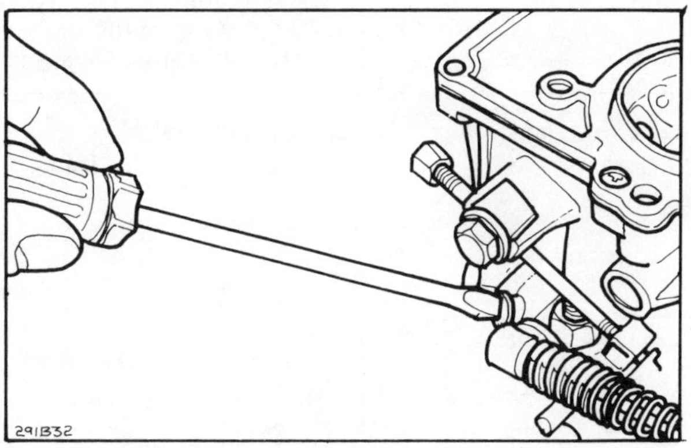

Fig. B:32 CVH engine idle mixture adjustment

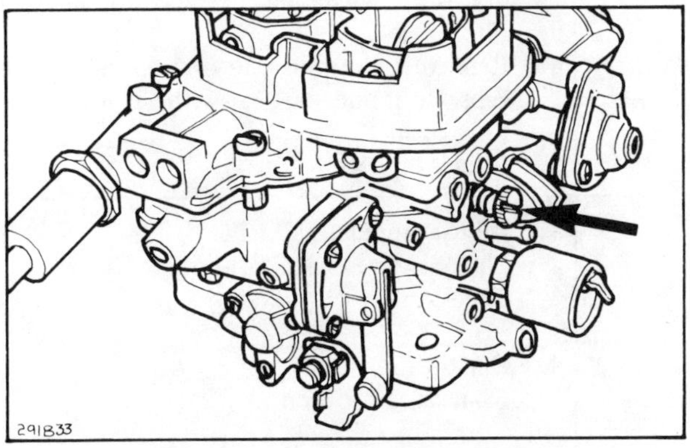

Fig. B:33 XR3 engine idle speed adjustment

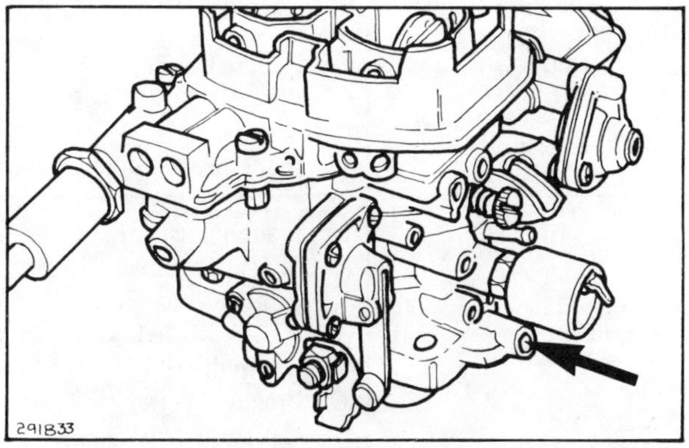

Fig. B:34 XR3 engine idle mixture adjustment

NON-START
Trouble Shooter

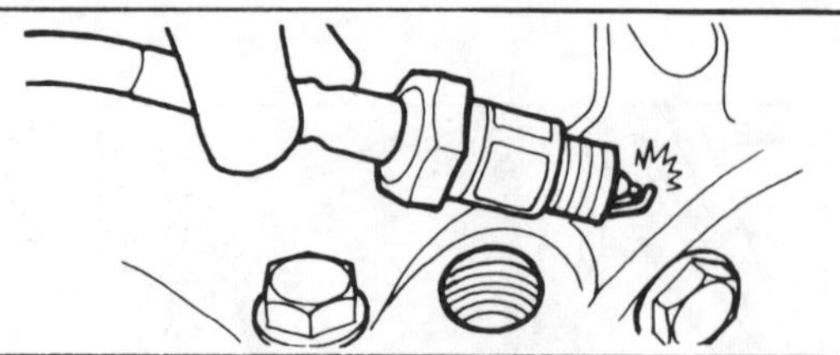

FAULT	CAUSE	CURE
Starter will not turn engine (Headlamps become dim after a minute	1. Battery discharged. 2. Battery faulty.	1. Check charging system and charge battery. 2. Fit new battery.
Starter will not turn engine (Headlamps dim when starter switch operated)	1. Bad connection on battery terminal 2. Starter jammed. 3. Seized engine.	1. Clean and tighten battery terminals. 2. Rock car in gear (Manual). 3. Establish cause of seized engine and rectify.
Starter will not turn engine (Headlamps stay bright)	1. Disconnected or loose engine earth. 2. Starter solenoid faulty. 3. Starter pinion not engaging. 4. Starter pinion/ring gear teeth damaged. 5. Starter motor faulty. 6. Starter solenoid connections loose. 7. Faulty starter/ignition switch.	1. Reconnect or tighten earth. 2. Fit new solenoid. 3. Remove starter and free off pinion. 4. Replace pinion/ring gear. 5. Fit new starter motor. 6. Tighten connections. 7. Fit new switch.
Engine turns slowly but will not start	1. Battery partly discharged. 2. Faulty battery. 3. Loose or corroded battery terminals. 4. Loose or disconnected engine earth. 5. Faulty starter motor. 6. Partially seized engine.	1. Recharge battery and check charging system. 2. Fit new battery. 3. Clean and tighten terminals. 4. Reconnect or tighten earth. 5. Fit new starter motor. 6. Establish cause and rectify.
Engine turns but will not fire	1. Ignition fault. 2. No spark at plug lead. 3. Mixture not reaching cylinders. 4. Engine flooded (over-use of choke). 5. Inadequate tappet clearances.	1. Check for spark at plug lead. 2. Check coil output to confirm high or low tension fault. If spark from coil, check HT leads, distributor cap and rotor arm, particularly for cracks, tracking or dampness. If no spark from coil, check ignition-coil connections and contact-breaker points for short circuits or disconnection. 3. Check petrol feed to carburettor by disconnecting fuel feed pipe and actuating pump. If no petrol, look into carburettor intake, operate throttle, observe whether damp or dry. If dry, clean jets, needle valve, internal filters etc. 4. Remove spark plugs, dry, clean, and check gaps. Refit and start car with throttle held fully open. 5. Adjust tappets.
Engine backfires	1. Inadequate tappet clearances. 2. Ignition timing incorrect. 3. Contact breaker points dirty or incorrectly set. 4. Damp distributor cap or leads. 5. Cracked or tracking distributor cap/rotor arm. 6. Incorrect plug lead positioning.	1. Adjust tappets. 2. Check and reset ignition timing. 3. Check and reset points. 4. Dry thoroughly. 5. Fit new cap/rotor arm. 6. Check and re-position leads.

Engine – ohv

INTRODUCTION . . . . [1]

The 1117 cc OHV engine fitted to the Escort is the same unit as fitted to the Fiesta model and is known as the 'L' engine.

With plenty of working room inside the engine compartment around the engine, it is quite easy for the DIY mechanic to carry out most of the normal overhaul or replacement jobs with the engine/gearbox unit in the car.

The cylinder head, oil pump, timing chain, and both front and rear crankshaft oil seals are easy to get at for replacement. The engine sump pan can be removed quite easily to replace the crankshaft big-end bearings, and for piston removal.

In fact, the engine will only need to be removed for major work such as crankshaft replacement, to remove the camshaft or for any work on the cylinder block such as a rebore.

The Escort uses metric fastenings and a comprehensive metric socket and spanner set will be needed for any overhaul work. In addition, a number of easily obtainable special tools will be needed for more ambitious work, such as a valve spring compressor, piston ring clamp and possibly a universal puller tool. All of these can usually be bought or even hired locally for the job - see Yellow Pages under 'Tools and Equipment Hire'.

CYLINDER HEAD . . . . [2]

Removal

1. Disconnect the battery earth strap. Remove the two crosshead screws (Fig. C:1) retaining the air cleaner to the carburettor.
2. Disconnect the upper breather hose from the oil filler cap and detach the vacuum pipe from the inlet manifold.
3. Remove the air cleaner assembly from the carburettor.
4. Drain the contents of the cooling system by disconnecting the bottom hose from the radiator outlet.
5. Disconnect the top radiator hose and overflow tank pipe from the thermostat housing (Fig. C:2).
6. Disconnect the (upper) heater hose from the automatic choke housing (Fig. C:3).
7. Pull back the clip over the end of the throttle cable and disconnect the inner cable from the carburettor throttle lever. Undo the bracket bolt (Fig. C:4) and detach the throttle cable assembly from the carburettor.
8. Make a note of the positions of the various pipes and wires around the carburettor and inlet manifold before disconnecting the brake servo vacuum pipe (if fitted), the fuel and vacuum pipes from the carburettor, and the breather hose from the inlet manifold. Plug the fuel pipes to prevent fuel loss and the ingress of dirt.
9. Disconnect the HT leads from the spark plugs by pulling on the terminal insulating sleeve, not on the wire itself. Disconnect the wire from the temperature gauge sender unit, the anti-dieselling valve on the side of the carburettor, and the cooling fan thermal switch (if fitted).
10. Lift off the hot air box and undo the bolts securing the exhaust down pipe from the manifold (Fig. C:5). Detach the pipe from the manifold and support it with wire to prevent it dropping.
11. Remove the oil filler cap with its breather hose and place it to one side. Undo the crosshead screws and carefully detach the rocker cover from the cylinder head. Place the cover upside down to prevent oil draining out.
12. Undo the four rocker shaft pedestal bolts evenly, a little at a time until the shaft is free from valve spring compression (Fig. C:6). Carefully lift off the shaft and place it inside the removed rocker cover.
13. Starting from the thermostat housing end of the engine, remove each push rod in turn, and place them through numbered holes in a piece of card.

NOTE: The push rods must be refitted in their original positions on reassembly.

14. Remove the spark plugs.
15. Release the cylinder head bolts evenly, in the reverse order of tightening shown in Fig. C:7.

16. Lift off the cylinder head complete with carburettor, inlet and exhaust manifolds. If the head will not free due to the adhesion of the gasket, tap around the side of the head with a wood, plastic or hard rubber mallet to break the seal. Do not try to insert a screwdriver between the gasket faces.

Dismantling

1. Undo the bolts and remove the inlet manifold complete with the carburettor and the exhaust manifold from the cylinder head (Fig. C:8).
2. Support the head upside down on wood blocks and remove all carbon deposits from the combustion chambers, valve heads and valve ports with a suitable scraper and wire brush. Take great care not to damage the machined face of the cylinder head.
3. Similarly clean all deposits from the cylinder block face and piston crowns, but ensure that carbon particles are not allowed to enter oil or waterways in the cylinder block. These can be covered with pieces of cloth or masking tape.
4. At each valve in turn: Compress the valve spring using a suitable compressor (Fig. C:9) and extract the two split tapered collets from around the valve stem. Release the compressor and remove the spring retainer and valve spring. Remove the oil seal fitted to the valve stem and withdraw the valve. Discard the oil seal. Mate each valve to its spring components and place to one side in removal order.
5. To dismantle the rocker shaft assembly (Fig. C:11), remove the split pin from one end of the rocker shaft and detach the flat washer, crimped spring washer and second flat washer which bear against the end rocker arm (Fig. C:10). The rocker arms, rocker shaft support pillars and springs can now be removed from the shaft. If necessary tap jammed rocker supports with a soft-faced hammer to free them. The plugs can be removed from each end of the rocker shaft to facilitate cleaning by piercing a hole in the plug and levering the plug out of the end of the shaft.

Cylinder Head Inspection

Refer to CVH ENGINE chapter for details of cylinder head, valves, guides and springs inspection.

Pushrods & Rocker Shaft Assembly

Clean the rods in solvent. Check the ends of the rods for nicks, grooves, roughness or excessive wear. Check each rod for straightness. Replace as necessary. Do NOT attempt to straighten a bent pushrod.

NOTE: The push rods can be checked visually for straightness by rolling them across a flat surface.

Inspect the shaft and rocker arm bores for nicks, scratches, scores or scuffs. Inspect the pad at the valve end of each rocker arm for indication of scuffing or abnormal wear. If the pad is grooved, replace the arm. Do NOT attempt to true the surface by grinding.

Check the clearance between each rocker arm and the shaft by measuring the inside diameter of the arm bore and the outside diameter of the shaft. If the clearance is excessive, the shaft and/or rocker arm should be replaced.

Check that all oil passages are clear. Replace any damaged adjusting screws.

Cylinder Head Reassembly

Reassemble the cylinder head in the reverse order of dismantling, with special attention to the following points:

a) Lap in each valve in turn using coarse, followed by fine, grinding paste until a gas-tight seal is obtained at the seat. This will be indicated by a continuous matt-grey ring around the valve face and seat. When this has been achieved, clean all traces of paste from the cylinder head and valves - this is most important.

b) Lubricate the valve guides and valves with SAE 90 Hypoid Oil before installing the valves.

c) The valve stem seal must always be renewed. Cover the collet grooves in the end of the valve stem with adhesive foil or tape to avoid damaging the seal during installation (Fig. C:12). Lubricate the seal with oil to make fitting easier. Remove the foil once the seal is in position.

d) Ensure that the valve stem is not damaged by the spring retainer when compressing the valve spring, and that the split tapered collets engage correctly in the valve stem and spring retainer when the spring is released.

e) If the rocker assembly was dismantled, reassemble it in reverse order. If the end plugs were removed, install new plugs in each end of the shaft. The shaft must be positioned so that the oil feed holes in the shaft for rocker lubrication point downwards to the front of the cylinder head. The bolt holes in the rocker shaft support pillars must be located on the same side as the adjusting screws in the rocker arms. Install the split pins at the ends of the shaft with their heads upwards, and bend over their legs to secure.

f) When refitting the inlet manifold assembly, apply sealing compound to both sides of the manifold gasket at the central waterway.

Cylinder Head Installation

Installation is a reversal of the removal procedure, with special attention to the following points:

1. Make sure all matings faces, especially those of the cylinder head and block are perfectly clean and free from old gasket material.
2. If the cylinder head was removed to change the cylinder head gasket only, then check the flatness of the cylinder head and block before reassembly.
3. Do not use sealing compound of any type on the cylinder head gasket.
4. The cylinder head bolts are of different lengths, two being shorter than the others according to the varying thickness of the cylinder head casting. Make sure the bolts are fitted in their correct respective positions.
5. Tighten the cylinder head bolts evenly in the sequence shown in Fig. C:7, and in the progressive stages given

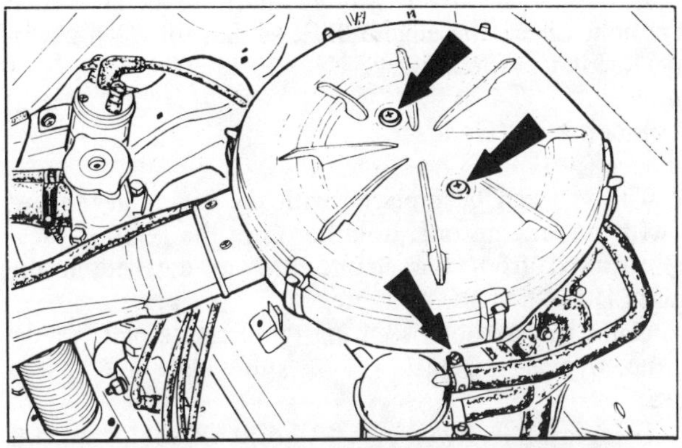

Fig. C:1 Air cleaner attachment screws

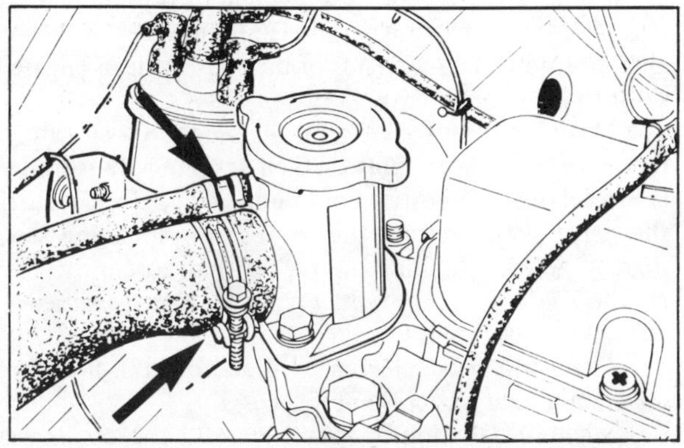

Fig. C:2 Thermostat housing hose connections

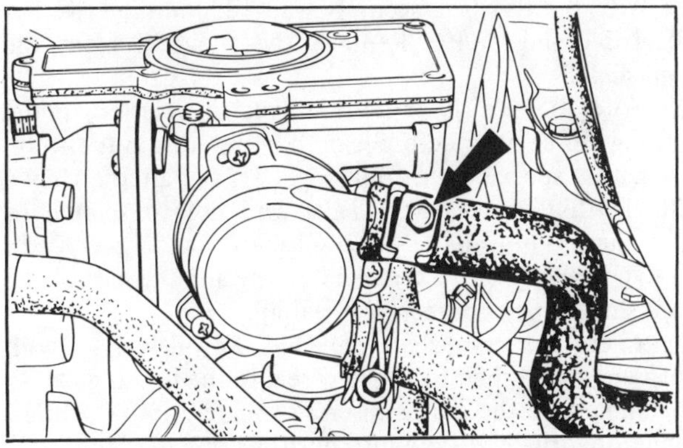

Fig. C:3 Heater hose connections at auto - choke

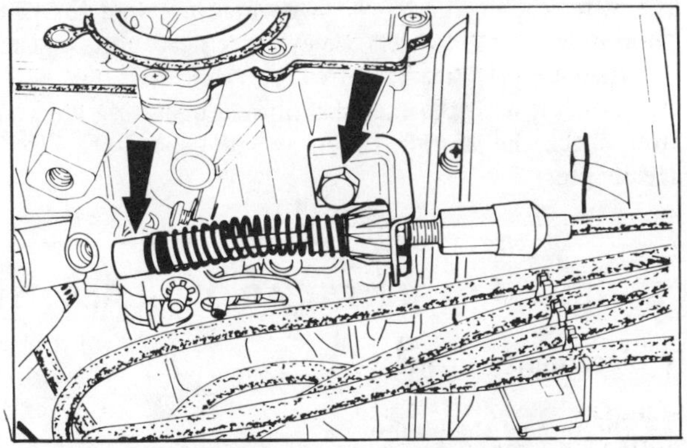

Fig. C:4 Throttle cable connections

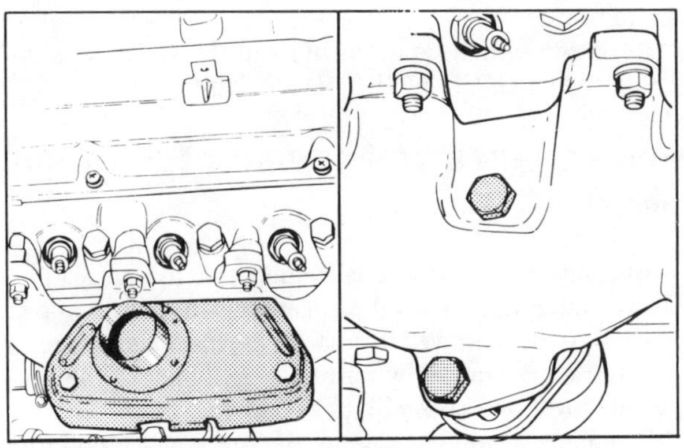

Fig. C:5 Hot air box and exhaust pipe fitting

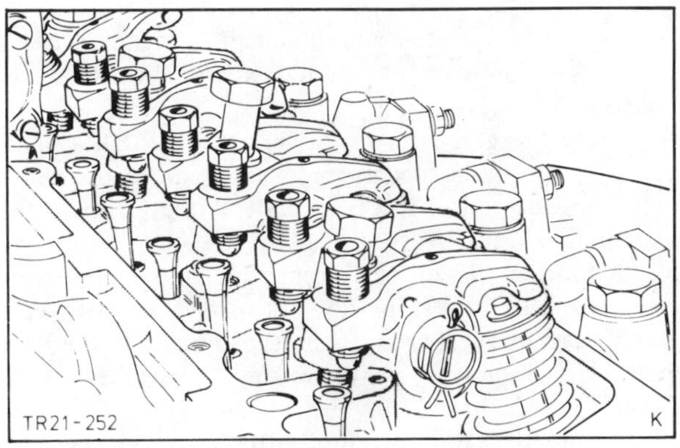

Fig. C:6 Removing rocker shaft from head

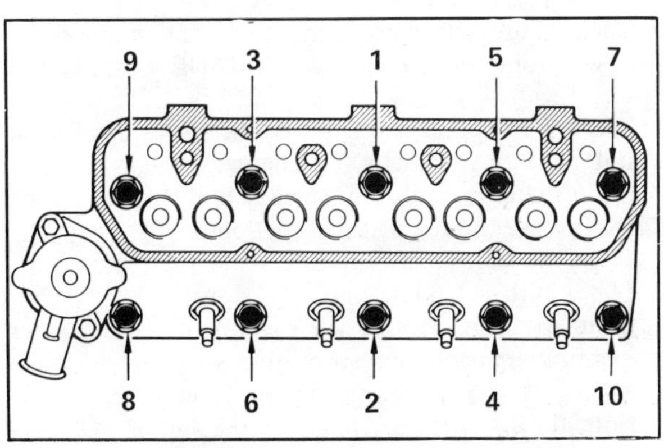

Fig. C:7 Cylinder head bolt tightening sequence

Fig. C:8 Removing inlet manifold from head

in the TECHNICAL DATA at the end of this manual.
6. Make sure that the pushrods are installed in their original positions. Dip the ends of the rods in clean engine oil prior to installing them.
7. Make sure the rocker arm adjusting screws locate correctly in the cupped ends of their respective pushrods. If any work has been carried out on the valves (e.g. recutting the valve seats) release the rocker arm adjusting screws slightly before installing the rocker shaft assembly. Tighten the shaft retaining bolts evenly to their specified torque.
8. Check the valve clearances as described in the TUNE-UP chapter and adjust them if necessary.
9. When installation is complete refill the cooling system, then start the engine and check for oil, water or exhaust system leaks.
10. After the engine has been warmed up, wait for 15 to 20 minutes and retighten the cylinder head bolts to their specified torque, - see TECHNICAL DATA.
11. Finally, with the engine at normal operating temperature, check the ignition timing and idle speed see TUNE-UP chapter.

CRANKSHAFT FRONT/REAR OIL SEALS . [3]

Replacement - Front

1. Disconnect the battery.
2. Slacken the alternator mounting and pivot bolts and push the alternator towards the engine. Remove the drive belt.
3. Jack up the vehicle and support it on stands. Refer to the jacking information contained in the ROUTINE MAINTENANCE chapter.
4. Place the gear lever in 1st gear and with an assistant applying the brake pedal, undo and remove the centre bolt from the end of the crankshaft. Withdraw the crankshaft pulley, using levers or a suitable puller.
5. Withdraw the oil seal from the front cover either using a special tool (Fig. C:13) or using a screwdriver to gently lever the seal from its bore. However, if the latter method is employed, great care should be taken not to damage the seal bore in any way.
6. Ensure that the bore in the front cover is clean and free from old sealer, etc.
7. Lubricate sealing lips of the new seal and the timing cover bore.
8. Press the new seal into the cover bore with the open side facing inwards, this can be done by inserting a spacer ring between the seal and the pulley and drifting on the pulley to drive the seal home (Fig. C:14). Alternatively, it is possible to fit the seal by gently tapping it into place, using a large socket as a drift. Whichever fitting method is used, it is important the seal lips are not damaged and that the seal is fitted square in its housing.
9. Refit the crankshaft pulley, making sure that the pulley sleeve is well lubricated to prevent damage to the oil seal lips, then refit the pulley bolt.
10. Tighten the pulley bolt to the correct torque using the same method as used to undo it.
11. Lower the car to the ground and refit the alternator drive belt. Check the engine oil level and top-up if necessary. Reconnect the battery.

Replacement - Rear

1. The seal can be replaced with the engine in the car, or with the engine out. In either case, the gearbox must be separated from the engine first as detailed in the CLUTCH & GEARBOX chapter.
2. Undo the six bolts securing the pressure plate flange to the flywheel. Detach the pressure plate and clutch plate.
3. Using a screwdriver or other suitable tool, wedge it between the engine block and flywheel ring gear teeth to prevent the crankshaft from turning.
4. With the flywheel securely wedged, undo the five flywheel securing bolts. Remove the flywheel from the crankshaft.
5. Note the position of the oil seal in relation to the housing bore edge before carefully levering out the oil seal with a special Ford tool (Fig. C:15) or alternatively a suitable thin, screwdriver. Take care not to scratch or score the seal housing or the crankshaft.
6. Thoroughly clean the seal housing and the part of the crankshaft which contacts the seal lip.
7. Smear the crankshaft end and the oil seal lip with clean engine oil then fit the new seal to the housing.
8. Make sure the seal is inserted squarely into the housing before driving it fully into position.
9. Refit the flywheel and tighten the bolts to the correct torque using the same method of wedging the flywheel when removing.
10 Refit the clutch components and the gearbox as detailed in the CLUTCH & GEARBOX chapter.

TIMING CHAIN & TENSIONER . . . . [4]

Removal

Although the timing chain and the tensioner can be replaced quite easily with the engine in-situ, the sump pan will have to be removed from the engine to gain access to the tensioner and allow correct positioning of the timing chain cover on reassembly.
1. Disconnect the battery earth strap and drain the engine oil from the sump pan as detailed in the ROUTINE MAINTENANCE chapter.
2. Slacken the alternator mountings and remove the drive belt from the pulleys, see ENGINE ELECTRICS chapter.
3. Raise and support the front of the car as detailed in the ROUTINE MAINTENANCE chapter.
4. Undo the crankshaft front pulley bolt and detach the pulley as detailed in the previous section.
5. Undo the sump pan bolts and remove the pan as detailed later in this chapter.
6. Undo the bolts around the edge of the timing cover and carefully remove the cover from the engine. Take care not to damage the oil seal lip on the crankshaft nose.
7. Note the fitted position of the oil slinger ring (concave side facing away from engine) before removing it

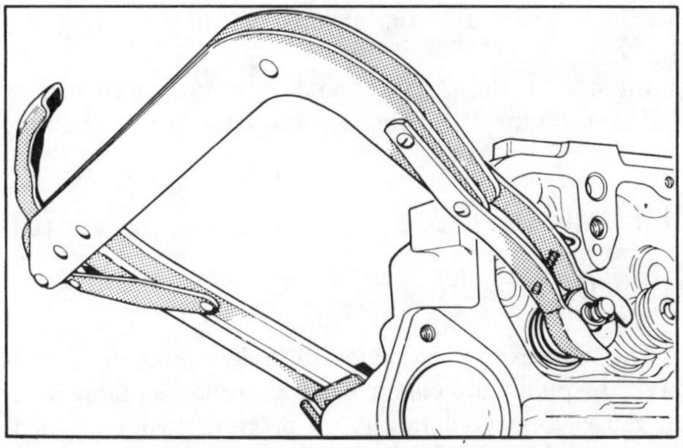

Fig. C:9 Using a valve spring compressor tool

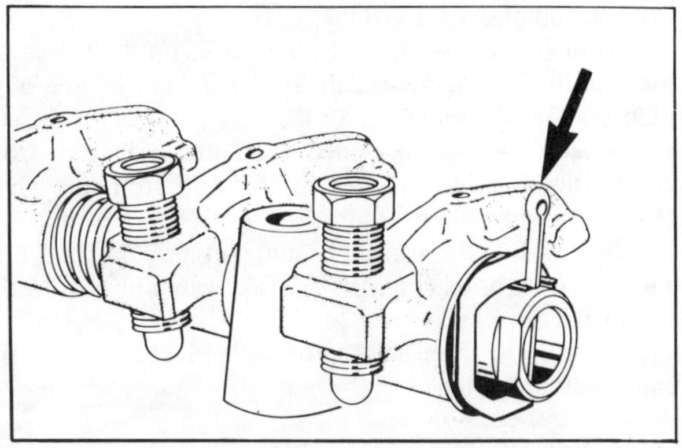

Fig. C:10 Rocker shaft securing pin

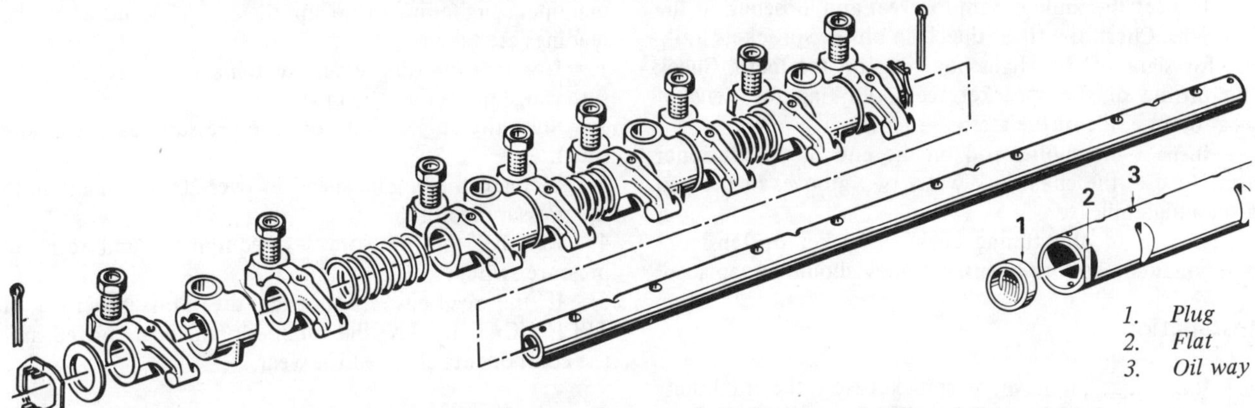

Fig. C:11 Exploded view of rocker shaft components (left). Details of shaft end fitting (right)

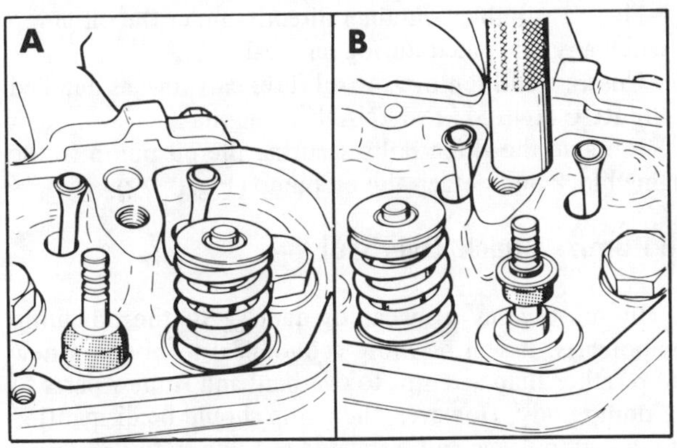

Fig. C:12 Valve stem oil seals, exhaust (A), inlet (B)

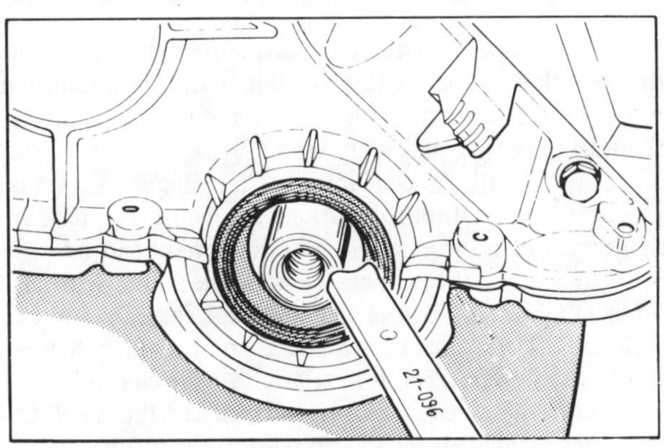

Fig. C:13 Removing crankshaft front oil seal

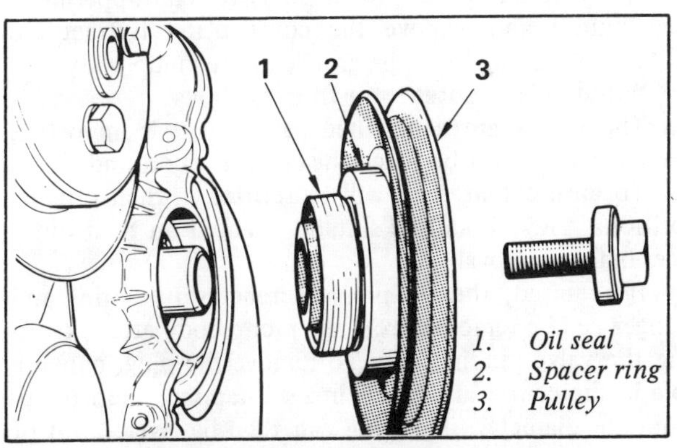

Fig. C:14 Fitting oil seal using spacer ring

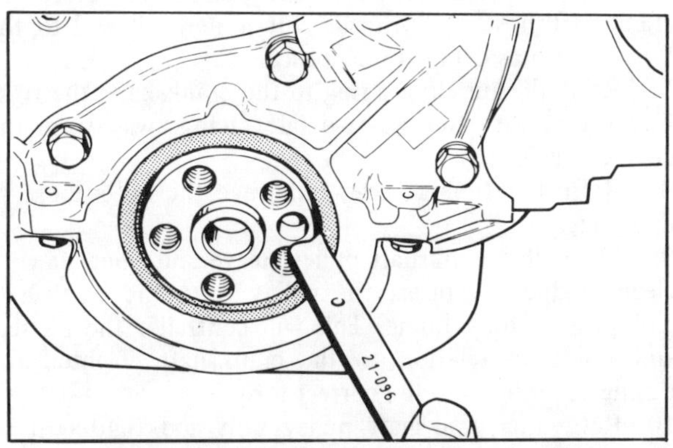

Fig. C:15 Removing crankshaft rear oil seal

from the crankshaft nose (Fig. C:16).
8. Turn the crankshaft until the marks on the flanges of the camshaft and crankshaft sprockets are in line with each other as shown in Fig. C:17.
9. Release the spring-loaded tensioner snail cam from the tensioner arm, and detach the arm from the pivot pin on the end main bearing cap (Fig. C:18).
10. Bend back the camshaft sprocket bolt locking tabs and undo the bolts (Fig. C:19). Take care not to turn the sprocket.
11. Detach the camshaft sprocket and chain from the crankshaft sprocket.

Inspection

Inspect the timing chain for wear and looseness at the link pins. Check the fit of the chain on the sprockets looking for signs of the chain bunching on the teeth. Check the profile of the sprocket teeth, looking for signs of wear on one side of the teeth.

Inspect the rubber pad on the end of the tensioner arm. In use, the chain link wears two grooves in the pad after a high mileage.

If any of the timing chain, sprocket or tensioner components are worn or suspect, they should be replaced.

Installation

1. If necessary, remove the sprocket from the crankshaft using a short universal puller (Fig. C:20). Fit a new sprocket, making sure that the woodruff key is correctly fitted.
2. Fit the chain to the camshaft sprocket and position the sprocket mark so it is in line with the crankshaft sprocket flange mark (Fig. C:17) before attaching the chain to the crankshaft sprocket.
3. Carefully fit the cam sprocket, aligning the dowel pin and bolt holes without disturbing the camshaft. Refit and tighten the cam sprocket bolts to the correct torque.
4. Check that the sprocket marks are in-line with each other (Fig. C:17), before bending over the cam sprocket lock tabs. If the marks are not aligned, detach the cam sprocket and reposition it in relation to the chain.
5. Refit the tensioner arm and release the snail cam tensioner spring (Fig. C:18), to tension the chain.
6. Clean the timing cover and engine block mating faces to remove all traces of old gasket. Also clean the crankshaft nose and threads. Fit a new oil seal to the timing cover as detailed previously.
7. Refit the oil slinger ring to the crankshaft, the right way round with the concave side facing away from the engine (Fig. C:16).
8. Refit the timing cover but leave the bolts loose at this stage.
9. Clean the crankshaft pulley sleeve and smear it with clean engine oil. Insert the pulley onto the crankshaft and push it fully home. This will centralise the oil seal and cover in relation to the crankshaft. Tighten the timing cover bolts to the correct torque.
10. Refit the crankshaft pulley bolt and tighten it to the correct torque. Use a wood block as a wedge to prevent the crankshaft from turning when tightening the bolt.
11. Fit a new sump gasket and refit the sump pan as detailed in 'Sump Pan - Installation' later in this chapter.

OIL PUMP. . . . . [5]

Testing Oil Pressure

The oil pressure can be tested to determine the condition of the pump and engine bearings using a suitable pressure gauge connected to the oil pressure switch connection on the side of the cylinder block. To get a reasonably accurate reading, it is important that the engine is at normal operating temperature and the correct speed when the readings are taken.

1. Unscrew the oil pressure switch and connect the pressure gauge pipe (Fig. C:21).
2. Start the engine and note the pressure reading at idle (hot).
3. Increase the engine speed to over 2000 rpm and note the pressure reading.
4. Remove the pressure test equipment and refit oil pressure switch.
5. If the readings are outside the limits specified in TECHNICAL DATA, the pump should be removed and the components checked for wear.

Removal

1. Place a suitable container directly under the oil pump to catch any oil spilled during removal.
2. Unscrew and remove the oil filter cartridge as detailed in the ROUTINE MAINTENANCE chapter.
3. Remove the three bolts securing the oil pump to the engine block and detach the oil pump (Fig. C:22).

Oil Pump - Check/Overhaul

In most cases of wear or damage to the oil pump components, it will be more economical to obtain a new pump rather than attempt to overhaul and fit new parts to the pump body. However, the pump should be dismantled after a high mileage and inspected as follows, to determine its condition.

1. Hold the pump body with the cover end uppermost, then unbolt and remove the cover plate. Extract the rubber sealing ring from the groove in the pump body.
2. Withdraw the outer rotor from the housing.
3. The rotors are a matched pair, and it is important that the outer rotor be fitted the correct way round.

To ensure that they will be refitted in their original positions, they should be suitably marked on their outer faces before removal.

4. If required, the oil pressure relief valve spring and plunger can be removed from the pump body after extracting the valve sealing plug. To remove the plug, punch a hole in the centre and screw in a self-tapping screw of appropriate diameter. The plug can then be levered out of the relief valve bore, and the spring and plunger removed.

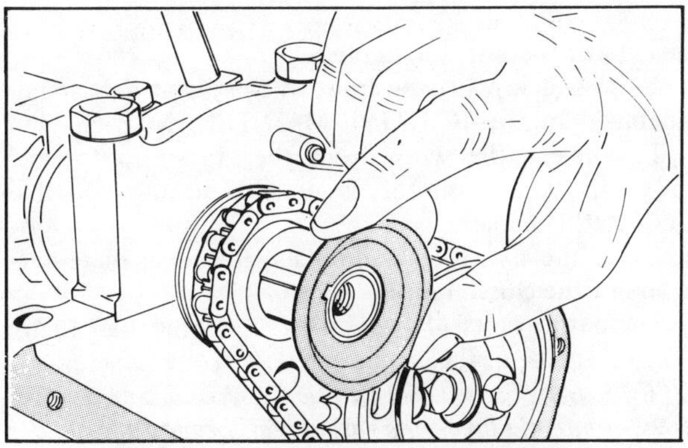

Fig. C:16 Oil slinger ring fitted position

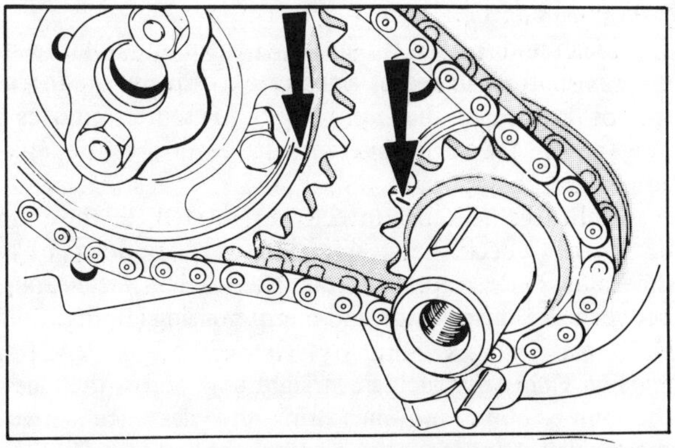

Fig. C:17 Timing sprocket alignment marks

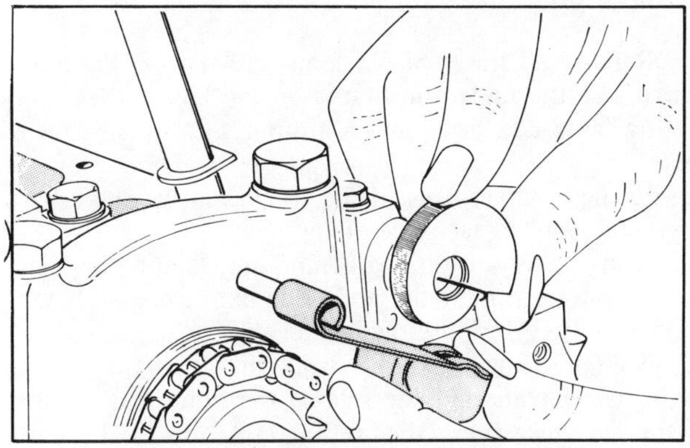

Fig. C:18 Chain tensioner arm and snail cam

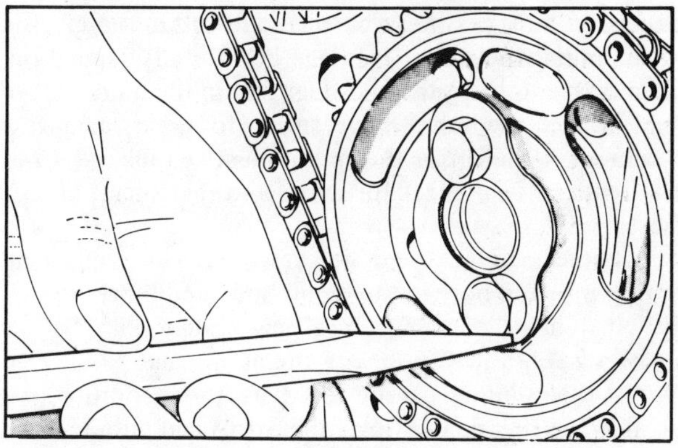

Fig. C:19 Releasing cam sprocket bolt tabs

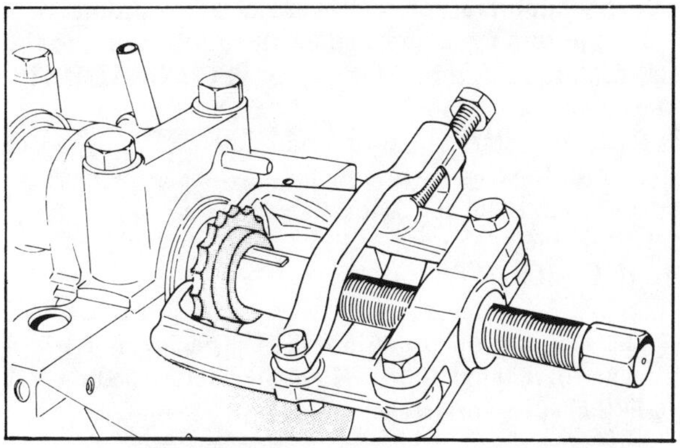

Fig. C:20 Removing chain sprocket from crankshaft

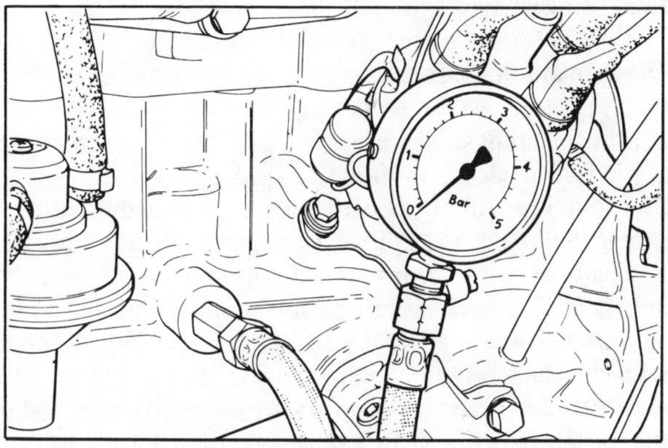

Fig. C:21 Oil pressure gauge connecting point

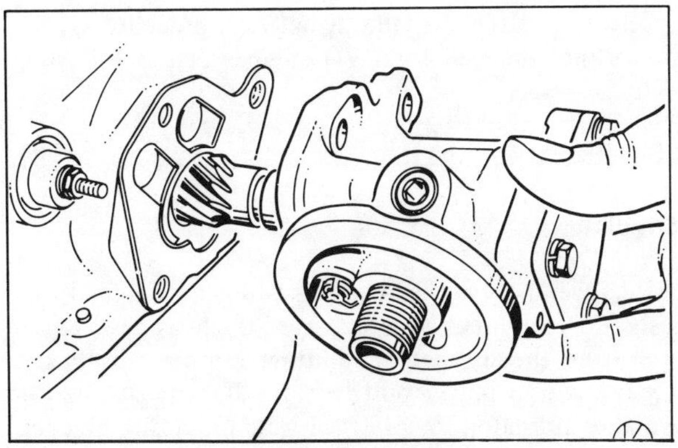

Fig. C:22 Removing oil pump from engine

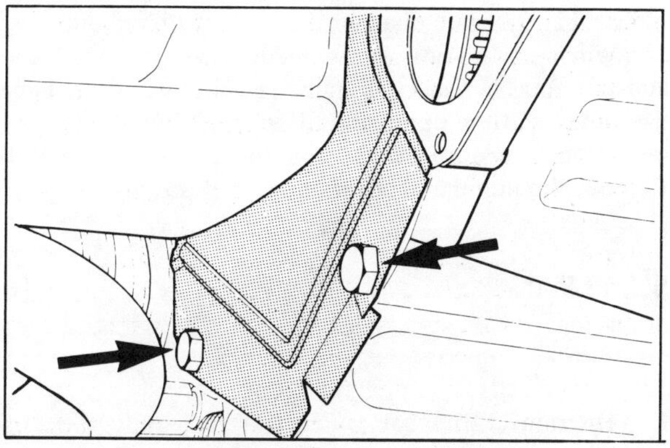

Fig. C:23 Clutch cover plate and bolts

5. Clean and inspect the inside of the pump housing and the rotors for damage or signs of excessive wear. Inspect the inside face of the pump cover for scores, grooves or other signs of wear - if present, the cover must be replaced.
6. Refit the rotors, correctly orientated, in the pump body, then check the clearance between the lobes of the inner and outer rotors, (A, Fig. C:24), then the clearance between the outer rotor and pump housing (B, Fig. C:24) with feeler gauges. Similarly, check the rotor end-float (C, Fig. C:24) by placing a straight-edge across the face of the pump body and measuring the clearance between the straight-edge and the end of the rotors. The wear limits for these clearances are given in TECHNICAL DATA at the end of the manual. If the rotor end-float is excessive, but the other clearances are within the specified wear limits, the pump face can be carefully lapped on a flat surface to bring the end-float within the limit.
7. Inspect the relief valve spring for wear, damage or weakness. If defective the spring must be replaced. Check the relief valve plunger for scores and free operation in its bore.
8. To reassemble the relief valve, fit the plunger and spring into the bore in the pump body and lubricate lightly with engine oil. Drive in a new sealing plug, flat side outwards, until it is flush with the pump face.
9. If the components are in satisfactory condition, reassemble the pump. Lubricate the rotors and pump housing liberally with engine oil before fitting the pump cover.
10. Finally, check that the rotors rotate freely by turning the drive gear shaft.

Oil Pump - Installation

1. Ensure that the joint surfaces on the pump and cylinder block are clean and free from old gasket material.
2. If a new or overhauled oil pump is being fitted, it should be filled with oil and turned by hand through a complete rotation prior to installation.
3. Locate a new gasket on the pump mounting flange and fit the pump assembly to the cylinder block. Secure with the three retaining bolts.
4. Fit a new filter cartridge of the pump. Wet the seal on the cartridge with clean engine oil, then screw the cartridge onto the threaded adaptor by hand until the seal just contacts the pump housing. Tighten the cartridge by a further half to three-quarters of a turn, by hand only.
5. When installation is complete, check the engine oil and top up as necessary. Finally, run the engine and check for oil leaks at the pump and filter.
6. If leakage does occur, stop the engine immediately. Remove the pump and check the gasket position.

SUMP PAN .... [6]

Removal

1. Disconnect the battery earth strap, undo the sump drain plug and drain the engine oil into a suitable container. Refit the sump drain plug.
2. Raise and support the front of the car on axle stands as detailed in the ROUTINE MAINTENANCE chapter.
3. Disconnect the wires from the starter motor and remove the starter motor as detailed in the ENGINE ELECTRICS chapter.
4. Undo the two bolts and detach the cover plate over the end of the clutch housing (Fig. C:23).
5. Undo the bolts (18) securing the sump pan flange to the underside of the engine and detach the sump.
NOTE: It may be necessary to use a screwdriver to lever off the sump and break the gasket seal. Leave a tray under the engine to catch any oil drips.

Installation

1. Remove all traces of old sump gasket from the sump flange and the underside of the engine block. Clean the mating faces and make sure oil pump pick-up strainer is clean.
2. Fit new rubber seals to the grooves in the crankshaft front and rear oil seal carriers (Fig. C:25).
3. Using grease or jointing compound, fit the new cork sump gasket strips to the underside of the engine block. Make sure that the bolt holes are all aligned.
4. On OHV engines only, check the gap between the sump baffle plate and the side of the sump (Fig. C:26). If the clearance is less than 0.080-0.091 in (2,0-3,8 mm) due to the sump pan having been damaged or dented, use a hammer and drift to obtain the correct dimension. This gap is very important for correct engine lubrication.
5. Fit the sump pan and tighten the bolts evenly to the specified torque in three stages - see TECHNICAL DATA at the end of this manual.
6. Refit the starter motor and clutch housing cover plate. Refill the engine with oil and reconnect the battery. Start the engine and check for oil leaks.

ENGINE MOUNTINGS................. [7]

The replacement procedure for the engine mountings is similar to that on the CVH engine. Refer to the CVH ENGINE chapter for details.

ENGINE REMOVAL & INSTALLATION ... [8]

The engine removal and installation procedure is the same as that detailed for CVH engines in the following chapter.

ENGINE OVERHAUL .... [9]

Dismantling

1. Remove the cylinder head assembly as detailed previously. Carefully remove and discard the head gasket.
2. Slacken the alternator mounting and pivot bolts and the water pump pulley bolt, remove the drive belt. Remove the alternator complete with its mounting bracket. Remove the water pump pulley.

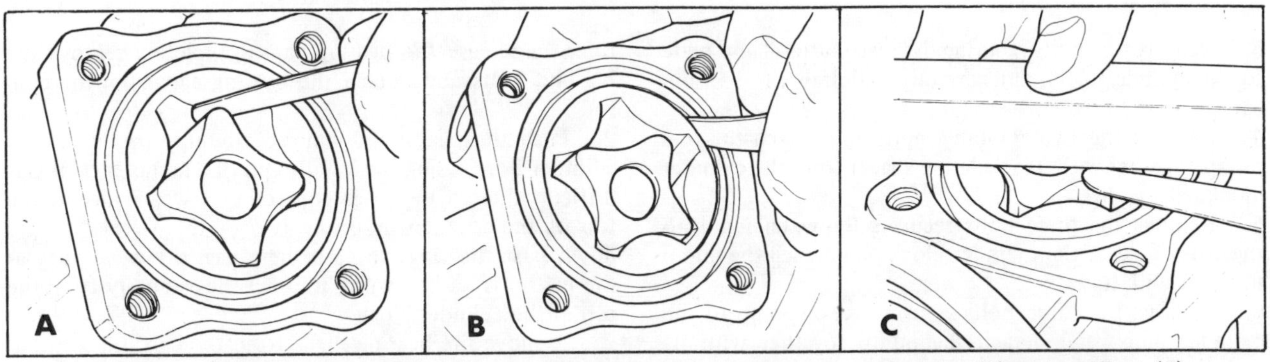

Fig. C:24 Checking oil pump for wear, at rotor lobe tip (A), outer rotor to housing (B) and endfloat (C)

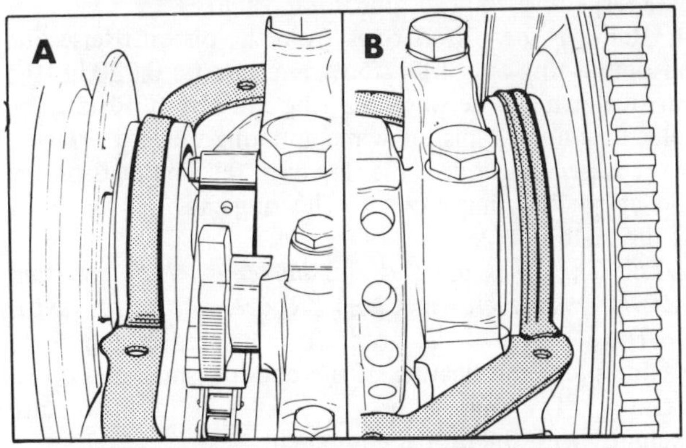

A. Timing cover end B. Flywheel end

Fig. C:25 Details of sump gasket at front and rear

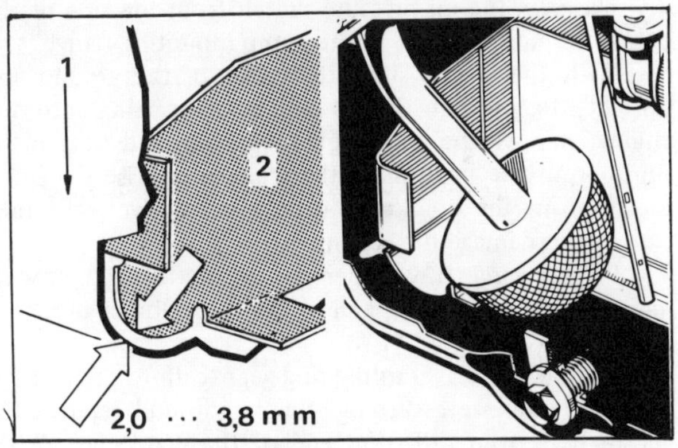

1. Sump pan 2. Baffle plate

Fig. C:26 Check sump pan baffle clearance

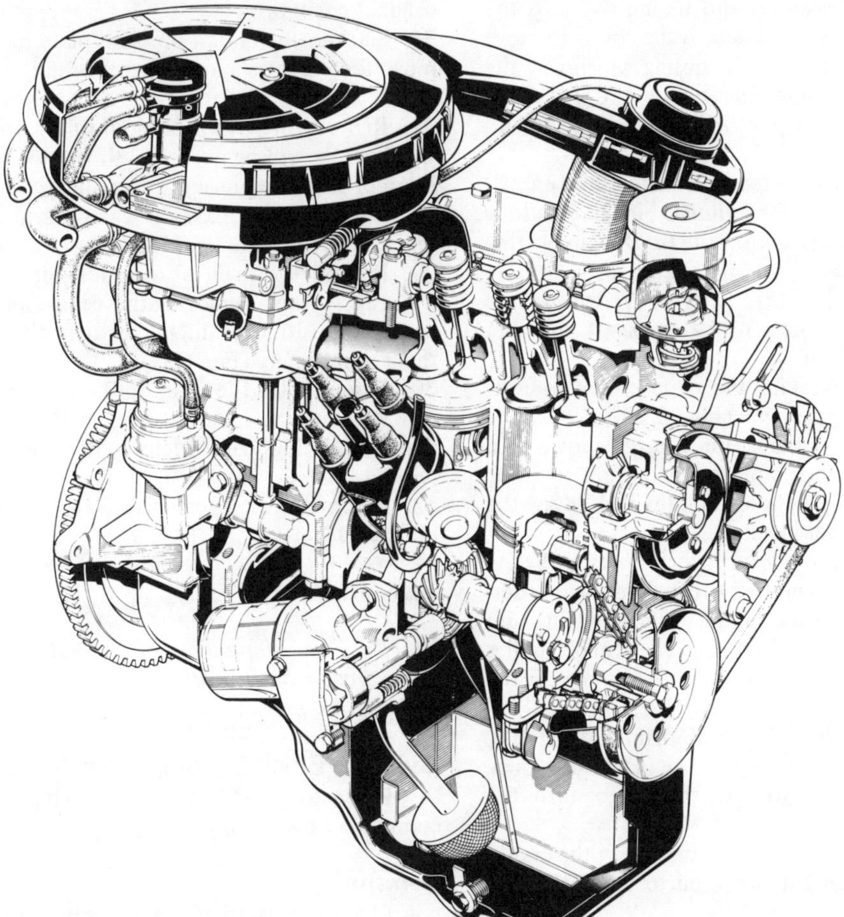

Fig. C:27 Cutaway view of Escort OHV 'L' engine showing component location

3. Remove the bolts securing the distributor clamp plate to the cylinder block and carefully withdraw the distributor.
4. Remove the two retaining bolts and detach the fuel pump together with its insulating spacer from the cylinder block (Fig. C:28).
5. Remove the three bolts securing the water pump to the front face of the cylinder block and detach the pump together with its gasket.
6. Remove the three bolts securing the oil pump to the cylinder block and detach the pump together with its gasket (Fig. C:22).
7. Unscrew the oil pressure switch from the side of the cylinder block adjacent to the pump mounting flange.
8. With the engine supported clear of the ground and upright, to prevent sludge and swarf getting into the engine, remove the sump retaining bolts and detach the sump from the bottom of the crankcase. Prise the sump flange from the crankcase with a screwdriver, but take care not to damage the mating surfaces.
9. Place a large tray under the engine to catch any remaining oil, water, etc., and invert the cylinder block assembly.
10. Use a screwdriver in the ring gear teeth to prevent the flywheel/crankshaft turning, then undo and remove the centre bolt from the front end of the crankshaft. Withdraw the pulley from the crankshaft, using levers or a suitable puller.
11. Remove the bolts securing the timing cover to the front of the cylinder block and detach the cover. Remove the oil slinger from the crankshaft, noting which way the dished side is facing for future reference (Fig. C:16).
12. Unbolt the oil pick-up pipe from the crankcase and remove it by moving it to and fro until it is free.
13. Unbolt the timing chain tensioner assembly (Fig. C:18) from the lower face of the cylinder block. Withdraw the tensioner arm from the pin on the front main bearing cap.
14. Release the lock plate tabs (Fig. C:19) remove the two retaining bolts and detach the camshaft sprocket complete with timing chain.
15. Release the lock plate tabs and remove the two bolts securing the camshaft thrust plate (Fig. C:29).
16. With the cylinder block positioned upside down turn the camshaft through one complete revolution to bring all the followers into the TDC position, then withdraw the camshaft from the cylinder block.
17. Extract the cam followers from their bores in the cylinder block, keeping them in their original installed order.
18. If required, withdraw the timing chain sprocket from the front end of the crankshaft using a suitable puller (Fig. C:20). Extract the Woodruff key from the slot in the crankshaft.
19. Check that all bearing caps, both main and big end, are suitably marked so that they can be installed in their original locations (Fig. C:30). At this stage, turn the engine on its side.
20. For each piston and con-rod in turn: Turn the crankshaft as necessary to bring the con-rod to the bottom of its travel.
21. Unscrew the big-end bolts two or three turns and tap them to release the bearing cap. Completely remove the big end bolts and detach the bearing cap from the con-rod.
22. Push the piston and con-rod assembly up out of the cylinder bore, using the handle end of a hammer. If this is difficult due to the rings contacting a wear ridge at the top of the bore, try inserting two or three feeler gauges down past the rings to ease them past the wear ridge as the piston is withdrawn. Withdraw the assembly from the top of the cylinder block.
23. Remove the bearing shells from the cap and con-rod. Identify the shells if they are to be re-used. Refit each cap to its respective con-rod, this is important.
24. Remove the piston rings from the piston. Raise one end of the ring out of its groove and insert a 0.020 in (0,5 mm) feeler blade between the ring and piston. Rotate the feeler around the piston while applying upward pressure to the raised portion of the ring until the ring rests on the land above the ring grooves. The ring can then be eased off the piston.
NOTE: The rings must always be removed and refitted over the piston crown - NEVER over the piston skirt.
25. The removal of the piston from its con-rod requires a suitable press and heating of the con-rod small end on reassembly. This job is best entrusted to a local auto machinist with the proper equipment.

Suitably identify the various components of the assembly so that they may be refitted, if necessary, in their original positions.
26. Slacken the clutch pressure plate bolts evenly and remove the pressure plate together with the clutch disc from the flywheel.
27. Remove the flywheel mounting bolts and remove the flywheel. It will be necessary to jam the crankshaft with a lump of wood to prevent the crankshaft turning.
28. Detach the crankshaft rear oil seal carrier from the rear face of the cylinder block (Fig. C:31).
29. Release the main bearing retaining bolts and detach the caps. When removing the centre main bearing cap, note the position of the two thrust half washers and mark them accordingly (Fig. C32).
30. Carefully lift the crankshaft out of the crankcase.
31. Remove the main bearing shells from the crankcase and the bearings caps. Identify the shells if they are to be re-used.
32. Remove the oil seals from the timing cover and rear oil seal carrier.

Inspection

NOTE: The components of the cylinder head assembly have already been dealt with under the heading CYLINDER HEAD previously.

Cylinder Block & Component Inspection

Refer to the CVH ENGINE chapter for details of component inspection.

Camshaft

Clean the camshaft thoroughly with petrol or other suitable solvent, then wipe it dry. Inspect the cam lobes

Fig. C.28 Removing fuel pump from engine

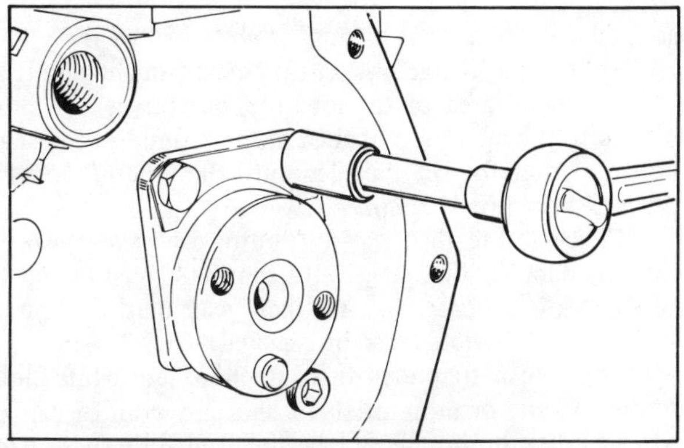

Fig. C:29 Removing camshaft thrust plate

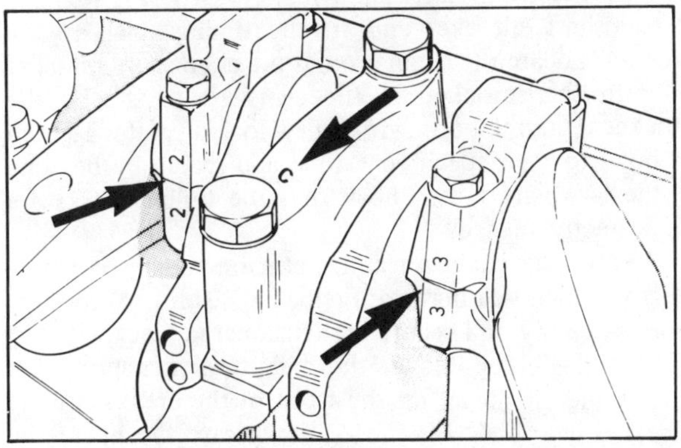

Fig. C:30 Main and big-end bearing cap marks

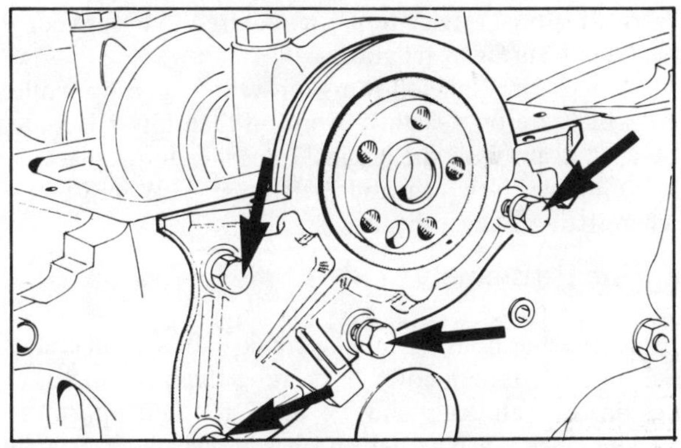

Fig. C:31 Crankshaft rear oil seal carrier bolts

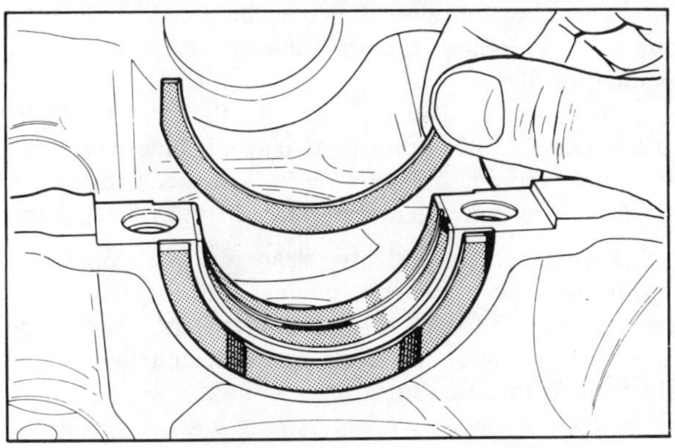

Fig. C:32 Crankshaft thrust washer position

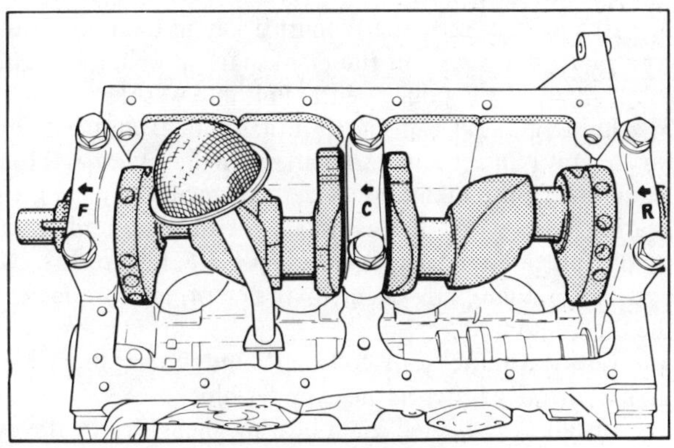

Fig. C.33 Main bearing cap position markings

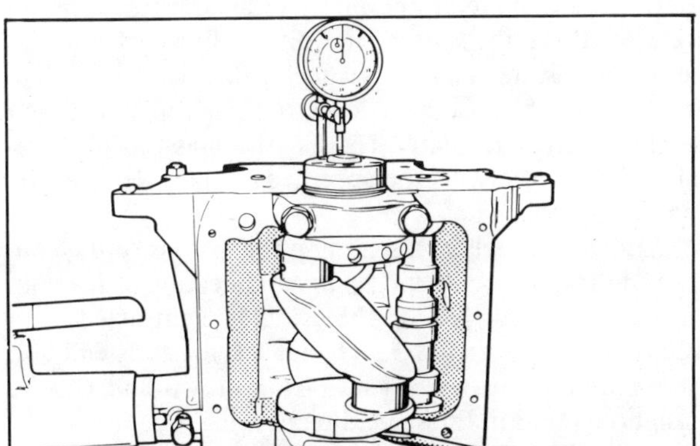

Fig. C:34 Checking crankshaft endfloat

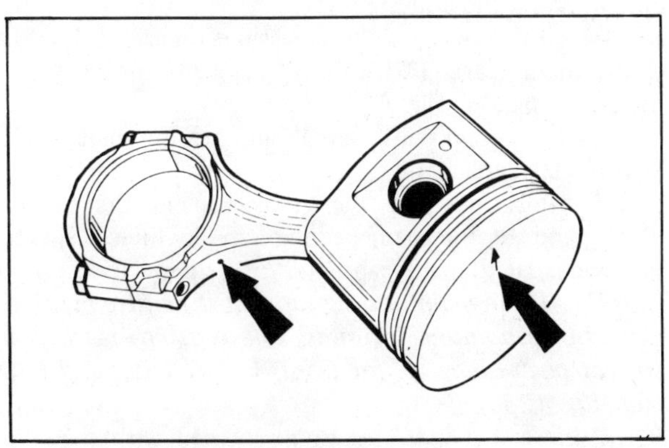

Fig. C:35 Correct relationship of piston to conrod

and journals for scoring or signs of abnormal wear.

Lobe wear characteristics may result in slight pitting in the general area of the lobe toe, but this is not detrimental to the operation of the camshaft, and the camshaft need not therefore be replaced until the lobe lift loss exceeds 0.005 in (0,127 mm).

Check the teeth of the distributor drive skew gear for wear or damage. If damaged, the camshaft must be replaced. If this is the case, then the skew gears on the oil pump and distributor should also be checked.

The camshaft runs in three steel-backed, white metal bushes. If one or more of the bushes are worn or damaged, all three bushes should be replaced, otherwise camshaft alignment may be affected. It is recommended that bush replacement be entrusted to an authorised Ford agent as special service tools are required to ensure correct alignment during installation.

Check the cam followers for wear or scoring. Followers which are only slightly worn on their lower faces may be used again with the original camshaft, but replacement is preferable. New cam followers must ALWAYS be used when fitting a new camshaft.

Engine Reassembly

1. Clean the housings of the crankshaft rear oil seal carrier and the timing cover, inspecting them for nicks, etc. Lubricate both seals and fit them to their appropriate housings using a hammer and hardwood block or a suitable size socket or tubular drift. Drive the seals in squarely up to their stops.
2. If removed, refit the Woodruff key in the nose of the crankshaft and press on the crankshaft sprocket, securing it by means of the pulley centre bolt and washer.
3. Lubricate the camshaft journals, lobes and bearings thoroughly. Insert the camshaft into the cylinder block from the front, taking great care to avoid damaging the bearing bushes. Fit the camshaft thrust plate and secure to the block face using a new lockplate. Secure the two bolts after tightening them to their correct torque with the lockplate tabs.
4. Lubricate the cam followers and insert them into their respective bores in the cylinder block.
5. Install the selected main bearing shells dry in the cylinder block and main bearing caps. Ensure that the locating tag on each shell correctly engages the corresponding notch in the housing.
6. Install the two half thrust washers on each side of the centre main bearing. With the oil groove facing away from the centre bearing Fig. C32.
7. Coat the crankshaft journals and bearing shells liberally with clean engine oil.
8. Carefully position the crankshaft in the cylinder block, and fit the main bearing caps in their respective locations. Tighten the cap bolts to their specified torque.

NOTE: All the main bearing caps must be fitted with the cast arrows pointing forwards. The caps are marked for correct positioning: F for front, C for centre and R for rear (Fig. C:33).

9. Using a dial indicator gauge located with the stylus against the crankshaft flange, as shown in Fig. C:34, check the crankshaft endfloat. Push the crankshaft towards the rear of the engine by levering on the crankshaft web with a screwdriver, and note the reading on the indicator dial. If the endfloat exceeds the specified limit, the thrust washers must be replaced with ones of suitable oversize to bring the figure within specification.

   If a dial gauge is not available, the endfloat can be measured using feeler gauges between the face of the thrust washer and the crankshaft web.
10. Turn the crankshaft until the timing mark on the sprocket is pointing towards the camshaft centre. Temporarily fit the camshaft sprocket and turn the camshaft until the timing mark on the sprocket is adjacent to and in line with the crankshaft timing mark (Fig. C:17). Remove the camshaft sprocket and fit the timing chain to the sprocket. Locate the chain around the crankshaft sprocket and refit the sprocket on the camshaft flange. Fit the sprocket retaining bolts with a new locking plate. Tighten the bolts to their specified torque and secure by bending up the lockplate tabs. Check that the timing marks are still correctly aligned.
11. Locate the chain tensioner arm on the pivot pin at the front main bearing cap. Preload the spring of the tensioner assembly and secure the tensioner in place with the two retaining bolts. Release the chain tensioner spring.
12. Fit the oil slinger on the front of the crankshaft the correct way round (Fig. C:16). Lubricate the lip of the seal in the front cover, slide the crankshaft pulley sleeve through the cover oil seal and fit the front cover. Use the pulley sleeve to centralise the cover before tightening the front cover bolts and the crankshaft centre bolt to their specified torques.
13. Slide the rear oil seal carrier over the crankshaft, the seal lips being lightly lubricated to avoid damage. Tighten the carrier retaining bolts evenly to their specified torque.
14. Make sure that the mating faces of the flywheel and crankshaft flange are perfectly clean and free from burrs, then locate the flywheel squarely on the crankshaft flange, and tap into place. Apply oil resistant sealer to the threads of the retaining bolts, then fit and tighten the bolts to their specified torque.

NOTE: Only re-use flywheel retaining bolts if they are in good condition.

15. Make sure the friction surfaces of the flywheel and clutch pressure plate are perfectly clean and free from oil or grease. Place the clutch disc on the flywheel, the flat side of the disc facing the flywheel. Align the driven plate with a mandrel or home-made centralising tool and attach the clutch pressure plate. Tighten the pressure plate retaining bolts evenly to the specified torque. Remove the centralising tool.
16. As detailed earlier, assembling the pistons to the conrods requires the use of special jigs and heating of the conrod small end. This is a job that should be entrusted to an authorised Ford agent. On receipt of these assemblies, make sure the arrows on the piston crown and conrod splash bore point in the same direction Fig. C:35.
17. For each piston and connecting rod in turn:

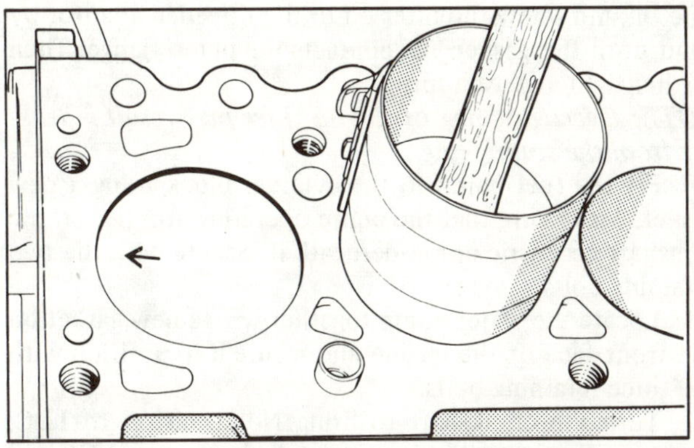

Fig. C:36 Fitting pistons to bores with tool

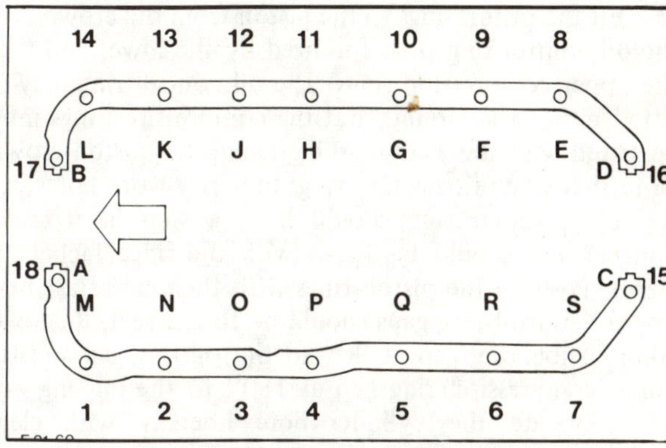

Fig. C:37 Sump bolt tightening sequence

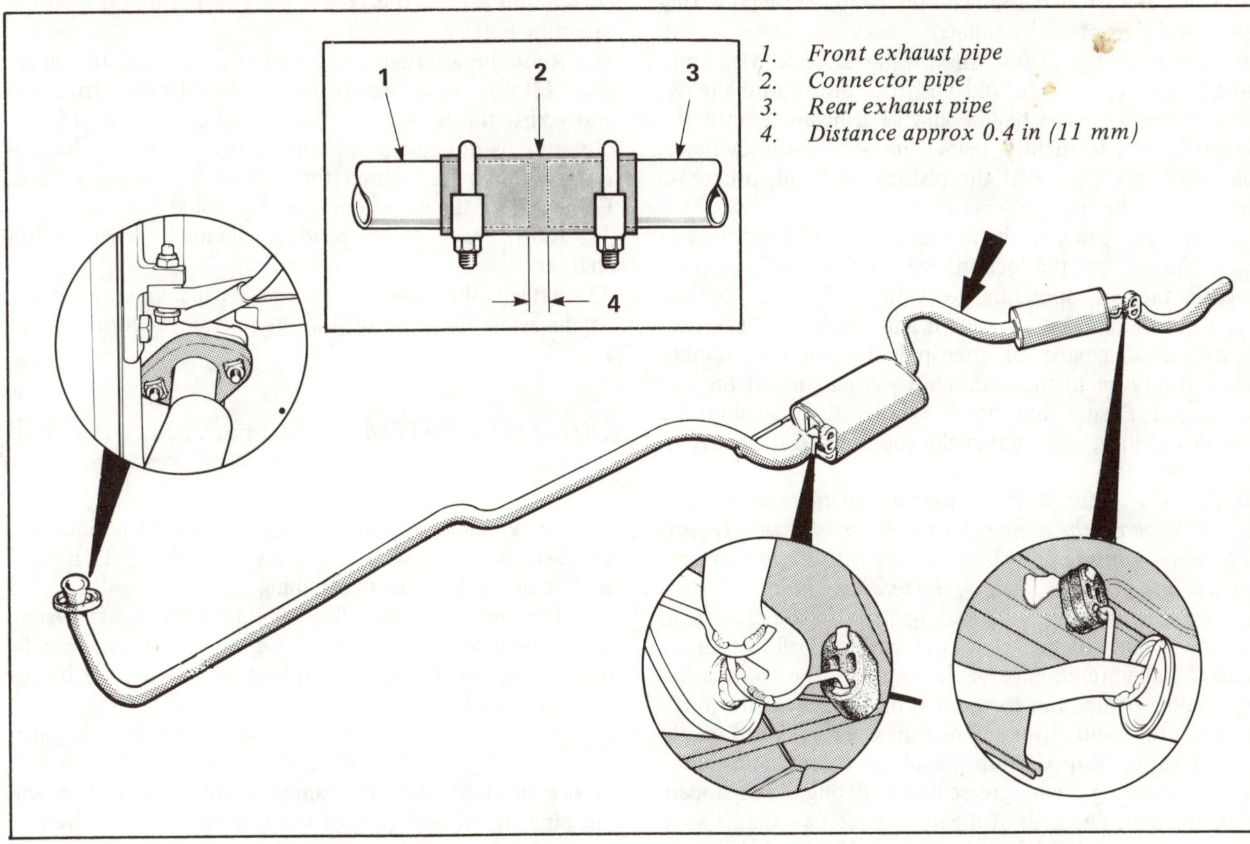

Fig. C:38 OHV exhaust system and mountings, details of connector sleeve inset

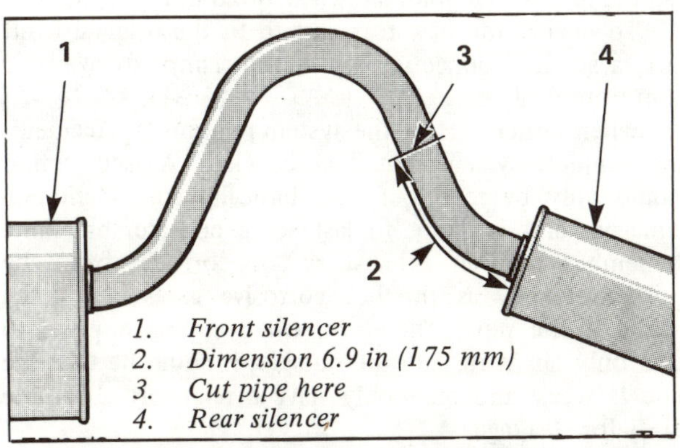

Fig. C:39 Position of hacksaw cut

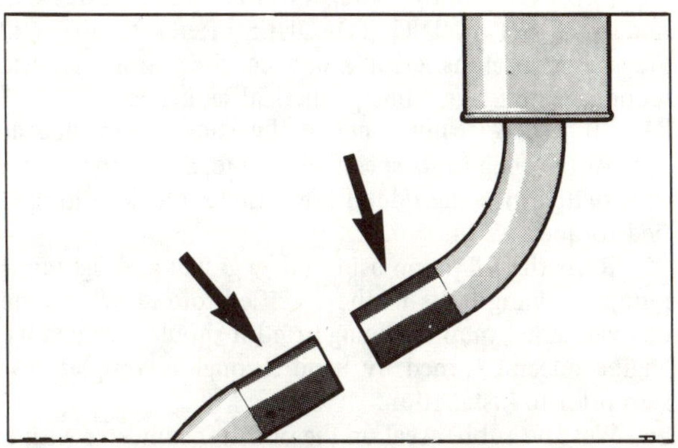

Fig. C.40 Mark cut pipe ends at positions shown

Fit the piston rings to the piston from the crown side, the oil control ring first, followed by the lower and then the upper compression rings. The oil control ring may be fitted either way round, but the compression rings must be fitted with the TOP marking facing the piston crown. Note that when fitting new rings to partly worn bores, the top compression ring should have a step in it (ridge dodger) and should be fitted with the ridge facing upwards. Position the piston rings with their gaps staggered. The oil control ring gaps should be to the rear, the lower compression ring gap at 90° to the oil ring gap and the upper compression ring gap at 180° to the oil ring gap.

Lubricate the cylinder bore liberally with clean engine oil.

18. Install the piston assembly in its correct respective bore with the arrow marking on the piston crown pointing towards the front of the engine.

19. Compress the piston rings, using a proper ring compressor tool (Fig. C:36 and push the piston into the cylinder bore with the handle end of a hammer until the piston crown it slightly below the top of the cylinder. Do NOT attempt to fit the pistons by hand, otherwise breakage of the rings may result.

20. Fit the bearing shells, dry, in the con-rod and big end caps. Ensure that the locating tag on each shell correctly engages the corresponding notch in the bearing housing.

21. Coat the crankshaft journal and bearing shells liberally with clean engine oil, then pull the con-rod assembly down firmly on to the crankshaft journal and fit the cap to the rod. Ensure that the cap is correctly located on the two dowel pins, and tighten the cap bolts to their specified torque.

NOTE: The identification numbers on the con-rod and cap must be on the same side of the assembly. Apply suitable metal jointing compound to the oil pickup pipe and press it into position. Tighten the retaining bolt.

22. Fit new rubber sump sealing strips in the grooves in the lower faces of the front cover and rear oil seal carrier with the chamfered edge of the seal into the groove. Apply sealing compound to the mating surfaces at the joints between the front cover and rear oil seal carrier and crankcase. Position a new sump gasket on the cylinder block flange, smearing it with grease before fitting to help locate it in position. The ends of the sump gasket should be located under the ends of the sealing strips (Fig. C:25).

23. Fit the sump and install the retaining bolts. Tighten the bolts evenly in three stages to the torque figures specified in TIGHTENING TORQUES. Refer to Fig. C:37 Stage 1 is in alphabetical sequence, Stage 2 in numerical sequence and Stage 3 in alphabetical sequence.

24. Fit a new sealing ring in the sump drain plug and tighten the plug to its specified torque. Screw the oil pressure switch into the side of the cylinder block to its specified torque.

25. Refit the oil pump using a new gasket and tighten the pump retaining bolts to the specified torque. When a new or overhauled pump is being fitted it should be filled with engine oil and turned by hand through a complete rotation prior to installation.

26. Wet the rubber seal on the oil filter cannister with engine oil and screw the filter on to the threaded adaptor by hand until the gasket just contacts the pump flange. Then further tighten ¾ of a turn.

NOTE: Overtightening of the oil filter may result in leakage from the sealing ring.

27. Fit the fuel pump to the cylinder block using a new gasket. Make sure that the pump operating arm lies on the camshaft eccentric not underneath it. Secure with the two retaining bolts.

28. Locate the water pump together with a new gasket on the front face of the engine and secure it in position with the three retaining bolts.

29. Turn the crankshaft to bring No. 1 piston to TDC. Position the distributor as detailed in the ENGINE ELECTRICS chapter, then fit the distributor into the engine and secure the clamp plate in the block with the retaining bolt.

30. Refit the alternator and mounting bracket to the engine. Fit the water pump pulley, then fit the drive belt and adjust the tension so that a total deflection of 0.5 in (13 mm) is possible at a point midway along the longest run of the belt. Tighten the alternator mounting bolts. Fully tighten the water pump pulley bolts.

31. Refit the cylinder head as detailed earlier in this chapter.

32. Attach the gearbox to the engine and install the engine assembly as detailed in the previous section.

EXHAUST SYSTEM....................[10]

The exhaust system should be examined periodically to check for corrosion and leaks - see PASS THE MoT section at the front of this manual.

The original production equipment exhaust system is of one-piece design, and the complete system can be removed as one piece without any unrelated parts having to be removed first.

A replacement exhaust system comes in two parts, the front pipe from the manifold through to the rear of the front silencer box, and the rear silencer box and tail pipe. If the rear part of the original system - the rear silencer and tailpipe - needs replacing but the front part is fit for further service, it can be replaced separately by cutting the pipe at the point shown arrowed in Fig. C:38.

To enable the new rear part to fit the original front part, a service connecting sleeve and clamps are available from Ford dealers.

When either part of the system requires replacement, the complete system should be checked. A used section should only be reused if it is basically sound and undamaged, and is likely to last some considerable time. Remember that an exhaust system corrodes from the inside outwards as the hot corrosive gases attack the inside of the pipe. Therefore, when a pipe appears to have only slight rust on the outside, it could be that the pipe is weak and may only have a relatively short life left before leakage.

As corrosion takes place throughout the entire length of the exhaust system, examine the pipe and silencers very carefully, giving each part a light hammer tap to check which parts are weak and which parts are sound. After examination, it may well prove that fitting a complete system will be the greater economy in the long run.

Replacement - Rear Silencer & Tailpipe

1. Raise and support the rear of the car on ramps or axle stands as detailed in the ROUTINE MAINTENANCE chapter.
2. Referring to Fig. C:39, measure a distance of 6.5 in (165 mm) from the front face of the rear silencer along the pipe. Using a hacksaw, cut the pipe squarely through at this point.
3. Detach the rear part of the exhaust system.
4. Clean the manifold pipe end with a file and use emery cloth to smooth the metal and remove any burrs for a distance of 3 in (80 mm) to allow the connector sleeve to slide over the pipe.
5. Measure back 1.9 in (50 mm) from the cut end of the manifold pipe and the new rear silencer pipe and scribe a line on the pipe. (Fig. C:40).
6. Smear grease or exhaust assembly paste over the two pipe joint ends then slide the connector onto the manifold pipe end until it is in line with the scribed line.
7. Fit the new rear exhaust section to the connector sleeve sliding it up the scribed line. (Fig. C:38 inset).
8. Fit the two 'U' clamps but leave the nuts loose at this stage.
9. Align the rear exhaust system with the front so that none of the mounting insulator rubbers are distorted or under strain. Check that the minimum distance between any part of the exhaust system and the car body or other components is 1.0 in (25 mm).
10. Tighten the connecting sleeve 'U' clamp nuts to a torque of 28 - 33 lb.ft (3,6 - 4,1 kg m).
11. Run the engine and check that the connector sleeve is free from leaks before lowering the rear of the car.

Replacement - Complete Exhaust System

1. Place the car over a pit, or raise and support the car so that all four wheels are off the ground as detailed in the ROUTINE MAINTENANCE chapter.
2. From inside the engine compartment, undo the two bolts and remove the hot air box over the top of the exhaust manifold.
3. Undo the two bolts and nuts securing the exhaust downpipe to the manifold. Disconnect the pipe from the manifold. Discard the nuts and bolts. New nuts and bolts should be fitted.
4. Unhook the exhaust system from the mounting rubber insulators and remove the complete system from under the car. Detach the mounting rubbers.
5. Clean the rear end of the manifold pipe and the front end of the rear silencer pipe with emery cloth and a file to remove any paint and burrs for a distance of 3 in (80 mm). This will allow the connector sleeve to slide over the pipe ends.
6. Measure back 1.9 in (50 mm) from the connector ends of both pipes and make a scribed line on each pipe.
7. Join the two pipe sections together with the connector sleeve so that the scribed lines are level with the connector sleeve ends. Fit the two 'U' clamps but leave the nuts loose at this stage.
8. Position the exhaust system under the car and connect it to the mounting rubber insulators.
9. Reconnect the manifold down pipe to the manifold using two new nuts and bolts, tightened loosely at this stage.
10. Align the exhaust system so that none of the mounting insulator rubbers are distorted or under strain. Check that the minimum distance between any part of the exhaust system and the car body or other component is 1.0 in (25 mm).
11. Tighten the manifold clamp nuts to a torque of 26 - 30 lb. ft (3,6 - 4,1 kg m), then tighten the connecting sleeve 'U' clamp nuts to a torque of 28 - 33 lb ft (3,6 - 4,1 kg m).
12. Run the engine and check for leaks before lowering the car to the ground (if raised on stands).

Engine – cvh

INTRODUCTION . . . . [1]

The CVH (Compound Valve angle, Hemispherical combustion chamber) engine is produced in three versions, 1117 cc, 1296 cc and 1597 cc, and has an alloy cylinder head with overhead camshaft and hydraulic tappets. This engine is known as the 'P' engine.

Although the engine block, crankshaft and pistons are of conventional Ford design, the remainder of the basic engine components differ from previous Ford practice. The oil pump is of the eccentric gear type and is fitted at the front of, and driven by the crankshaft. The only way that the oil pump can be checked for wear is by testing the pump pressure and by examination of the moving parts. As no spare parts are supplied, a new pump assembly should be fitted if the old one is worn or suspect.

The camshaft is driven by the crankshaft using a toothed belt which also drives the water pump at the front of the engine. Water pump replacement is therefore slightly more difficult than normal as the timing cover and belt must be removed before access can be gained to the pump.

The cylinder head, as mentioned previously, is made from alloy and it is important that it is only removed and refitted to the engine block at room temperature (approx. 68 deg F, 20 deg C). Two guide bushes are pressed into the engine block to ensure correct alignment of the cylinder head and gasket, thus avoiding the need for guide studs screwed into the head bolt holes in the block. Both the camshaft and valve tappets run directly in bearing bores in the cylinder head without replaceable bearing bushes. If either the camshaft or tappet bores are worn or damaged, a new cylinder head should be fitted.

It is very important that the cylinder head bolts are never reused once unscrewed, similarly they must not be retorqued after the initial tightening sequence. The bolts are designed to stretch on initial tightening and may break if retorqued.

Although the engine compartment looks crowded like a plumber's nightmare with pipes and wires sprouting from the engine, there is a reasonable amount of room for the DIY mechanic to carry out the normal service and overhaul or replacement jobs with the engine in the car.

With the timing belt cover removed, there is adequate access to replace the camshaft timing belt, crankshaft front oil seal and the water pump. At the rear (gearbox) end of the head, the camshaft can be removed with the head on the engine after detaching the camshaft belt and sprocket, valve tappets, fuel pump and distributor.

Under the engine, the sump pan can be removed with the engine in-situ for crankshaft big-end bearing access and to remove the pistons.

All nuts and bolts on the CVH engine are metric and a comprehensive socket and spanner set will be needed for any overhaul work. In addition, a small number of special tools will be needed for more ambitious work, such as a universal type valve spring compressor, piston ring clamp and a short reach, half-cup type puller for the crankshaft pulley. The latter is a Ford tool No 21-098 although all these tools and the equivalent Ford one should be available at a good car accessory shop, or even hired for the occasion - see Yellow Pages under 'Hire Services Tools and Equipment'.

CRANKSHAFT FRONT OIL SEAL . . . . [2]

Replacement

1. Replacing the oil seal is quite easy and there is a reasonable amount of room to do the job. You will need a short-reach, half-cup type puller tool similar to the Ford one shown in the illustrations to draw off the crankshaft timing belt pulley.
2. Disconnect the battery earth strap and place the gear lever in 1st gear with the handbrake fully applied.
3. Undo the bolt securing the crankshaft belt pulley.
4. Slacken the alternator adjuster arm securing bolt and slide the alternator towards the engine. Detach the alternator belt from the crankshaft pulley and remove the belt.
5. Carefully lever the crankshaft 'V' belt pulley off the end of the crankshaft, using a suitable puller if necessary.
6. Undo the four bolts and detach the timing belt cover (Fig. D:1).
7. Place the gear lever in 4th gear and release the handbrake. Push or pull the car to turn the engine until the camshaft belt pulley mark lines up with the TDC mark on the cylinder head (Fig. D:2). Place the gear lever in the

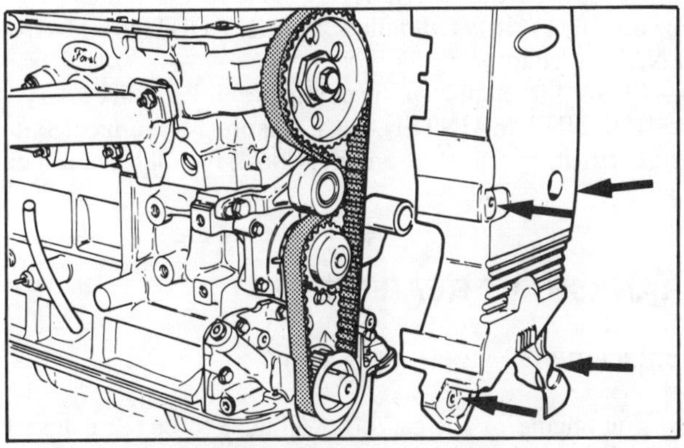
Fig. D:1 Timing belt cover attachment bolts

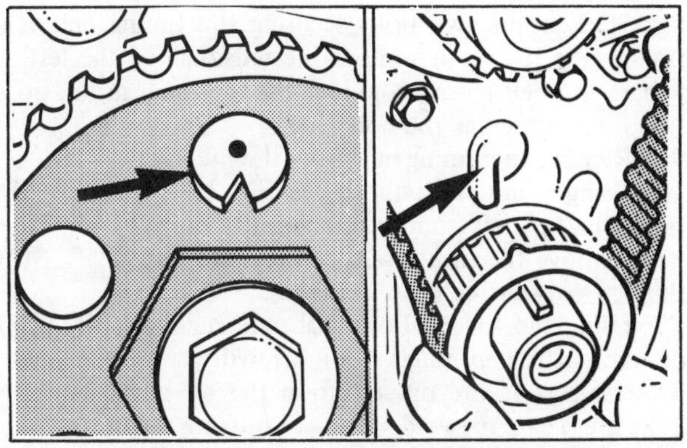
Fig. D:2 Camshaft & crankshaft pulley TDC marks

Fig. D:3 Timing belt tensioner and bolts arrowed

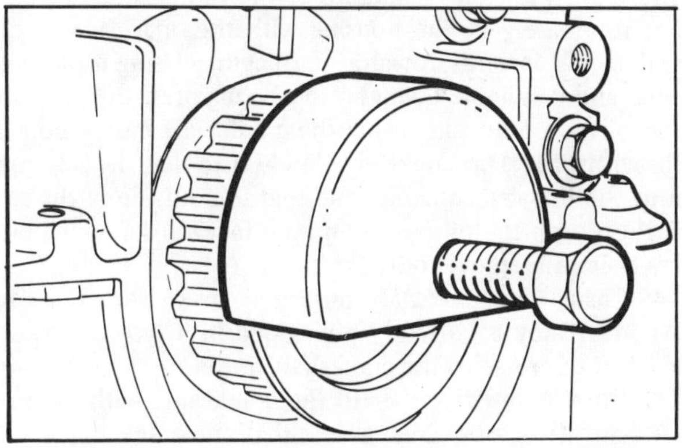
Fig. D:4 Removing timing belt pulley with puller

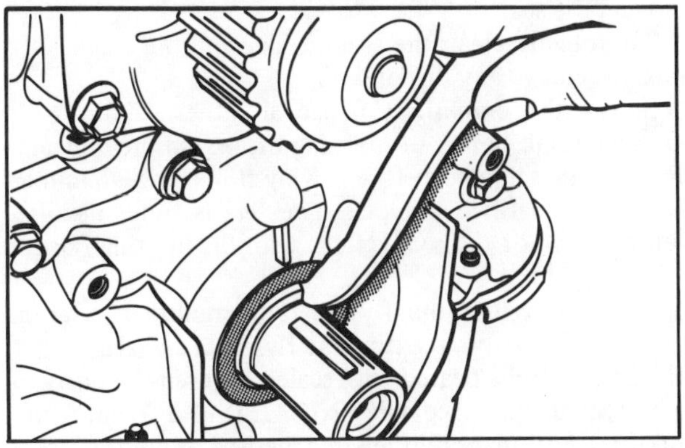
Fig. D:5 Removing crankshaft front oil seal

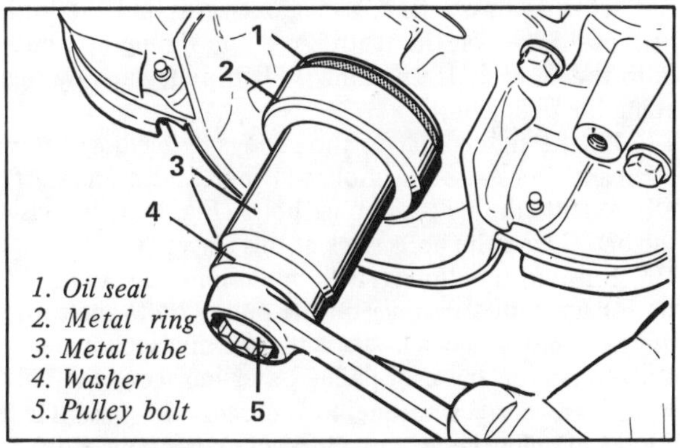

Fig. D:6 Method of fitting new front oil seal

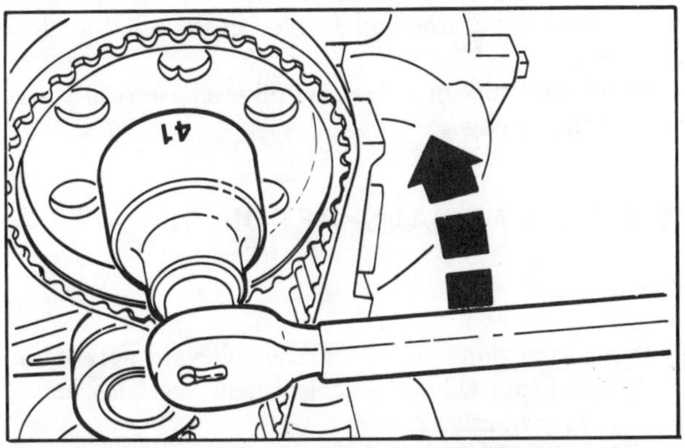

Fig. D:7 Tensioning timing belt with 41 mm socket

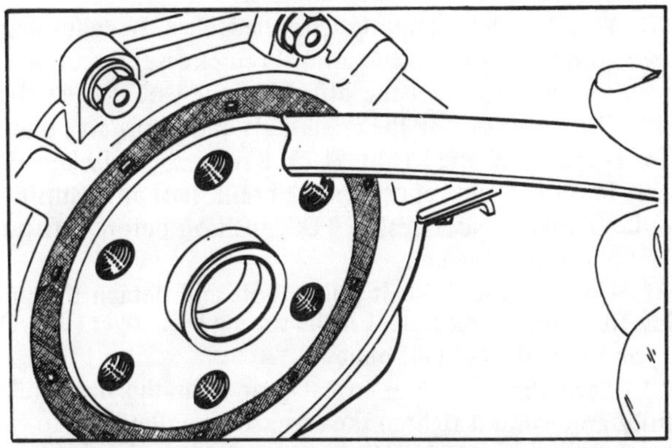
Fig. D:8 Removing crankshaft rear oil seal with tool

neutral position.

8. Slacken the two bolts securing the timing belt tensioner (Fig. D:3), then move the tensioner to the left to release the belt tension and lock the tensioner in this position by retightening the bolts.

9. Remove the timing belt from the pulleys.

10. Using a suitable puller - see Fig. D:4, withdraw the timing belt pulley from the crankshaft.

11. Remove the thrust washer from the crankshaft. Note that the curved side faces outwards.

12. Using the special Ford seal remover tool (Fig. D:5), or alternatively a longish thin screwdriver will do, carefully lever out the oil seal from the oil pump housing. Take great care to avoid scoring or burring the metal when doing this. Clean the housing and the crankshaft thoroughly afterwards.

13. Lightly oil the sealing lip of the new seal, and position it squarely in the housing with the open side of the seal facing inwards towards the engine. Using a piece of tube and a suitable diameter metal ring press the seal into the oil pump housing using the crankshaft pulley bolt as shown in Fig. D:6. Make absolutely sure that the seal goes into the housing squarely, and that the outside of the seal is flush with the oil pump housing face. Unscrew the pulley bolt and remove tools.

14. Clean the crankshaft nose and make sure that the Woodruff key is correctly positioned in its groove. Apply a smear of grease to the crankshaft nose.

15. Fit the thrust washer to the crankshaft with its curved side facing outwards. Check that the crank keyway is aligned with the TDC mark (Fig. D:2).

16. Loosely fit the timing belt pulley to the crankshaft end with its thrust face, facing outwards and the timing belt positioned on the teeth. Press the timing belt pulley onto the crankshaft using the 'V' belt pulley and by tightening the pulley bolt.

17. Fit the timing belt to the pulleys by feeding it round in an anti-clockwise direction, starting at the crankshaft.

18. Release the belt tensioner bolts (Fig. D:3) to tension the belt. Leave the bolts slack at this stage.

19. Tension the timing belt by holding the crankshaft stationary with a spanner on the pulley bolt, and turning the camshaft anti-clockwsie with a 41 mm socket and torque wrench on the cam pulley outer hexagon (Fig. D:7). Apply the correct torque to the camshaft - see TECHNICAL DATA in an anti-clockwise direction. Hold the wrench at the correct torque and tighten the tensioner bolts, right-hand bolt tightened first.

20. With the belt correctly tensioned, turn the crankshaft two complete turns in the normal (clockwise) direction of rotation, viewed from the offside of the car. Check that the TDC positions of the crankshaft and camshaft pulleys are correctly aligned (Fig. D:2). If not, then slacken the belt tensioner and reposition the crankshaft and camshaft pulleys to the No 1 piston TDC position before refitting and tensioning the belt.

21. Unscrew the 'V' belt pulley bolt and detach the pulley from the crankshaft. Fit the timing belt cover (Fig. D: 1) and refit the 'V' belt pulley.

22. Place the gear lever in 1st gear with the handbrake fully applied and tighten the crankshaft pulley bolt to the correct torque - TECHNICAL DATA. Place the gear lever in neutral position.

23. Reconnect the alternator 'V' belt to the pulleys and tension the belt as detailed in the ROUTINE MAINTENANCE chapter.

24. Check the engine oil level and top it up if necessary - see ROUTINE MAINTENANCE chapter. Reconnect battery earth strap and start engine. Check for oil leaks under engine.

CRANKSHAFT REAR OIL SEAL . . . . [3]

Replacement

1. The engine or the gearbox will have to be removed to gain access to the clutch/flywheel and the oil seal behind. The engine removal procedure is detailed later in this chapter. Gearbox removal is detailed in the CLUTCH & GEARBOX chapter, depending on which unit is to be removed. Note that six new flywheel to crankshaft bolts will be needed on reassembly.

2. Remove the clutch pressure plate and driven plate (disc) from the flywheel as detailed in the CLUTCH & GEARBOX chapter.

3. Undo the flywheel bolts and detach the flywheel from the crankshaft. In order to prevent the crankshaft from turning when undoing the bolts, the flywheel should be held stationary using a stout screwdriver engaged between the flywheel ring gear teeth and the engine block.

4. Using the special Ford seal remover tool (Fig. D:8) or, alternatively a thin screwdriver, carefully lever the oil seal out of the housing. Take great care not to score or damage the seal housing or the crankshaft surface when removing the seal.

5. Thoroughly clean the seal housing and the crankshaft flange to remove any dirt and oil.

6. Smear the crankshaft flange and the seal lip with clean engine oil before positioning the seal in the housing. Make sure it is square before driving it in using a hammer and smooth hardwood block. The seal is fully installed when the face of the seal is flush with the housing flange, all the way round.

7. Make sure that the flywheel to crankshaft mating faces are clean before refitting the flywheel and securing it with six new bolts coated with sealer. Tighten the bolts to the correct torque - see TECHNICAL DATA, keeping the crankshaft stationary by the same method as used to undo it.

8. Refit the clutch assembly to the flywheel and refit the engine or the gearbox as detailed in the relevant 'Removal' sections.

9. Check and top-up the engine oil level afterwards before starting the engine.

TIMING BELT & CAMSHAFT OIL SEAL . . . [4]

Replacement

1. Detach the timing belt from the pulleys as detailed in 'Crankshaft Front Oil Seal - Replacement' section, operations 1 to 9 previously.

2. Hold the camshaft pulley stationary using a screwdriver through the pulley holes and resting on the cylinder

Fig. D:9 Method of undoing cam belt pulley bolt

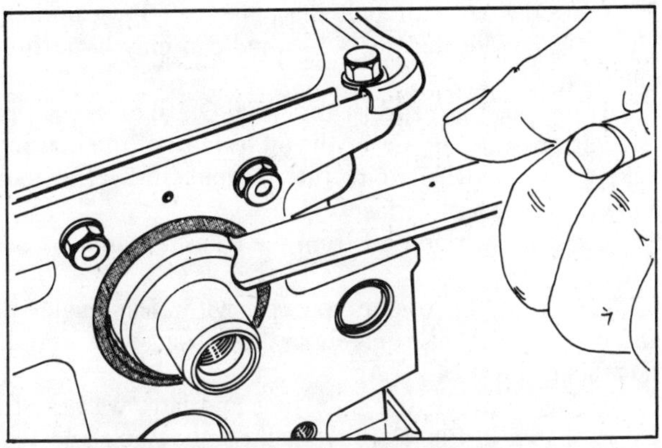

Fig. D:10 Removing camshaft oil seal with tool

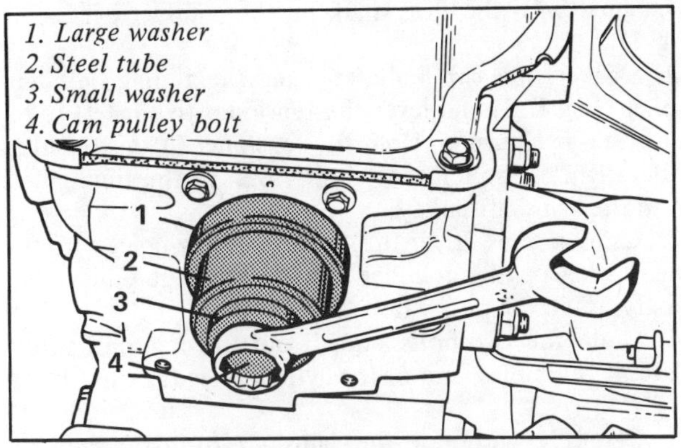

Fig. D:11 Method of fitting camshaft oil seal

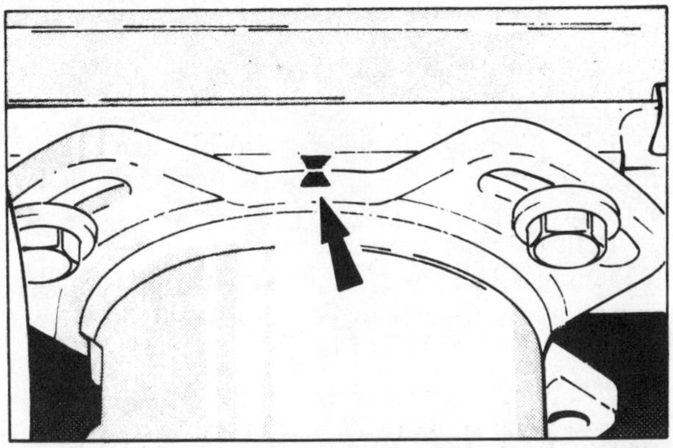

Fig. D:12 Location mark on distributor and head

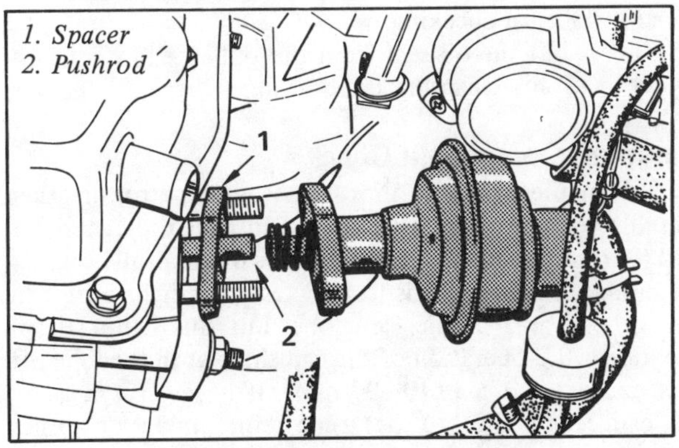

Fig. D:13 Removing fuel pump from cylinder head

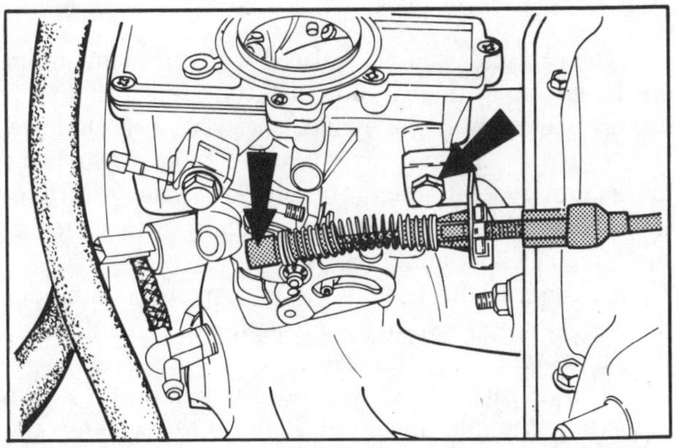

Fig. D:14 Throttle cable connection and bracket

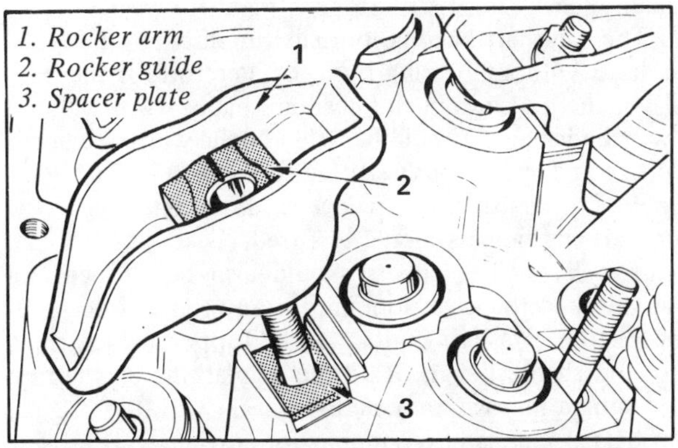

Fig. D:15 Valve rocker arm and components

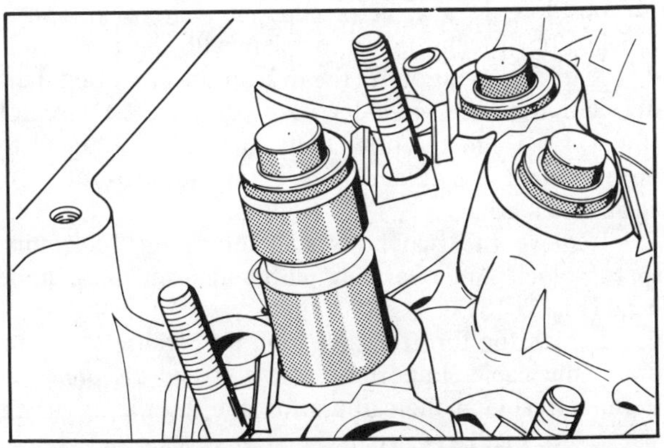

Fig. D:16 Withdraw hydraulic tappets from bores

head (Fig. D:9), and undo the cam pulley bolt.
3. Carefully lever or pull the pulley off the camshaft. The pulley is located by a key and can only be refitted one way.
4. Using the Ford special tool (Fig. D:10) or a long thin screwdriver, carefully lever the oil seal out of the housing. Take great care when doing this to avoid scoring the housing or the camshaft surface.
5. Clean all dirt and oil from the camshaft and the seal housing.
6. Smear the lip of the new seal with clean engine oil and position the seal squarely in the housing with its open side facing inwards.
7. Using a suitable diameter tube and large washers, fitted as shown in Fig. D:11, refit the pulley bolt and tighten it to press the seal fully into the housing until the outer face is flush with the cylinder head face. Make absolutely sure that the seal is fitted square and not misaligned in the housing.
8. Smear the camshaft nose with oil and check that the Woodruff key is correctly located in its groove.
9. Refit the camshaft pulley and tighten the bolt to the correct torque using the screwdriver to hold the pulley as on removal.
10. Refit the timing belt and tension it correctly, then refit the timing belt cover and crankshaft 'V' belt pulley as detailed under 'Crankshaft Front Oil Seal - Replacement' section - operations 17 to 24.

CAMSHAFT REPLACEMENT (IN-SITU). . . . [5]

Removal

The camshaft can be replaced with the cylinder head on the engine in the car, although only some of the operations will apply if the cylinder head has been removed. It is essential that no dirt is allowed to enter the engine lubrication system, therefore, all parts and your hands should be clean when dismantling and refitting the camshaft and ancillary components.
1. Disconnect the battery earth strap, then disconnect the crankcase ventilation hose from the inlet manifold and the cam cover.
2. Remove the air cleaner and disconnect the vacuum hose from the inlet manifold. Remove the screen washer reservoir.
3. If the spark plug leads are not numbered, mark them for position, 1 to 4, before disconnecting them and removing the distributor cap complete with leads.
4. Mark the position of the distributor mounting flange in relation to the cylinder head for correct reassembly (Fig. D:12). Undo the three bolts and withdraw the distributor from the cylinder head. Place the distributor in a clean safe place.
5. Remove the two bolts and detach the fuel pump, spacer block and operating push rod from the cylinder head (Fig. D:13).
6. Detach the throttle cable from the carburettor by undoing the cable support bracket bolt and disconnecting the inner cable at the carburettor lever by sliding off the sleeve - see Fig. D:14.
7. Disconnect the crankcase ventilation hose from the cylinder block and then remove the cam cover from the cylinder head (secured by nine bolts) to expose the valve rocker arms and hydraulic tappets.
8. Unscrew the nut securing each rocker arm to the stud and remove the arms, complete with guide block - see Fig. D:15. The arms with their respective guide blocks should be placed on a clean surface in removal order. Discard the nuts. New nuts should be fitted on reassembly.
9. Lift out the hydraulic tappets from the cylinder head bores (Fig. D:16) and place each one in its removal order alongside the rockers. If necessary, mark the removal order for all components from 1 to 8 for refitting.
10. Undo the four bolts and detach the timing belt cover from the engine block (Fig. D:1).
11. Using a suitable spanner on the crankshaft pulley turn the engine in the normal direction of rotation (clockwise as viewed from the front) until the camshaft pulley mark is aligned with the TDC mark on the cylinder head - see Fig. D:2.
12. Slacken the two bolts securing the timing belt tensioner (Fig. D:3) and lever the tensioner to the left to release the belt tension. Hold the tensioner in this position and retighten the tensioner bolts. Detach the timing belt from the camshaft pulley.
13. Using a screwdriver through the pulley holes with the end resting on the cylinder head to hold the pulley stationary, undo the pulley bolt (Fig. D:9).
14. Undo the two bolts and lift out the camshaft thrust plate at the pulley end of the cylinder head. (A in Fig. D: 17).
15. Carefully withdraw the camshaft from the cylinder head (B, Fig. D:17) making sure that the cam lobes do not score the cam bearing housings. Place the camshaft on a clean surface for checking.
16. Lever out the oil seal from the pulley end of the cylinder head, and clean the seal housing.

Camshaft Component Check

Clean the camshaft thoroughly with petrol or other suitable solvent, then wipe it dry. Inspect the cam lobes and journals for scoring or other signs of abnormal wear.

Lobe wear characteristics may result in slight pitting in the general area of the cam lobe, but this is not detrimental to the operation of the camshaft until the lobe lift loss exceeds 0.005 in (0,127 mm). If necessary, measure the cam length (heel to toe) using a micrometer or vernier caliper.

Check the distributor drive cog for wear or damage similarly, check the distributor shaft slot for wear.

The camshaft bearing journals run direct in the cylinder head without bearing bushes. Therefore, if the journals or the housings in the head are badly worn, either a new cylinder head should be fitted or the worn one bored oversize and special bushes fitted to the cam journal housings by specialist auto engineers. In either case, a new camshaft and tappets should be fitted.

The hydraulic tappets should also be cleaned and then checked for wear, looking for any signs of deep pitting on the camshaft contact face. Tappets which show only the slightest signs of pitting or marking on the face after a high mileage, may be used again with the original camshaft if this is also suitable for further service. New tappets must ALWAYS be fitted with a new camshaft.

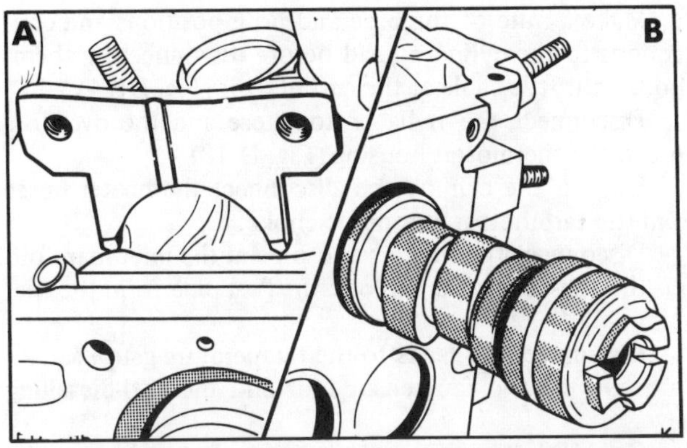

Fig. D:17 Thrust plate (A) and camshaft removal (B)

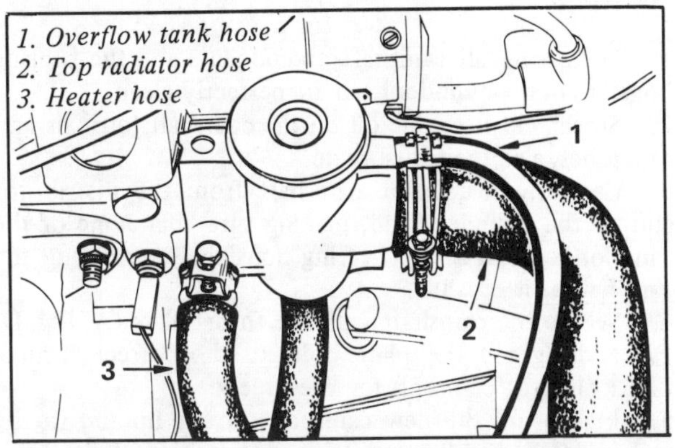

Fig. D:18 Hose connections at thermostat housing

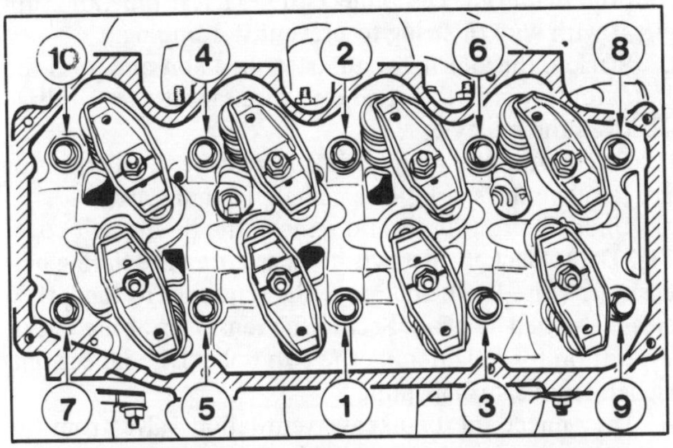

Fig. D:19 Cylinder head bolt tightening sequence

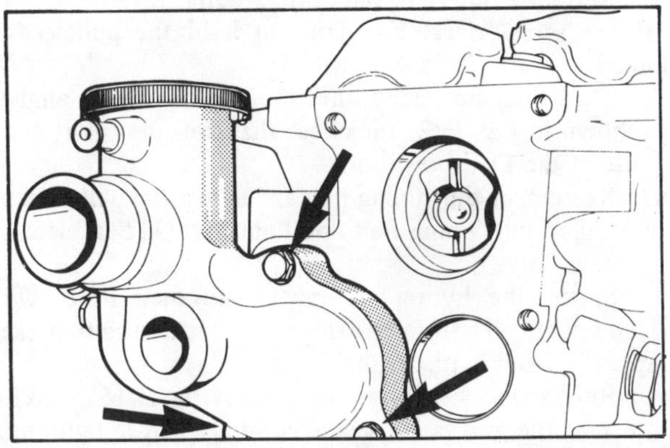

Fig. D:20 Thermostat housing bolts arrowed

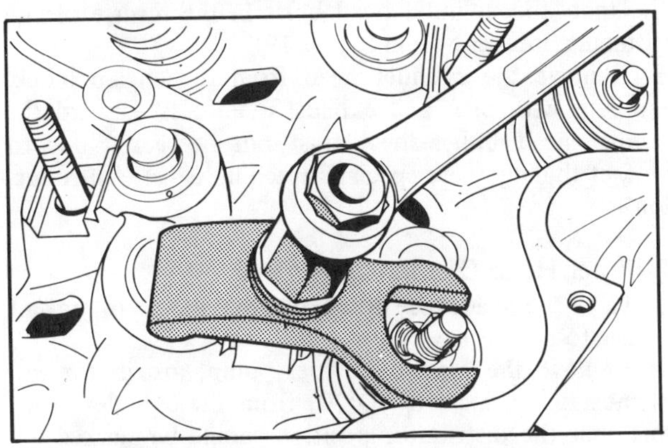

Fig. D:21 Using Ford valve compressor tool

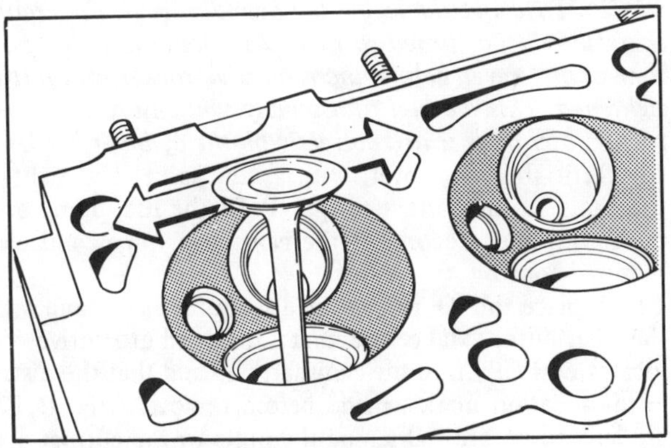

Fig. D:22 Measuring valve guide for wear

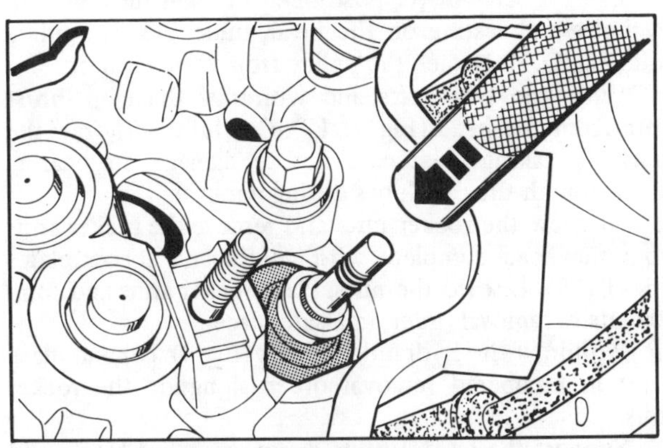

Fig. D:23 Installing valve stem oil seals

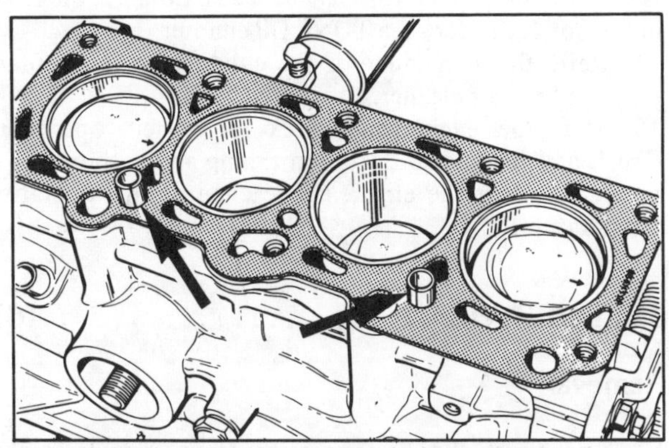

Fig. D:24 Head gasket showing guide bushes arrowed

Refitting

1. Make sure all camshaft components and the bearing housings in the cylinder head are perfectly clean.
2. Smear clean engine oil on the camshaft journals and cam lobes, also the thrust plate.
3. Carefully install the camshaft from the distributor end of the cylinder head, making sure that none of the cam lobes contact the bearing housings by rotating the camshaft as it goes in.
4. Secure the camshaft with the thrust plate (A, Fig. D: 17) and tighten the plate bolts to the correct torque. Check that the camshaft revolves freely.
5. Lightly oil the new camshaft oil seal lip and fit the seal as detailed in 'Timing Belt and Camshaft Oil Seal' section (4) previously.
6. Refit the pulley to the camshaft and tighten the bolt, using a screwdriver (Fig. D:9) to hold the pulley stationary.
7. Turn the cam pulley until the TDC marks are aligned as shown in Fig. D:2, and check that the crankshaft mark is also at the TDC position.
8. Reconnect the timing belt and adjust the belt tension as detailed in 'Timing Belt and Camshaft Oil Seal' section (4) previously.
9. Smear the hydraulic tappets and their bores with clean engine oil before refitting them. Make sure that each tappet is fitted in its original bore.
10. Refit the rockers arms and guides (Fig. D:15), making sure that the guides are seated correctly before tightening the nuts to the correct torque. Note that new nuts must be fitted.

NOTE: To avoid damaging the cylinder head, the crankshaft should be turned so that each tappet is at the bottom of its travel before the respective rocker arm nut is tightened. This method of assembly will ensure that each rocker fulcrum is seated correctly on the cylinder head.

11. Refit the cam cover, with a new gasket and tighten the bolts to the correct torque. Refit the fuel pump and push rod, then reconnect the cam cover pipes and carburettor controls.
12. Replace the 'O' ring in the distributor casing and refit the distributor. Make sure that the distributor drive cog locates correctly into the camshaft end and that the distributor location marks, made before removal (Fig. D:12) are aligned before tightening the bolts to the correct torque.
13. Refit the distributor cap and leads. Connect the leads in the correct order - see TUNE-UP chapter.
14. Refit the remainder of the cylinder head ancillary pipes and the air cleaner.
15. Start the engine and check for correct operation. Check and readjust the ignition timing and carburettor if necessary when the engine has reached normal operating temperature.

CYLINDER HEAD .... [6]

Removal

1. Disconnect the battery earth strap and drain the coolant into a suitable container by disconnecting the bottom hose at the radiator.
2. Make a note of the pipe and hose positions and connections to the cylinder head before disconnecting them, also the throttle cable at the carburettor (Fig. D:14).
3. Disconnect the radiator top hose and the overflow hose at the thermostat housing (Fig. D:18).
4. Slacken the clamps and disconnect the heater hoses from the carburettor automatic choke.
5. Disconnect the brake servo pipe at the inlet manifold and then disconnect and block the fuel pipe from the fuel pump.
6. Disconnect the leads to the temperature gauge sender unit, the electric fan sensor unit and the anti-dieselling valve.
7. On 1100cc variants only, remove the shield over the exhaust manifold. On all models, unbolt the exhaust pipe from the manifold. Disconnect the exhaust pipe and support it with wire or string to prevent it dropping.
8. Slacken the alternator adjuster slide bolt and push the alternator towards the engine to disconnect the 'V' drive belt from the pulleys.
9. Disconnect the plug leads and remove the distributor cap. Remove the spark plugs using a 16 mm socket.
10. Undo the four bolts and remove the timing belt cover (Fig. D:1). Slacken the two bolts securing the timing belt tensioner (Fig. D:3), and move the tensioner to the left to release the belt tension. Secure the tensioner in the released position by retightening the two bolts. Disconnect timing belt from camshaft pulley.
11. Disconnect the crankcase ventilation valve from the cam cover, then undo the nine bolts and detach the cam cover from the cylinder head.
12. Undo the cylinder head bolts in the reverse of the tightening order shown in Fig. D:19.
13. Remove the cylinder head from the engine block, complete with inlet and exhaust manifolds. Discard the cylinder head bolts, they must not be reused due to the fact that they are special stretch bolts - see 'Introduction'.

Cylinder Head Dismantling

1. Place the head, face down on wood blocks to prevent damage to the valves.
2. Remove the manifolds, fuel pump, distributor and thermostat housing (Fig. D:20) from the cylinder head. Note that the distributor position should be marked before removal (Fig. D:12).
3. Using a screwdriver positioned through the cam pulley holes and resting on the head, undo the cam pulley bolt (Fig. D:9). Detach the pulley from the camshaft.
4. Undo the two bolts and withdraw the cam thrust plate from the head (Fig. D:17). Carefully withdraw the camshaft, taking great care not to damage the bearing housings with the cam lobes during removal.
5. Unscrew the rocker nuts and remove the rocker arms from the head complete with guides and spacer plates (Fig. D:15). Discard the nuts. Place rocker arms and components in removal order, in a safe place.
6. Withdraw the hydraulic tappets (Fig. D:16) and place them in numbered removal order alongside the rocker arms.
7. Support the head, upside down, on wood blocks and remove all carbon deposits from the hemispherical com-

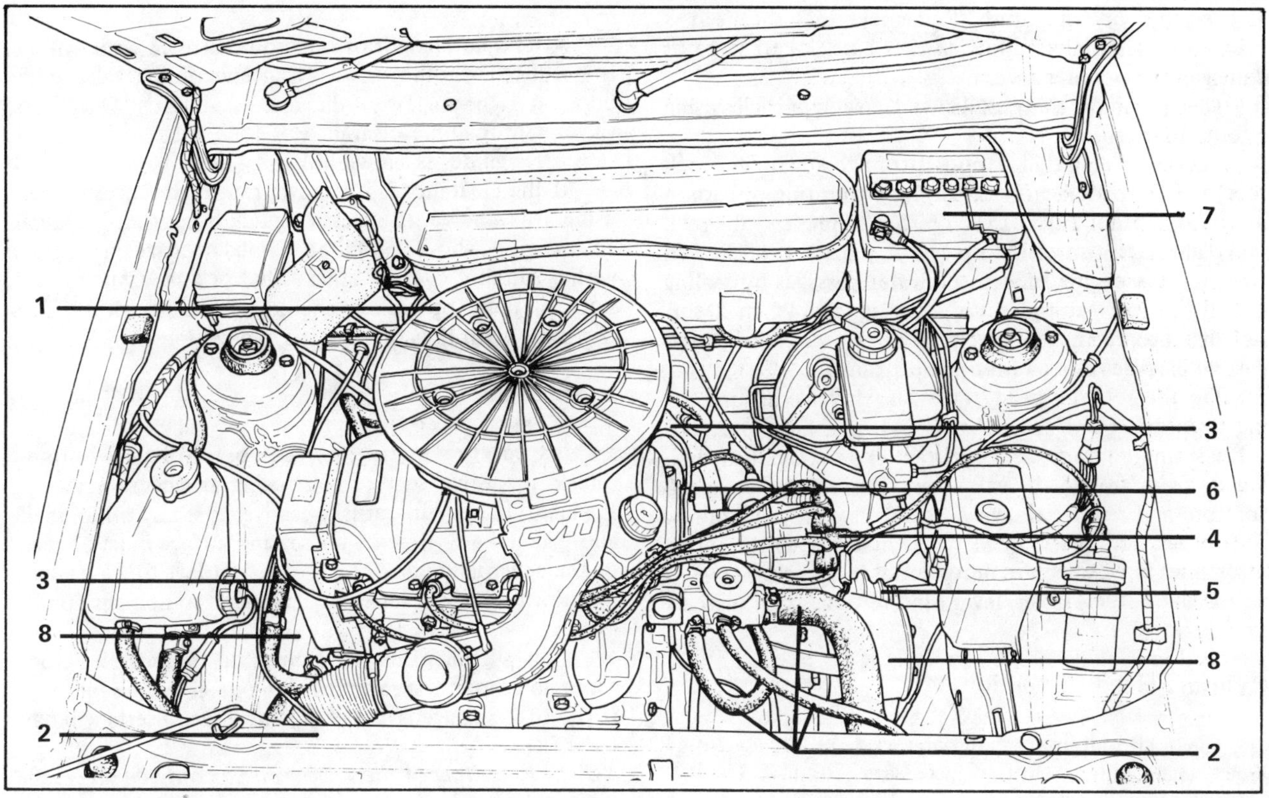

1. *Air cleaner assembly* 2. *Radiator hoses* 3. *Heater hoses* 4. *Coil lead & LT lead* 5. *Driveshafts* 6. *Gearshift rod* 7. *Battery* 8. *Engine mountings*

Fig. D:25 Components to be removed or disconnected when removing the engine unit (CVH type shown)

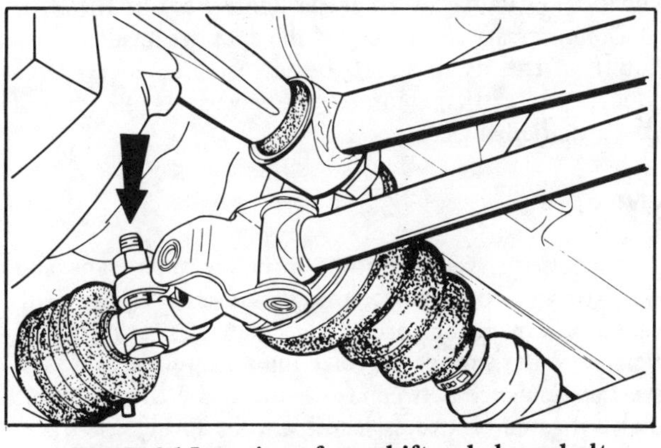

Fig. D:26 Location of gearshift rod clamp bolt

Fig. D:27 Engine stabiliser bar connection

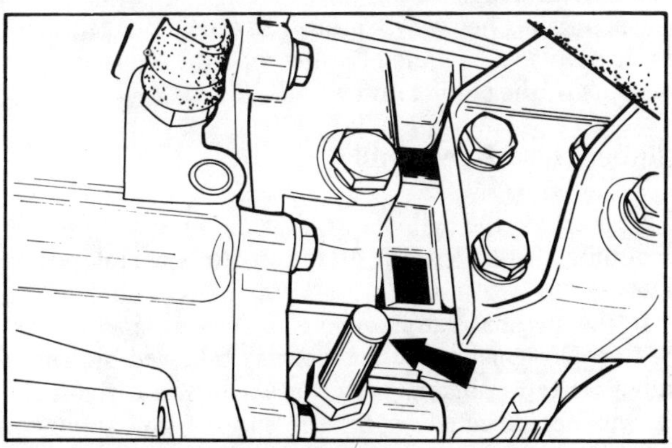

Fig. D:28 Selector shaft locking cap nut

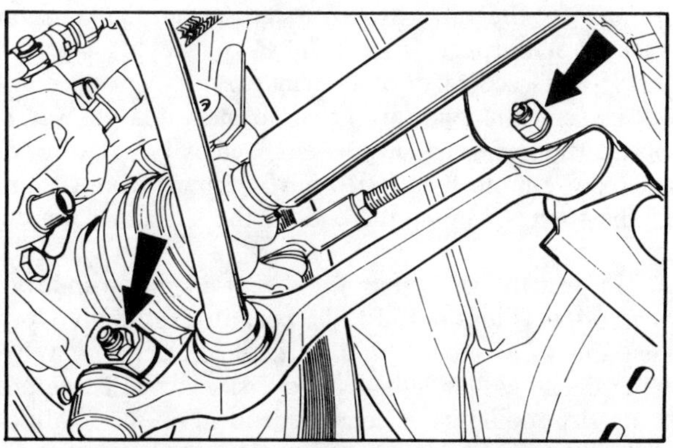

Fig. D:29 Suspension arm and pivot securing nuts

bustion chambers, valve heads and valve ports with a suitable scraper or wire brush. Remember that the head is made from aluminium alloy, so take care not to score or damage the machined face.

8. Clean or blow away all dirt and carbon particles using a foot pump air line.

9. Similarly, clean all deposits from the cylinder block face and the piston crowns, leaving a thin ring of carbon around the outer edge of each piston to preserve the seal. Make absolutely sure that no carbon particles are allowed to enter the engine oilways or water passages by sealing off all the holes with pieces of masking tape. When all carbon has been removed, clean or blow away all dirt and carbon particles using a foot pump air line.

10. On the cylinder head, remove each valve in turn, using a special Ford tool (Fig. D:21) or a large universal valve spring compressor, to extract the two split tapered collets from around the valve stem. Release the compressor tool and remove the valve spring assembly. Withdraw the valves and place them, and their respective spring assemblies in numbered removal order in a safe place.

11. Using a screwdriver, lever off the valve stem oil seals.

Cylinder Head Inspection

Clean all old gasket material, etc., from the machined faces, using emery cloth if necessary. Remove all dirt, grease, etc., with solvent.

Blow out all oil and water passages, particularly the oil transfer passage with compressed air. Probe all bolt holes to ensure they are clear. If necessary, the holes can be cleaned out with a suitable size of tap.

Inspect the head for cracks and all machined surfaces for burrs, scratches, or nicks. Remove any burrs with an oil stone.

Check the cylinder head gasket surface for flatness using a straight edge. If it is found to be distorted, the surface must be planed or ground true, or the head exchanged.

Check the sparking plug holes for stripped or damaged threads. If present, it may be possible to rectify such damage with a suitable size of tap.

Check the core plugs for leakage and replace as necessary.

Valve Inspection

Inspect the valve head for signs of burning, erosion, cracking or warpage. Inspect the valve face and edges for pits, grooves, scores, or other damage.

Inspect the end of the valve stem for grooves or scores. Minor damage may be removed by grinding but do not remove more than 0.010 in (0,25 mm) from the end of the stem. Chamfer the end of the stem as necessary afterwards.

Inspect the valve stem for a bent condition and for a wear ridge. If it is found to be bent or excessively worn, then the valve should be discarded. Replacement valves are available with standard and oversize stems. The appropriate size should be selected dependent on the condition of the valve guide.

Discard valves which are severely damaged.

Valve Guides

After some length of operation an oval wear pattern is produced on the valve spring end of the valve guide bore, at right-angles to the crankshaft axis. Check the valve stem-to-guide clearance of each valve in turn in its respective guide, as shown in Fig. D:22.

If the clearance is excessive, repeat the measurement using a new valve. If excessive guide bore wear is indicated, the guide should either be reamed oversize, and a valve with a suitably oversize stem fitted or a new guide fitted. Both of these operations should be left to local auto engineers with the correct tools and equipment.

Valve Seats

Inspect the seating surface for pitting or burning. Check the valve seat run-out using an accurate gauge, if available. If the run-out is excessive, or if any other of the above faults are present, the seating surface must be recut to the specified angle. This job is best left to a local auto machinist who will have the necessary equipment for the job.

Only sufficient metal to clean up the pits and grooves, or to correct the valve seat run-out should be removed. If the seat damage is severe then new seat inserts will have to be fitted.

NOTE: Recutting of the valve seats can only be successful if the valve guide bore is not worn.

Where inserts have not been installed previously, it will be necessary to machine an appropriate recess in the cylinder head. If for any reason an existing insert has become loose or damaged, an oversize insert can be fitted.

The necessary machining of the cylinder head and installation of the inserts should be entrusted to local auto engineers. After fitting inserts, the valve seats must be cut to the specified angle.

Valve Springs

Inspect the valve springs, spring retainer caps and collets for wear or damage, and discard any visually damaged parts. Check the springs for squareness, using a steel square on a flat surface. Revolve the spring slowly and observe the clearance between the top coil and the square. If the out of square is excessive, discard the spring.

Replace the springs as a matter of course if the car has completed a high mileage.

When fitting new valve springs it is recommended that they be replaced in sets. Ensure that the replacement springs are of the correct rating.

Cylinder Head Reassembly

Reassemble the cylinder head in the reverse order of dismantling, with special attention to the following points:

a) If the original valves are being reused, lap in each valve to its original seat using 'coarse', followed by 'fine' grinding paste to obtain a perfect gas-tight seal at the seat. This will be indicated by the presence of a continuous matt grey ring around the valve face and the seat. Note that it may take more than five minutes of continuous grinding to obtain the correct finish. The seal can be checked by cleaning all paste from the valve and seat, then refitting the valves and spark plugs, and filling the combustion chambers with paraffin. If the seal is good, there should be no detectable leakage of paraffin through the ports for at least 20 minutes.

b) When valve grinding is complete, remove all the valves (keeping them in their removal order at all times) and thoroughly wash the cylinder head and valves with paraffin to blow all components dry with compressed air from the foot pump. Blow through all oilways and water passages in the cylinder head several times in both directions. This is MOST important.

c) The valve stem oil seals must always be replaced, and great care must be used when fitting the new seals. Cover the valve stem collet grooves with adhesive foil and smear engine oil over the foil and valve stem before sliding the seals into position over the stem. This will avoid damaging the seal lip. Use a tubular drift to press the seals into position (Fig. D:23).

d) Refit each valve spring assembly and make sure that the split tapered collets engage correctly with the valve stem (Fig. D:21) as the valve spring compressor is released.

e) Oil the camshaft, the bearing housings in the head, and the thrust plate, before inserting the camshaft into the head. Take care not to damage the bearings with the cam lobes when refitting. Lock the camshaft in position with the thrust plate.

f) Fit a new oil seal to the front of the camshaft as detailed in 'Timing Belt & Camshaft Oil Seal', section (4) previously.

g) Oil the hydraulic valve tappets and their bores in the cylinder head before refitting the tappets in their original positions.

h) Refit the rockers arms and guides, making sure that the guides are correctly seated before tightening the nuts to the correct torque. Note that new nuts must be fitted.

NOTE: The camshaft should be turned so that each tappet is at the bottom of its travel before the respective rocker arm nut is tightened. This method of assembly will ensure that each rocker fulcrum is correctly seated on the cylinder head.

i) Refit the cam belt pulley and tighten the bolt by the method used to undo it, refer to Fig. D:9. Position the camshaft pulley at the TDC mark (Fig. D:2).

j) Refit the manifolds, fuel pump and distributor (Fig. D:12) to the cylinder head.

k) Make sure that the mating faces of both cylinder head and engine block are perfectly clean and that the head guide bushes are correctly positioned (Fig. D:24). Fit the new head gasket, ensuring that it is the correct type and has a part No as follows:

1100 cc. 81SM-6051-N2D/81SM-6051-N3D
1300/1600 cc 81SM-6051-R2D/81SM-6051-R3D

l) NOTE, before fitting the head, that the crankshaft should be turned anti-clockwise so that No. 1 piston is about 0.8 in (20 mm) before the TDC position. This will avoid any damage to the valves and pistons.

m) Always fit new cylinder head bolts and tighten them in the sequence shown in Fig. D:19, to the correct torque in four stages - see TECHNICAL DATA at the end of this manual.

n) Set the crankshaft and the camshaft at the TDC positions (Fig. D:2) before refitting and tensioning the belt as detailed in 'Timing Belt & Camshaft Oil Seal', section (4) previously.

SUMP PAN . . . . [7]

The sump pan removal and installation procedure on the CVH engine is identical to that on the OHV engines, and reference should be made to the 'OHV Engine' section for details.

ENGINE MOUNTINGS. . . . . [8]

Replacement

Engine mounting replacement is quite straightforward once the weight of the engine has been taken off the mounting rubbers. To do this, you will have to raise the engine slightly, either by attaching lifting equipment to the engine and gearbox, or by placing a jack with suitable wood spacer under the engine sump pan. The latter method is probably the easier and simpler way.

Having raised the engine slightly, the rear mountings can be replaced quite easily once the bolts to the engine, the body panel and the mounting centre bolt have been undone.

The engine rear or front nearside mounting under the gearbox is removed by the same method as for the rear mountings except that the centre bolt is undone first and then the engine lifted to allow the mounting bolts to be undone. Then the mounting rubber can be removed.

The mountings are installed in the reverse order of removal, making sure that none of the mounting rubbers are misaligned before refitting and tightening the fixing bolts.

ENGINE REMOVAL & INSTALLATION . . . [9]

Removal

The engine and gearbox is removed as one unit from under the car using a suitable hoist or block and tackle to take the combined weight which is around 350 lb (160 kg). On 4-speed gearbox models, select 4th gear to ensure correct adjustment of the gear shift mechanism when refitting. On 5-speed gearbox models, select reverse gear for the same reason.

1. Disconnect the battery earth strap and remove the bonnet panel as detailed in the BODY & FITTINGS chapter.
2. Remove the air cleaner assembly (Fig. D:25), and associated pipes from the engine as detailed in the TUNE-UP chapter.
3. Drain the cooling system, disconnect the top radiator hose and overflow tank pipe from the thermostat housing (Fig. D:25) and disconnect the heater hoses from the lateral water pipe and automatic choke by referring to the COOLING SYSTEM chapter.
4. Disconnect the throttle cable (Fig. D:14) and fuel pipe at the carburettor as detailed in the FUEL SYSTEM

chapter. Plug the fuel pipe to prevent fuel loss and the ingress of dirt. On models fitted with a brake servo, disconnect the servo vacuum pipe from the inlet manifold.

5. Disconnect the wires from the rear of the alternator, the temperature gauge sender unit, the oil pressure switch, coil, reversing light switch and the anti-dieselling valve. If necessary, make a note of where each wire connects to before disconnecting (Fig. D:25).

6. Unscrew the speedometer drive cable at the gearbox end, and disconnect the clutch cable from the clutch release arm as detailed in the CLUTCH & GEARBOX chapter.

7. Remove the hot air box over the exhaust manifold (OHV only) and undo the bolts securing the exhaust pipe to the manifold. If necessary, apply penetrating oil to the bolts if they are reluctant to move.

Detach the pipe flange from the manifold and suspend the pipe on a piece of wire to prevent it from dropping.

8. Raise and support the car on axle stands as detailed in the ROUTINE MAINTENANCE chapter, so that all four wheels are off the ground and that it is raised equally at the front and rear. The car should be raised at least 2 ft (61 cm) to allow the engine to be removed from under the front panel. Drain the engine oil.

9. Disconnect the exhaust pipe mountings and remove the complete exhaust system as detailed later in this chapter.

10. Disconnect the wires to the starter motor and the engine earth strap from the engine block.

11. Undo the clamp bolt and withdraw the gearshift rod from the gearbox selector shaft (Fig. D:26). Tie the gearshift rod to the stabiliser rod alongside, and unhook the spring.

12. Undo the bolt and disconnect the stabiliser rod from the gearbox (Fig. D:27). Note that there is a washer (arrowed) fitted between the rod boss and the gearbox.

13. Drain the gearbox oil fully by undoing the cap nut on the selector shaft locking assembly (Fig. D:28).

NOTE: Take care to collect the locking pin and spring fitted under the cap nut as it is removed. Refit the cap nut and components after the gearbox has drained.

14. Undo the nut and remove the bolt securing the offside lower suspension arm pivot to the body frame (Fig. D:29).

15. Disconnect the lower suspension arm ball joint from the hub spindle carrier by holding the Torx bolt and undoing the securing nut (Fig. D:30) then withdrawing the Torx bolt.

16. Using a screwdriver as shown in Fig. D:31, disconnect the offside driveshaft from the gearbox by levering it, and, pushing the road wheel outwards at the same time until the CV joint splines are clear of the gearbox. Suspend the driveshaft on the steering rack as shown in Fig. D:32 to avoid straining the CV joints.

NOTE: To prevent the differential pinions from being displaced once the drive shaft is removed, insert a length of clean, smooth dowel (a bit of broom handle is ideal) into the shaft hole to keep the pinions in place.

17. Disconnect the nearside lower suspension arm and ball joint and disconnect the nearside driveshaft from the gearbox, following the same procedure as detailed for the offside in operations 14 - 16 previously.

18. Attach lifting equipment to the engine and gearbox, then raise the engine assembly sufficient to take the weight off the mountings (Fig. D:33).

19. Undo the centre bolts from the front and rear engine/gearbox mountings.

20. Undo the bolts and detach the tie-rod or stabiliser mounting plates to the body on either side (Fig. D:34). Detach the assembly complete with lower suspension arms.

21. Make sure that all engine mounting bolts are removed and that there are no engine connections left to undo.

22. Carefully lower the engine and gearbox assembly down through the engine compartment to the ground. Detach the lifting equipment and manoeuvre the engine out from under the car.

NOTE: If the car has not been raised high enough, the engine can be carefully tilted on its side and then removed. Take care to avoid damaging any of the engine ancilliaries, oil filter and carburettor etc when doing this.

Installation

The engine is installed in the reverse order of removal, but the following points should be noted:

a) Make sure the engine and gearbox is positioned on the ground as near in-line with the engine mounting holes as possible, and the hoist also correctly positioned. This will allow the engine to be fitted easily and with as little manoevrering as possible.

b) Fit new snap rings to the ends of the drive shaft inner CV joint shafts (Fig. D:35). Press each drive shaft firmly into the gearbox by applying pressure to the outside of each front wheel. Take care not to rock the car on the axle stands when doing this.

c) Note that there is a washer fitted between the gearshift stabiliser rod and the gearbox casing when refitting.

d) On 4-speed gearboxes, reconnect the gearshift linkage as follows:

- Check that the gear lever is in the 4th gear position.
- From the underside of the selector housing, insert a 3,5 mm drift pin into the housing, see Fig. D:36 to lock the gear lever in 4th gear. Reconnect shaft spring.
- Using a suitable pin, as shown in Fig. D:37, turn the gearbox selector shaft in the clockwise direction as far as the stop will allow, then slide it fully into the gearbox. Attach a rubber band to the pin to hold the shaft in position to equalise any bearing play.
- Reconnect the gearshift rod and tighten the clamp bolt to the correct torque. Remove drift pin and pin from selector shaft.

e) On 5-speed gearboxes, reconnect the gearshift linkage as follows:

- Check that the gear lever is in the reverse gear position.
- Using a suitable pin, as shown in Fig. D:37, turn the gearbox selector shaft in a clockwise direction from the neutral position as far as the stop will allow, then push the selector rod fully into the gearbox to engage reverse gear. Hold the selector shaft in this position by attaching a rubber band to the pin.
- Taking great care not to disengage the gearshift rod from the selector shaft, push the gear lever down-

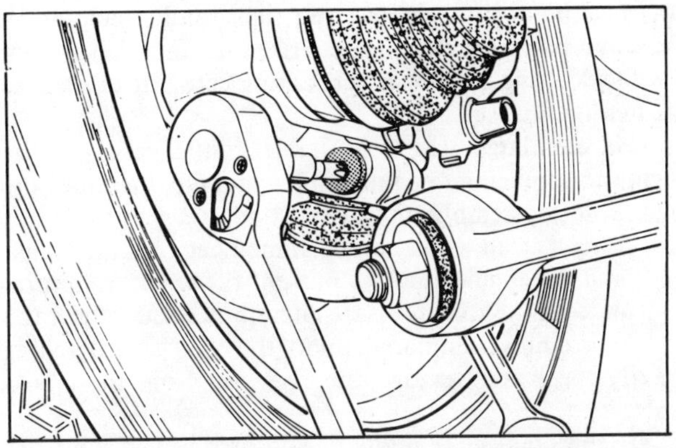

Fig. D:30 Detaching ball from hub carrier

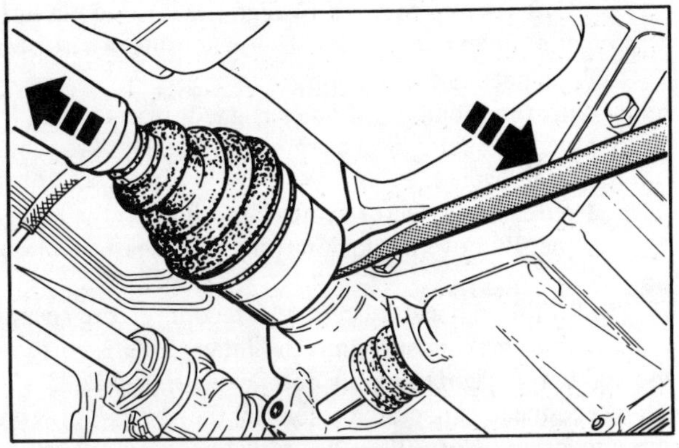

Fig. D:31 Method of disconnecting driveshaft

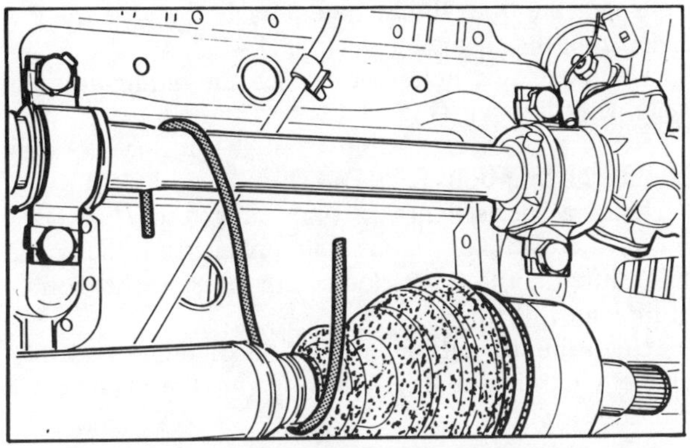

Fig. D:32 Suspending disconnected driveshaft

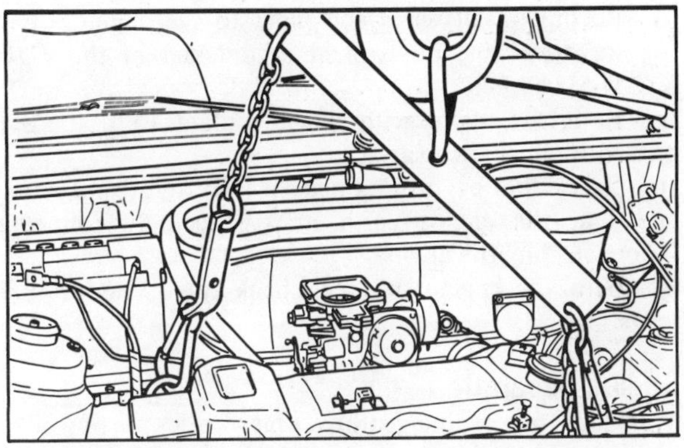

Fig. D:33 Engine lifting bracket attachment points

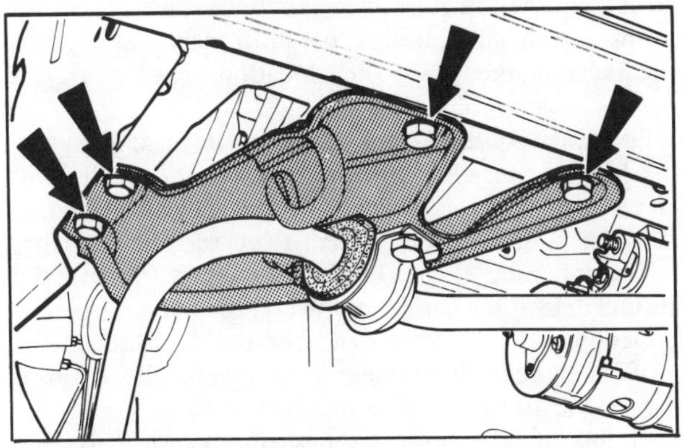

Fig. D:34 Tie rod/stabiliser mounting plate

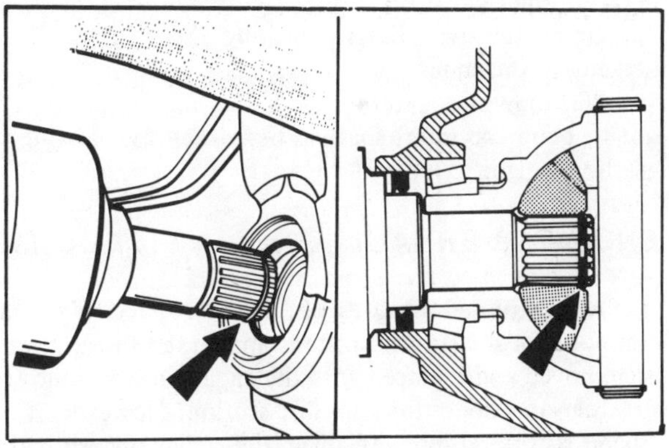

Fig. D:35 Snap ring position on inner driveshaft

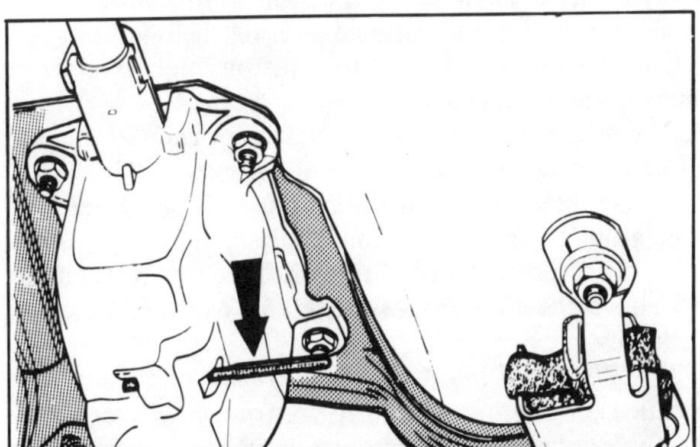

Fig. D:36 Locking gear lever-4-speed gearboxes

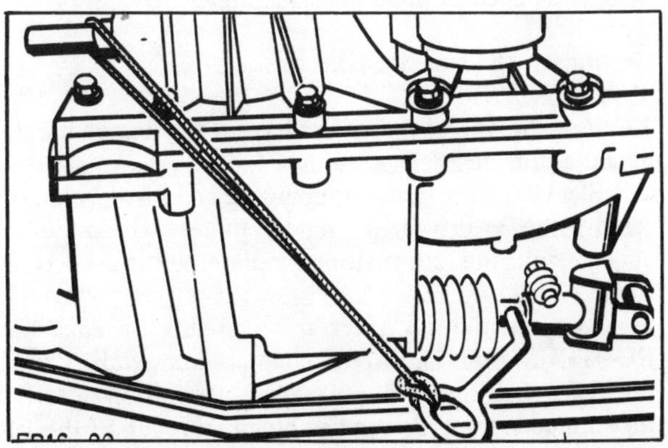

Fig. D:37 Method of positioning selector shaft

wards in reverse gear position and then move it gently back and forward to align the locking pin holes on the underside of the selector housing. An assistant will be needed for this job.

- With the locking pin holes aligned, insert a mandrel or special locking tool as shown in Fig. D:38 to secure the gear lever.

- Tighten the selector shaft clamp bolt to the correct torque. Remove the locking pins from shaft and housing.

f) Refit the exhaust system and reconnect the speedometer cable, making sure that the inner cable is correctly located before tightening the outer cable ring (Fig. D:39).

g) Reconnect the clutch cable in the reverse order of disconnecting. Note that no adjustment of the clutch cable is necessary as the automatic adjustment will take up any slack once the cable is operated.

h) Reconnect all wires and pipes to the engine. Check the throttle cable adjustment as detailed in the FUEL SYSTEM chapter.

i) Refill the engine with oil and coolant. Refill the gearbox with oil (if necessary).

j) Check that everything has been refitted to the engine and that the engine earth strap is securely connected before starting the engine.

k) With the engine running, check for oil and coolant leaks.

Engine Separation

1. Place the engine/gearbox unit on supporting wood blocks on level ground.
2. Undo the bolts and detach the front gearbox mounting from the gearbox casing.
3. Undo the two bolts attaching the gearbox clutch housing to the engine.
4. Withdraw the gearbox from the engine, taking care not to bend the input shaft in the clutch by allowing the gearbox weight to hang on the shaft.

ENGINE OVERHAUL ....[10]

As mentioned previously in the 'Introduction', certain components such as the pistons and big-end bearings can be removed and replaced with the engine in-situ, following the relevant operations in this section. However, if the engine is to be fully overhauled, the engine unit should be removed from the car and separated from the gearbox first, as detailed in the previous section.

Dismantling (Fig. D:40)

1. Remove the cylinder head from the engine block as detailed previously. Discard the head gasket.
2. Slacken the alternator mounting and pivot bolts, then remove the 'V' drive belt from the pulleys. Undo the bolts and detach the alternator complete with bracket and shield.
3. Remove the oil filter using a strap wrench - see ROUTINE MAINTENANCE chapter and withdraw the oil dipstick. Disconnect the crankcase ventilation hose from its connection on the engine block adjacent to the filter housing.
4. With the engine supported clear of the ground and upright to prevent any sludge or dirt in the sump from entering the engine, remove the flange bolts and detach the sump pan from the engine. Prise the sump flange from the block using a screwdriver, taking care not to damage the mating surfaces.
5. Place a large tray or suitable container under the engine to catch any remaining oil, water etc, then turn the engine block assembly on its side.
6. Using a clean wood block jammed between the crankshaft and the engine block to prevent the crank from turning, undo the clutch pressure plate bolts and detach the clutch assembly. Similarly, undo the six bolts attaching the flywheel and remove the flywheel from the crankshaft.

   If the crankshaft pulley is still attached at the front of the engine, undo the central bolt and detach the pulley at this stage.
7. Undo the four bolts and detach the rear oil seal carrier from the engine block (Fig. D:41).
8. Undo the two bolts and detach the timing belt tensioner with its spring (Fig. D:42).
9. Remove the water pump from the engine block by undoing the four bolts (Fig. D:43).
10. Using a special half-cup type puller tool (Fig. D:4), remove the camshaft timing belt pulley from the crankshaft. Note the fitted position of the thrust washer behind before removing it.
11. Undo the six bolts and detach the oil pump assembly, complete with intake filter bracket, from the engine block (Fig. D:44).
12. Turn the crankshaft so that all pistons are at the mid-position in their bores, then carefully scrape the carbon deposit from the top edge of each cylinder bore.
13. Check that all bearing caps, both main and big-end, are suitably marked with their location - see Fig. D:45 - for correct reassembly.
14. For each piston and con-rod in turn: Turn the crankshaft as necessary to bring the con-rod to the bottom of its travel. Unscrew the big-end cap bolts, two or three turns and gently tap them to release the cap from the locating pins (Fig. D:46). Completely remove the bolts and detach the cap from the con-rod.
15. Gently push the piston and con-rod out through the top of the cylinder bore using a hammer handle. If this is difficult due to the rings contacting a wear ridge at the top of the cylinder bore, try inserting two or three thin feeler gauges, equally spaced, down past the rings to ease them past the wear ridge as the piston is withdrawn.
16. Place each big-end cap and bearing shells with its respective con-rod. Arrange all four pistons and con-rods in numbered removal order - this is important.
17. Remove the piston rings from each piston. Raise one end of the ring out of its groove and insert a 0.020 in (0,5 mm) feeler blade between the ring and piston. Rotate the feeler around the piston while applying upward pressure to the raised portion of the ring until the ring rests on the land above the ring grooves. The ring can then be eased off the piston.

NOTE: The rings must always be removed and refitted over the piston crown - NEVER over the piston skirt.

18. The removal of the piston from its con-rod requires a suitable press and heating of the con-rod small end on reassembly. This job is best entrusted to a local Ford dealer

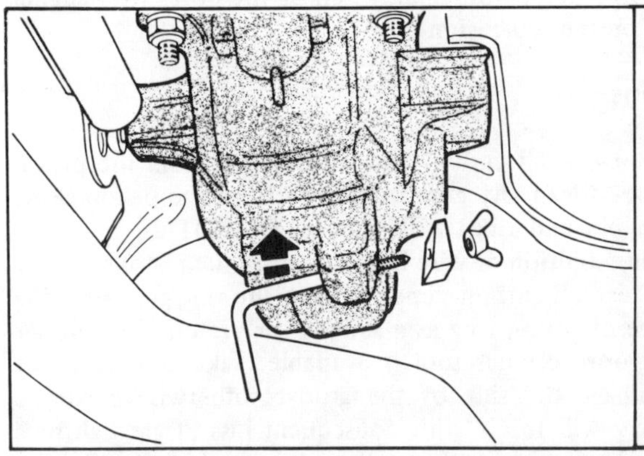

Fig. D:38 Locking gear lever - 5- speed gearboxes

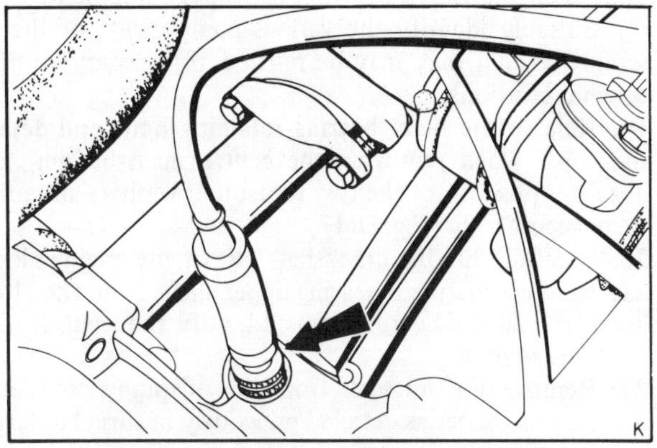

Fig. D:39 Speedometer cable connection

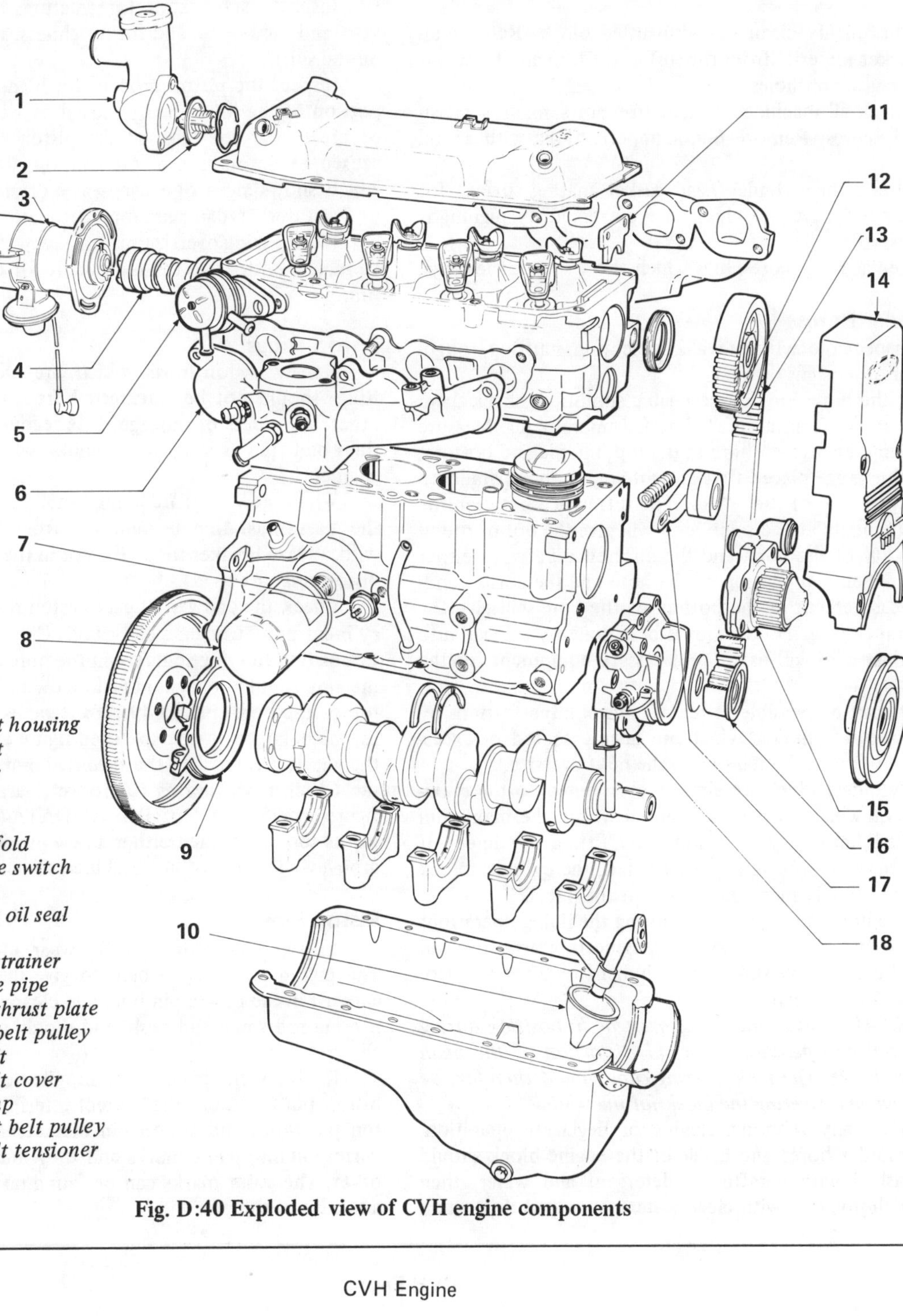

1. *Thermostat housing*
2. *Thermostat*
3. *Distributor*
4. *Camshaft*
5. *Fuel pump*
6. *Inlet manifold*
7. *Oil pressure switch*
8. *Oil filter*
9. *Crankshaft oil seal carrier*
10. *Oil pump strainer with intake pipe*
11. *Camshaft thrust plate*
12. *Camshaft belt pulley*
13. *Timing belt*
14. *Timing belt cover*
15. *Water pump*
16. *Crankshaft belt pulley*
17. *Timing belt tensioner*
18. *Oil pump*

Fig. D:40 Exploded view of CVH engine components

or auto engineering firm with the proper equipment.

Suitably identify the various components of the assembly so that they may be refitted, if necessary, in their original positions.

19. Release the main bearing retaining bolts and detach the caps. When removing the centre main bearing cap, note the position of the two thrust half washers and mark them accordingly (Fig. D:47).

20. Carefully lift the crankshaft out of the engine block.

21. Remove the main bearing upper shells from the block housings and the bearings caps. Identify the shells if they are to be re-used.

22. Remove the oil seals from the oil pump cover and rear oil seal carrier as detailed previously in this chapter.

Cylinder Block & Component Inspection

Thoroughly clean the dismantled block. Remove all old gasket material from the surfaces. Clean and blow out all passages, bolt holes, etc.

Check all machined surfaces for burrs, nicks, scratches and scores. Remove minor imperfections with an oil stone.

Check the cylinder head gasket mating surface for flatness and have the block face corrected by planing if necessary.

Replace any core plugs which show signs of leakage.

Cylinder Bores

Inspect the cylinder walls for scoring, scuffing, roughness or other signs of wear.

If the bores look in reasonable condition, check them for wear, using an accurate internal micrometer. Measure the diameter of each bore at the top, middle and bottom, with the gauge placed first at right angles, then parallel to the centre line of the engine. The difference between the measurements at the same level will give the out of round or ovality of the bore, and the difference between either measurements at the top of the bore and the corresponding measurement at the bottom of the bore will give the bore taper (Fig. D:48). This is best done by a local auto machinist who will have the necessary equipment for the job.

It may be possible to remove minor imperfections by honing, but cylinders which are deeply scored or excessively worn must be rebored to the next oversize.

Cylinders which are already over bored to the maximum, or which cannot be rebored within the maximum specified limit may be restored by fitting dry liners. If liners have not previously been fitted, the cylinder block must be suitably machined to accept the liners.

Needless to say, both the boring the lining operations must be entrusted to a specialist engineering firm who will have the necessary skill and equipment required to carry out work of this nature.

NOTE: All main bearing caps must be in position during the reboring operation to avoid distortion of the main bearing bores. The main bearing caps should therefore be fitted when delivering the block for machining.

After any reboring, honing or deglazing operation, the cylinder bores and inside of the engine block should be washed with paraffin or detergent and water, then rinsed thoroughly with clean water and wiped dry with a clean lint free cloth. Finally, smear the bores with engine oil to prevent corrosion.

Pistons

Remove all carbon and oil deposits from the piston surfaces. Clean any gum or varnish from the piston skirt, piston pin and rings with solvent. Do NOT use a caustic cleaning solution or wire brush to clean the pistons.

Clean all carbon deposits from the ring grooves using a piece of piston ring as a scraper, or preferably a proper ring groove cleaner tool if available. Take great care not to damage the side of the groove, otherwise excessive sideplay will result with consequent loss of gas tightness and excessive fuel consumption. Ensure that the oil drain holes in the oil control ring groove are clear.

Inspect each piston for fractures at the ring lands, skirt and pin bosses, and for roughness, scoring or scuffing on the skirt.

Inspect the piston grooves for high steps at the inner portion of the lower lands caused by wear. Spongy, eroded areas near the edge of the piston crown are usually caused by detonation or pre-ignition. Replace any piston which shows signs of wear, wavy or stepped ring lands, or fractures or damage from detonation or pre-ignition.

When fitting oversize pistons, each cylinder bore must be machined to give the correct fit for the individual piston.

Piston Rings

Renew piston rings which are worn or damaged. Rings should not be transferred from one piston to another, regardless of mileage. It is recommended that new rings be fitted as a matter of course when overhauling the engine.

Where new rings are being installed in a used cylinder, the glaze must first be removed from the cylinder walls with a glaze breaker tool. (Except in the case where Cords piston rings are used).

Check the end gap of each piston ring in its respective cylinder bore, using a feeler gauge. Use the inverted piston to position the ring squarely in the bore. With worn bores, the ring should be located just below the wear ridge at the top of the bore (Fig. D:49). If the gap is too small, then it can be enlarged to the specified figure by careful filing. If the gap is greater than the specified limit, try another ring set. Check that the ring to groove clearance is within the figure specified in TECHNICAL DATA (Fig. D:50). If the gap is too great, then either a new piston is needed or the grooves will have to be machined to accept ring 'inserts'.

Piston Pins

Inspect each piston pin for wear, etching or fracture. The piston pins are selected to give the correct interference fit in the piston pin bore and clearance fit in the connecting rod small end bush, and should not be interchanged.

Replacement pistons are supplied complete with the piston pin to ensure the correct interference fit. The piston pin bores and piston pin diameters are graded with corresponding paint marks and they must match one another. The paint marks can be found on the piston crown and edge of the pin.

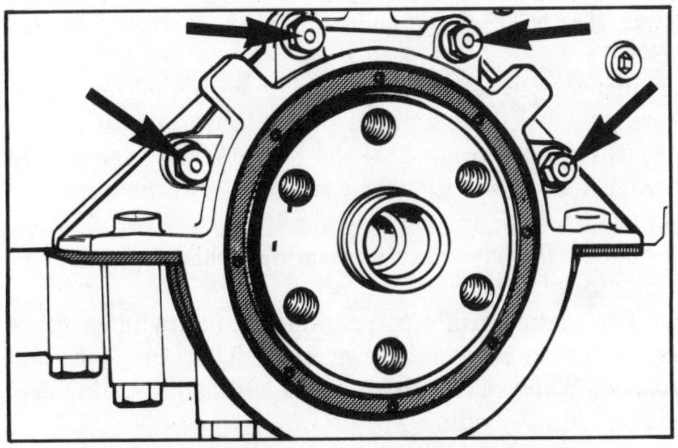

Fig. D:41 Crankshaft rear oil seal carrier bolts

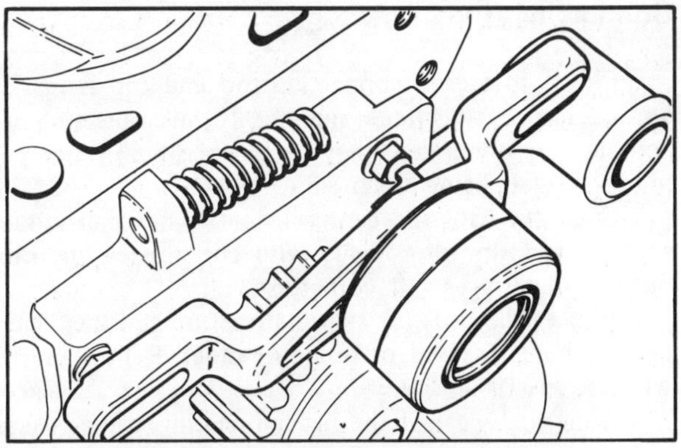

Fig. D:42 Timing belt tensioner and spring

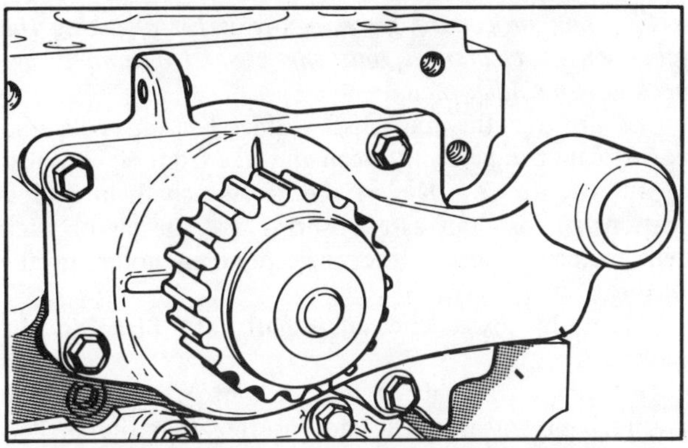

Fig. D:43 Water pump assembly fitting

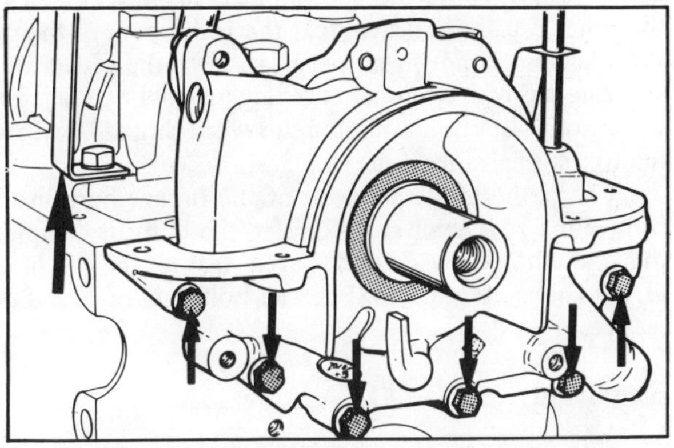

Fig. D:44 Oil pump bolts and filter bracket bolt

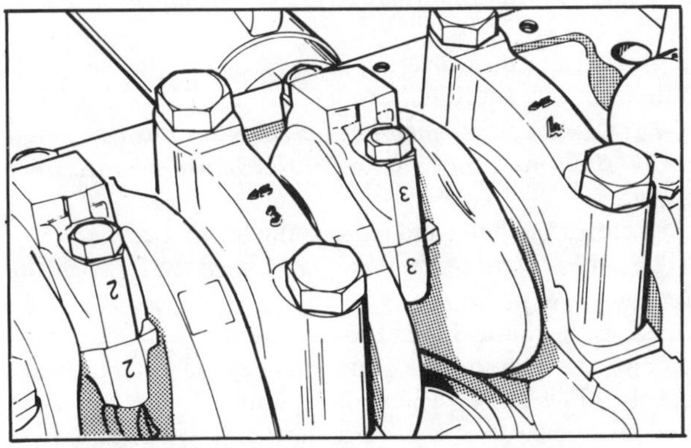

Fig. D:45 Big-end and main bearing cap marks

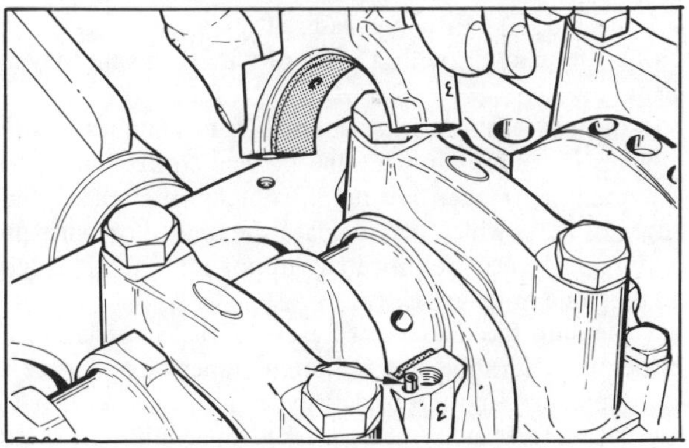

Fig. D:46 Removing big-end bearing caps

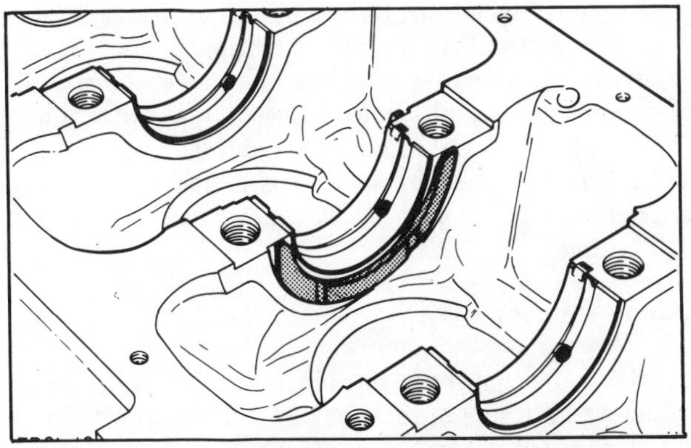

Fig. D:47 Correct position of crank thrust washers

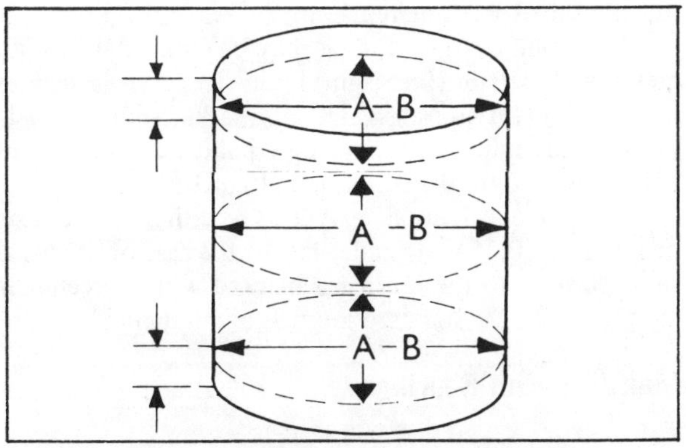

Fig. D:48 Measuring points for cylinder bore wear

Connecting Rods

Check that each connecting rod and cap is marked with its respective cylinder number. If the connecting rods and caps are unmarked, they should be suitably stamped, unless they are being discarded.

Clean the rods and caps with petrol or other suitable solvent, and dry thoroughly with compressed air. Blow out the oil passages with compressed air.

Inspect the rods for signs of fracture and the big end bearing bores for out of round or taper. Renew any rod which is unsatisfactory.

Abnormal wear of the big end bearings can be caused by either a bent rod, a worn or damaged crankpin or a tapered big end bore. A shiny surface on the pin boss side of the piston, above the pin on one side and below it on the other usually indicates that the rod is bent or the piston pin hole is not in proper relation to the piston skirt and ring grooves. A proper checking jig will be required to check for these conditions and this work should be entrusted to a specialist machine shop.

Check the inside diameter of the bronze bush in the connecting rod small end. Replace the bush or complete rod if the bore is worn or damaged. Inspect the condition of the connecting rod big end cap bolts and replace if defective.

Crankshaft

Wash the crankshaft with petrol or other suitable solvent and dry thoroughly. Take great care to avoid damaging the machined surfaces. Blow out the oil passages with compressed air. In some cases it may be necessary to probe the passages with a piece of wire to ensure they are clear.

Inspect each bearing journal for scratches, scores, grooves or cracks. Inspect the oil seal contact surface at the crankshaft rear journal for nicks, scratches, sharp edges or burrs which may damage the oil seal or cause premature seal wear. Remove any minor imperfections with an oil stone or emery cloth.

Measure the diameter of each journal in at least four places to determine out of round, taper or undersize. If the taper or out of round is excessive, or if any of the journal surfaces are severely marked the crankshaft should be reground or replaced. If any of the journals will not clean up within the minimum specified regrind diameter, the crankshaft must be replaced.

The main bearing journals may be either standard diameter or 0.010 in (0,25 mm) undersize, even in new engines. Standard journals are not marked but undersize journals are indicated by a yellow paint mark on the first balance weight on the crankshaft (Fig. D:51).

The big end journals may also be either standard size or 0.010 in (0,25 mm) undersize. In the case of undersize, the web next to the journal is marked with a green paint spot (Fig. D:52). Standard journals are unmarked.

Main/Big-End Bearings

Clean the bearing shells thoroughly with petrol or other suitable solvent and dry thoroughly with compressed air. Do NOT scrape gum or varnish deposits from the shells.

Inspect the bearing surface of each shell carefully. Bearings which have worn, scored, chipped or pitted surfaces should be replaced. If the bearing base is visible the bearing is worn and should be replaced, also the respective journal checked for ovality.

Check the clearance of bearings which appear to be satisfactory, as described below.

The main bearing parent bore in the cylinder block may be either standard diameter or 0.015 in (0,38 mm) oversize. Where an oversize exists the main bearing caps are marked with yellow paint spots.

Bearing Clearances

NOTE: The procedure given below is for checking the main bearing clearances, but the big-end clearances are checked in a similar manner.

1. Ensure that the crankshaft surface bearing shells and bearing caps are perfectly clean and free from oil and dirt.
2. Fit the upper halves of the bearing shells into their locations in the crankcase, ensuring that the tag on each shell correctly engages the corresponding notch in the housing.
3. Carefully lower the crankshaft into place in the crankcase.
4. Fit the lower half of the shell into the cap of the bearing being measured, again ensuring that the shell is correctly located.
5. Place a piece of Plastigage across the full width of the crankshaft journal, and about 0.25 in (6 mm) off centre (A, Fig. D:53).
6. Install the main bearing cap and tighten the cap bolts to their normal specified torque.

NOTE: None of the remaining bearing caps must be fitted during this procedure. Each bearing must be measured separately.

7. Release the bolts and remove the bearing cap.

NOTE: The crankshaft must not be moved while the Plastigage is in position.

8. Measure the width of the compressed plastic filament (B, Fig. D:54) using the scale provided on the Plastigage pack (C, Fig. D:54). The widest point will give the minimum clearance and the narrowest point the maximum clearance. The difference between the two readings will therefore give the taper on the journal.
9. To check the journal for out of round, clean all traces of Plastigage from the journal and bearing insert, rotate the crankshaft a quarter of a turn and repeat the measuring procedure. The difference between the two readings will indicate the out of round of the journal.

NOTE: When the measurement is completed, all Plastigage material must be cleaned off the bearing insert and crankshaft journal.

Engine Reassembly

1. Make sure that the engine block is thoroughly cleaned to remove all traces of dirt and swarf using an air line to blow through all oil and water passages.
2. Position the engine block upside down on wood

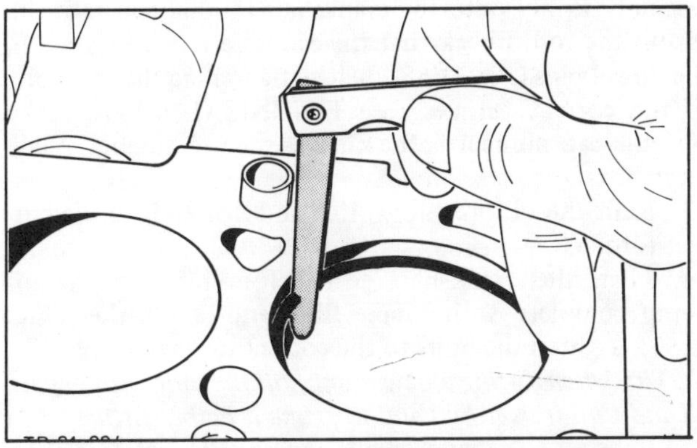

Fig. D:49 Ring position when checking ring gap

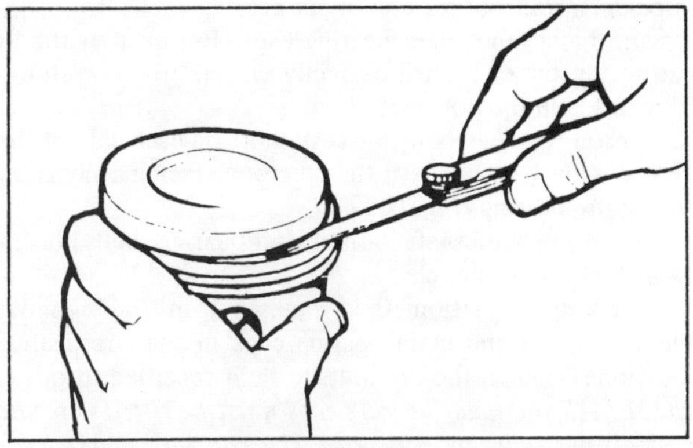

Fig. D:50 Measuring ring to piston groove clearance

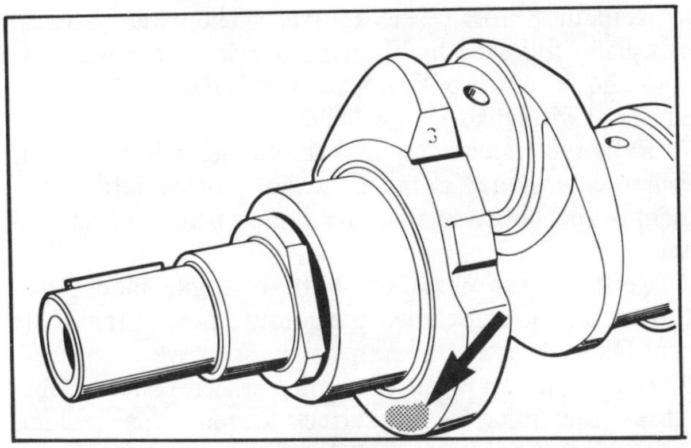

Fig. D:51 Main bearing undersize mark position

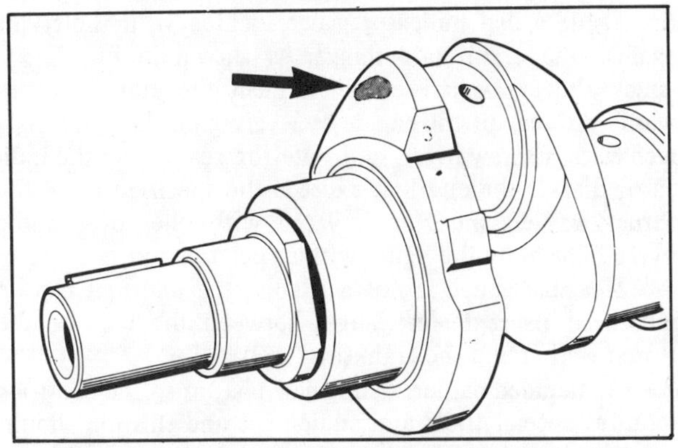

Fig. D:52 Big-end bearing undersize mark position

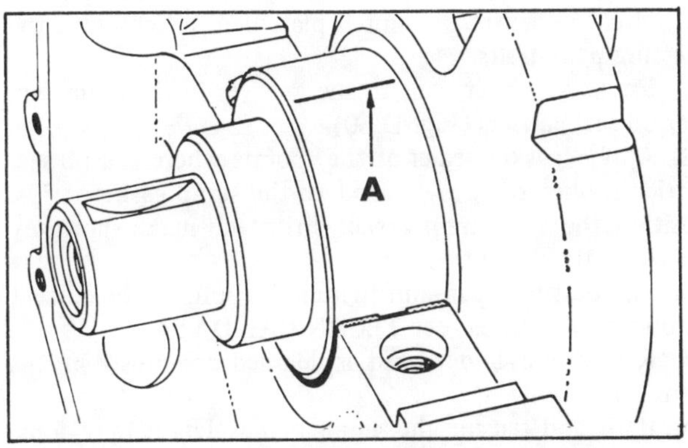

Fig. D:53 Position plastigage strip at point (A)

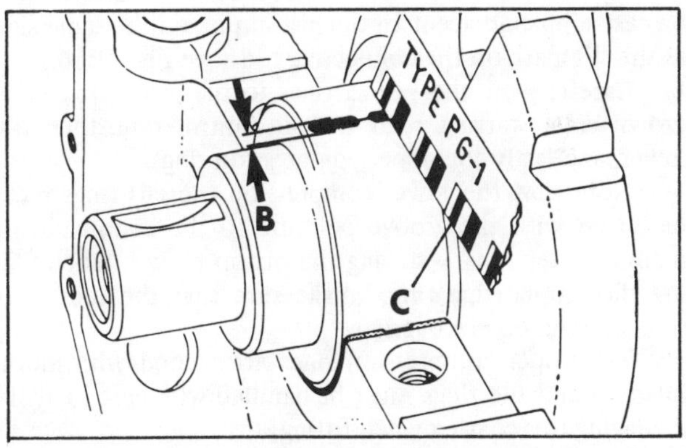

Fig. D:54 Measuring plastigage thickness with scale

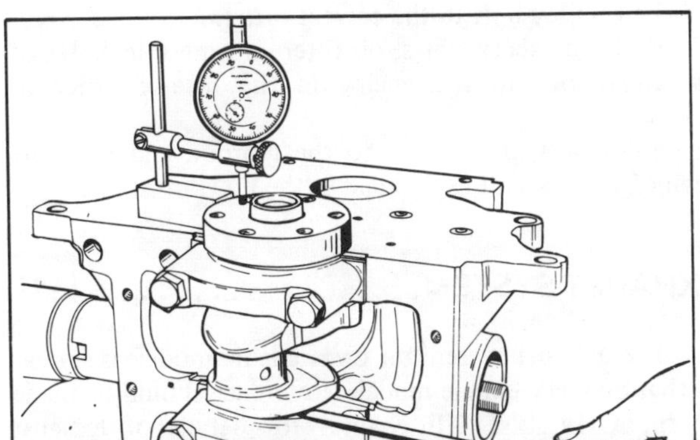

Fig. D:55 Measuring crankshaft endfloat

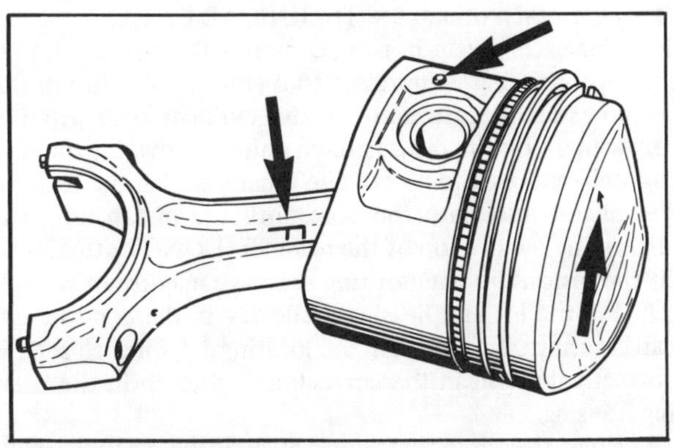

Fig. D:56 Correct relationship of piston to con rod

blocks. Install the selected main bearing shells dry in the engine block and main bearing caps. Ensure that the locating tag on each shell correctly engages the corresponding notch in the housing.

3. Install the two half thrust washers on each side of the centre main bearing with the oil groove facing away from the centre bearing (Fig. D:47).

4. Coat the crankshaft journals and bearing shells liberally with clean engine oil.

5. Carefully position the crankshaft in the cylinder block, and fit the main bearing caps in their respective locations. Tighten the cap bolts to their specified torque.

NOTE: All the main bearing caps must be fitted with the cast arrows pointing forwards. The caps are marked for correct positioning: F for front, C for centre and R for rear (Fig. D:45).

6. Using a dial indicator gauge located with the stylus against the crankshaft flange, as shown in Fig. D:55, check the crankshaft endfloat. Push the crankshaft towards the rear of the engine by levering on the crankshaft web with a screwdriver, and note the reading on the indicator dial. If the endfloat exceeds the specified limit, the thrust washers must be replaced with ones of suitable oversize to bring the figure within specification.

If a dial gauge is not available, the endfloat can be measured using feeler gauges between the face of the thrust washer and the crankshaft web.

7. As detailed earlier, fitting new pistons to the con-rods requires special tools and equipment and this job should be done by a local Ford dealer or auto engineers. If new pistons have been fitted, check that the con-rods have been correctly fitted with the arrow on the piston crown or cast nipple adjacent to the piston pin, on the same side as the 'F' mark on the connecting rod - see Fig. D:56.

8. Carefully fit the piston rings to the piston from the crown side, starting with the oil control ring, then the lower, and lastly the upper compression rings.

Note that the lower compression (nosed) rings must be fitted with the groove pointing to the piston pin or manufacturer's mark facing the piston crown. When fitting the supporting ring, make sure that the ends abut squarely and do not overlap.

The upper compression rings are coated with molybdenum and the rings must be handled with care to avoid damaging this coating when fitting.

9. Smear the pistons and cylinder bores with clean engine oil then position the rings correctly on the pistons relative to each other - see TECHNICAL DATA.

10. For each piston in turn, compress the piston rings using a proper ring compressor tool (Fig. D:57). Install the piston assembly into its respective cylinder bore with the arrow mark on the piston crown pointing towards the timing belt end of the block. Use a hammer handle to push the piston fully into the bore until the piston crown is slightly below the top of the cylinder. Do NOT attempt to fit the pistons by hand or ring breakage may result.

11. Fit the big-end bearing shells dry in the con-rod and cap housings. Ensure that the locating tag on each shell is correctly located in the corresponding notch in the bearing housing.

12. Coat the crankshaft journal and the bearing shells with clean engine oil, then pull the con-rod and piston assembly firmly onto the crankshaft journal and refit the cap to the rod. Check that the cap is correctly located on the dowel pins (Fig. D:58), before tightening the cap bolts to the correct torque - see TECHNICAL DATA. Make sure the cap number is the same as the rod number, made on removal.

13. Refit the oil pressure switch and tighten it to the correct torque.

14. Align the crankshaft gear before refitting the oil pump complete with intake filter and new gasket (Fig. D:59). Tighten the bolts to the correct torque.

NOTE: When fitting a new pump, fill the unit with engine oil and turn it over by hand to prime it before fitting.

15. Lubricate the crankshaft and seal lip before fitting the oil seal to the pump as detailed previously under 'Crankshaft Rear Oil Seal'.

16. Refit the thrust washer (curved side outwards) to the crankshaft, followed by the timing belt pulley with the aid of the 'V' - belt pulley. Make sure the crankshaft nose is smeared with grease beforehand.

17. Refit the water pump and the timing belt tensioner. Lever the tensioner as far as possible to the left before pinching the bolts, so that the timing belt can be fitted later.

18. Refit the rear oil seal carrier to the engine block, then fit the oil seal as detailed previously under 'Crankshaft Front Oil Seal'.

19. Refit the flywheel, securing it with new bolts. Tighten the bolts to the correct torque using a wood block to jam the crankshaft and prevent it from turning.

20. Make sure the flywheel face is clean and grease free before refitting the clutch assembly. Use an aligning tool (mandrel) to centre the clutch plate before tightening the pressure plate bolts.

21. Fit new rubber seals to the grooves in the front and rear oil seal carriers (Fig. D:60).

22. Apply gasket sealer at the points where the oil seal carrier rubber seals are joined to the sump gasket strips. Position the two sump gasket strips and make sure they are correctly aligned.

23. Fit the sump pan and tighten the bolts to the correct torque in two stages - see TECHNICAL DATA.

24. Refit the cylinder head as detailed previously in this chapter.

25. Refit and tension the timing belt as detailed previously under 'Timing Belt and Camshaft Oil Seal'.

26. Refit the timing belt cover and tighten the crankshaft 'V' - belt pulley bolt to the correct torque.

27. Refit the spark plugs, oil filter and the remainder of the ancilliaries to the engine in the reverse order of removal.

28. Reconnect the engine to the gearbox and refit the engine/gearbox unit as detailed in the previous section.

EXHAUST SYSTEM . . . . [11]

The exhaust system on CVH engine models is similar to that on OHV engine models. Reference should be made to the OHV ENGINE chapter for details of exhaust system replacement. Fig. D:61 shows the various system arrangements on CVH engine models.

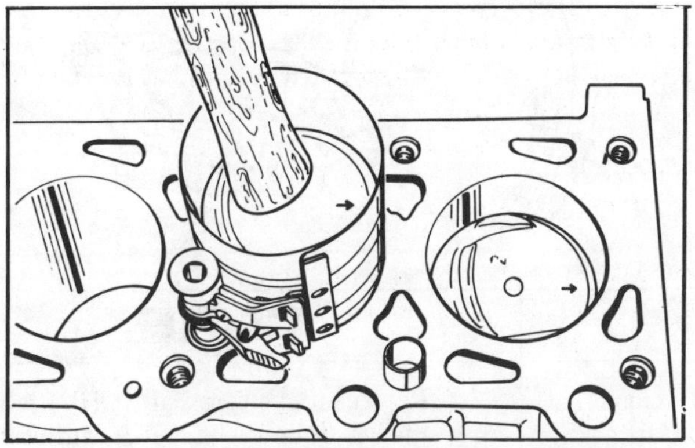

Fig. D:57 Method of installing piston using clamp

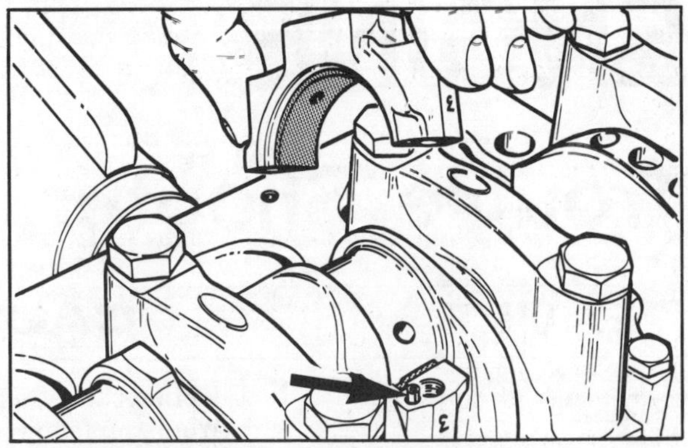

Fig. D:58 Big-end cap locating dowel pins

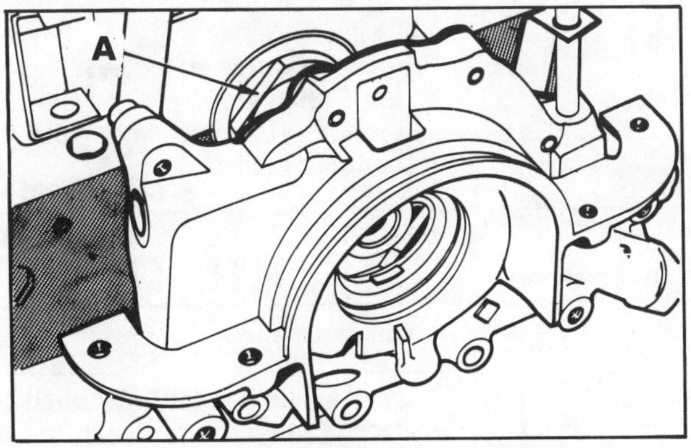

Fig. D:59 Oil pump installation, crankshaft gear (A)

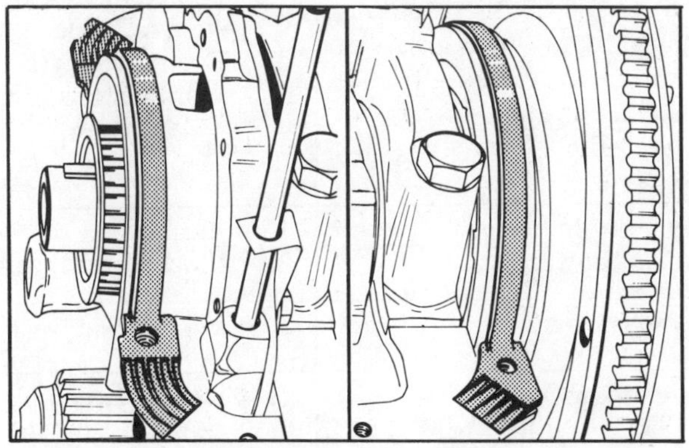

Fig. D:60 Fit new sump rubber seals to end grooves

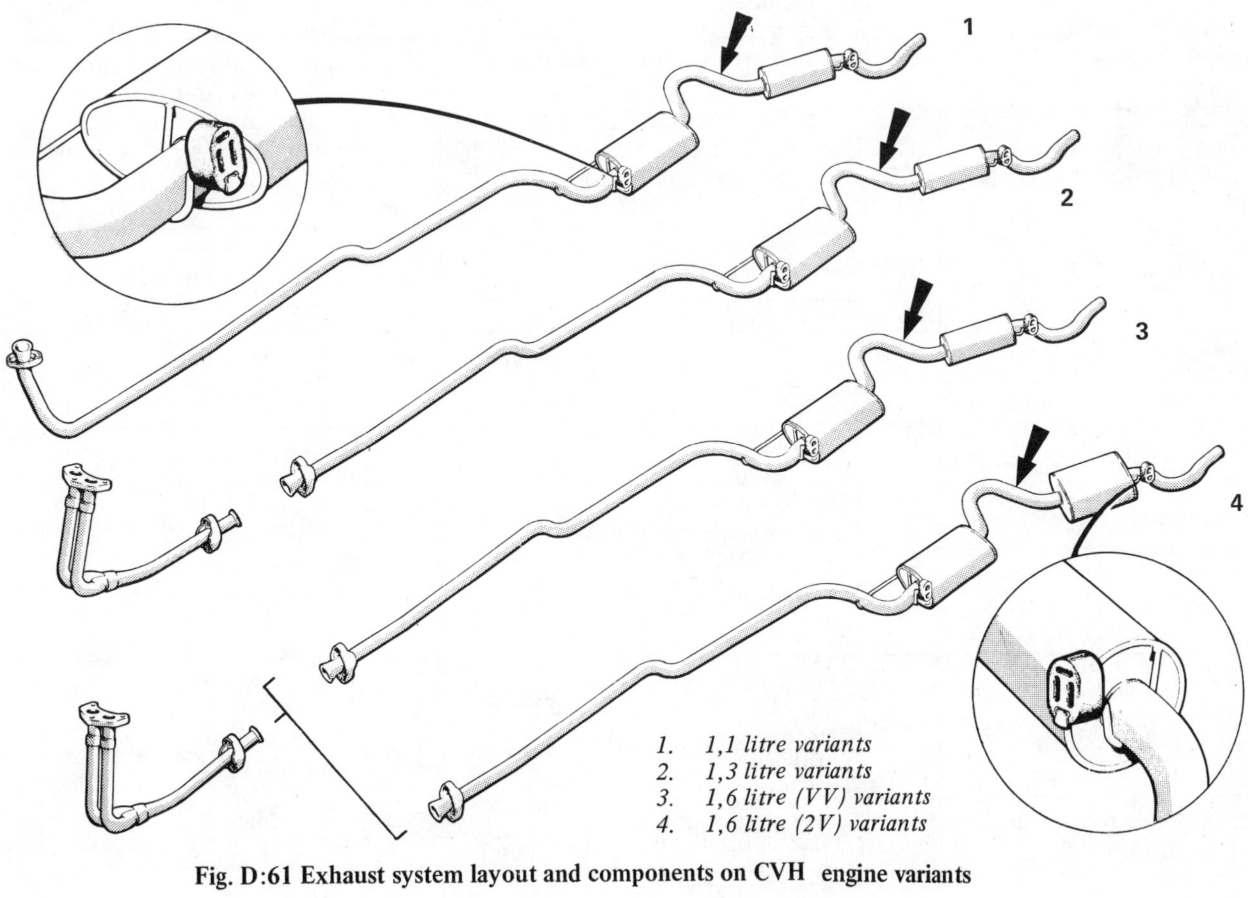

1. *1,1 litre variants*
2. *1,3 litre variants*
3. *1,6 litre (VV) variants*
4. *1,6 litre (2V) variants*

Fig. D:61 Exhaust system layout and components on CVH engine variants

ENGINE

Trouble Shooter

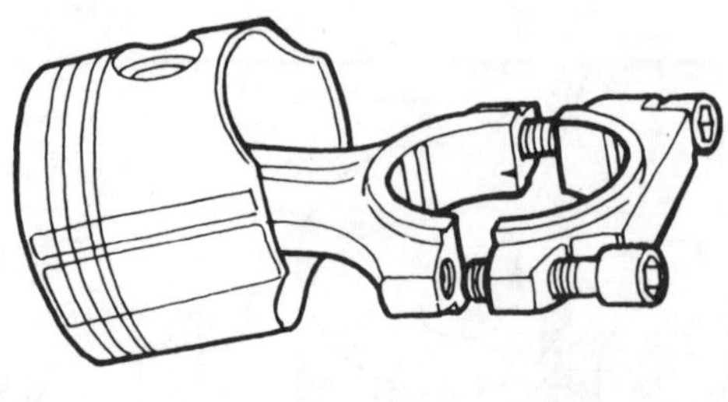

FAULT	CAUSE	CURE
Noisy tappet (with correct clearance	1. Worn rocker or shaft (OHV engine) 2. Worn cam follower tappet (OHC engine)	1. Replace rockers or shaft (OHV). 2. Fit new follower/tappet & camshaft.
Lack of compression	1. Faulty valve seat, excessive wear in stem or guide. 2. Faulty head gasket. 3. Worn pistons, rings and bores.	1. Recut seat and valve, fit new guide and valve. 2. Fit new gasket or reface head. 3. Either fit new rings, pistons and rings and rebore. If engine badly worn then recondition engine.
Smoke from exhaust. Lack of power	1. As above. 2. Blocked crankcase breather.	1. As above. 2. Check breathing apparatus as above.
Piston slap	1. Worn pistons, rings and bores.	1. Engine overhaul.
Big-end knock	1. Wear between big-end shell and crankshaft. Wrong torque on bolts.	1. Depending on wear, fit new shells, regrind crankshaft and check torque.
Mains rumble	1. Wear between main bearing shells and crankshaft.	1. As above.
Cam follower tap	1. Camshaft worn or follower dished.	1. Examine and replace followers or camshaft. Or both.
Knocking when clutch depressed. Movement at crank pulley	1. Excessive crankshaft end-float. Wear between crank and thrust washer.	1. Fit new thrust washers and recheck clearance.
Clattering from front of engine	1. Worn or slack timing chain, worn chain tensioner.	1. Fit new chain and tensioner. Adjust chain where necessary.
Small-end or gudgeon pin knock	1. Excessive wear between gudgeon pin and con-rod.	1. Check piston assemblies for wear.
Lack of oil pressure	1. Lack of oil. 2. Faulty oil pump. 3. Blocked oil pick-up strainer. 4. Faulty pressure-relief valve. 5. Blocked oil filter. 6. Excessive wear in crankshaft journals.	1. Install fresh oil. 2. Fit new pump. 3. Clean pick-up. 4. Fit new relief valve. 5. Fit new filter. 6. Overhaul engine.
Oil leaks	1. Sump gaskets or packings. 2. Front and rear crankshaft oil seal. 3. Rocker or camshaft gasket. 4. Oil filter.	1. Fit new gaskets. 2. Fit new seals. 3. Fit new gasket. 4. Check filter seal.
Lack of power (engine in good condition)	1. Faulty ignition timing. Faulty sparking plugs, points or condenser. Wrong valve clearance.	1. Tune engine.

Engine Electrics

INTRODUCTION .... [1]

The engine electrical system is divided into three main sections. These are the starting, ignition and charging circuits. Components such as the temperature sender unit, transmission electrics and the warning check system are dealt with in the relevant sections.

The procedure for charging the battery is given in the GENERAL ELECTRICS chapter. Many charging problems come from lack of battery maintenance and drive belt failure. The alternators fitted to the Escort are all similar in design and construction, although there are detail differences between different makes. The Charging Circuit Test in the Lucas Alternator section applies to all types.

To solve continuity and earthing problems a test lamp and a selection of jump leads should prove sufficient. However, to test some of the more complex assemblies a multi-meter will be required. This should have voltage, amperage and preferably, resistance ranges. The instrument may be combined with a dwell meter and tachometer or rev-counter. See the TUNE-UP chapter for details of the applications of these scales. Other tools that may prove necessary are electrician's screwdrivers, a selection of spanners, pliers and wire strippers. Note that the testing of electronic components such as the electronic ignition module, alternator diodes should be left to an auto electrician or Ford dealer. Special equipment is also necessary to test armature windings and the distributor advance characteristics, although the latter can be checked, see the TUNE-UP chapter.

LUCAS ALTERNATOR .... [2]

Precautions

It is essential that the following points be observed when carrying out any maintenance or repairs on a charging system which incorporates an alternator:

1. Never disconnect or reconnect the battery leads while the alternator is running, otherwise the rectifier pack and/or regulator unit may be damaged. It is likewise inadvisable to make or break any other connection in the charging circuit while the engine is running.
2. Never run the alternator with the main output cable disconnected either at the alternator or the battery while the field windings are energised, otherwise the rectifier diodes may be damaged.
3. When reconnecting the battery leads, always ensure that the correct polarity is observed - reversed battery or charger connections will damage the rectifier diodes. Connect the positive lead to the battery positive terminal first, then the negative lead to the negative terminal.
4. No attempt should be made to polarise the alternator. The alternator is for use on negative earth systems only and the polarity cannot be reversed by "flashing" the field terminals as with a DC generator. Any attempt to do so will damage it.
5. Always disconnect the battery earth lead before using a battery charger as this will isolate the alternator from the battery and external charging equipment. The charger must always be connected in parallel with the battery - positive to positive and negative to negative.

Charging Circuit Test

1. Check that the generator drive belt is correctly adjusted. The battery should be in good condition and its terminals clean and tight, also the connections and leads in the rest of the charging system should be clean and making good contact.
2. Disconnect the cable plug from the rear of the alternator, then connect the negative side of a voltmeter to a good earth and switch the ignition on. Connect the other voltmeter lead to each of the disconnected alternator cables in turn (Fig. E:1). If a battery voltage is not recorded at the blue cable from the dashboard, then check the warning light circuit and the bulb.
3. If a battery voltage is not recorded at the red main charging leads to the starter motor, check the wiring and the connections between the battery and the alternator.

4. If all is well with the wiring, test the alternator output as follows (Fig. E:2). Disconnect the multi-plug and connect a jump lead between the blue lead terminal and its socket on the alternator. Connect an ammeter between the output leads and multi-plug red lead terminals. Connect a voltmeter across the battery terminals. Fit a 30 amp resistor between the battery positive terminal and earth. With the engine running at 3,000 rpm with the headlamps, heater blower motor and heated rear window working, the alternator output should be as specified (See TECHNICAL DATA). The voltage across the battery should not drop to less than 13 volts. With the ignition turned off, disconnect the resistor between battery and earth.
5. If the voltage is less than specified, the output is less than prescribed. Check the voltage regulator as follows (Fig. E:3). With the engine running at 3,000 rpm with the headlamps, heater blower motor and heated rear window off, note the ammeter reading. The reading will drop until 10 amps is indicated. At this point the voltmeter across the battery terminals should read between 13,7 and 14,5 volts. If outside these limits, the voltage regulator is defective.
6. Connect the positive side voltage drop test (Fig. E:4). If the voltager reading is more than 0,5 volt, with the engine running at 3,000 rpm with the headlamps on, there is a fault on the positive side of the charging circuit wiring. If within specification, connect the negative side volt drop test (Fig. E:5). If the reading on the voltmeter is in excess of 0,25 volt, there is a fault in the negative side wiring of the charging circuit.

NOTE: Most faults allied to the charging system are usually due to either worn brushes or to a defective voltage regulator.

Replacement of both the brushes and the integral regulator unit on alternators is detailed in this chapter. If anything more involved such as the rectifier diodes, slip rings, bearings etc., are faulty then it is recommended that any repairs are given to an auto electrical specialist or an exchange unit obtained.

In most cases, the latter will be found to be the most economical and convenient solution rather than attempt to obtain parts and repair the alternator.

NOTE: It should be noted that with an alternator, the battery should ALWAYS be disconnected before starting work on the charging system to avoid any possibility of damage to the semi-conductor devices in the rectifier and diode units.

Removal and Installation

1. Disconnect the battery and remove the heat shield, held by two bolts. Pull the multi plug from the rear of the alternator.
2. Slacken the alternator mounting bolts (Figs. E:6 & 7) and push the alternator towards the engine until the drive belt can be detached from the pulley.
3. Remove the mounting bolts and detach the alternator from the engine.
4. Install the alternator in the reverse order of removal. Adjust the drive belt tension as detailed in the ROUTINE MAINTENANCE chapter.

Brush Replacement

The brushes can be easily checked and replaced after removal of the alternator. Remove the black plastic cover from the rear of the alternator. This is secured by two retaining screws with hexagon heads and a small box spanner will be required to release these (Fig. E:8).

Remove the two screws securing the two brush retaining plates and withdraw the brush assemblies from the brush box (Fig. E:9). Fit the new brush assemblies into the brush box and secure in position with the securing screws. Check that the various terminals are correctly reconnected before refitting the end cover.

Alternatively, the complete brush box assembly, can be removed from the slip ring end bracket. This will allow inspection of the slip rings and the brush holders. Remove the two screws securing the brush box flange to the slip ring end bracket and lift off the assembly.

Inspect the brushes for wear. With the brushes in the free position, measure the amount by which they protrude beyond the brush box moulding. The brush length when new is 0.5 in (13 mm). If the amount protruding is near the wear limit of 0.2 in (5 mm), then the brush assemblies should be renewed.

Remove the old brush assemblies from the brush box and fit the new ones, as detailed above.

Check the new brushes for freedom of movement in their holders. Clean any brushes which are stiff with a petrol-moistened cloth, or by lightly polishing the brush sides on a smooth file if necessary.

Inspect the surface of the slip rings on the end of the rotor. If there is any evidence of roughness or burning this can be cleaned off with very fine glass paper. On no account must emery cloth or similar abrasive be used.

Before refitting the brush box, clean off any dirt which may have collected around the slip ring end bracket or the apertures in the plastic end cover.

Check that the brushes are correctly positioned in their holders, then locate the brush box assembly on the alternator end bracket and secure with the two retaining screws. Then refit the end cover.

Voltage Regulator (Fig. E:10)

The voltage regulator can be removed and refitted after removal of the black plastic end cover, see Brush Replacement. The regulator wiring connections can then be disconnected, then the retaining screw removed. The regulator can then be withdrawn. Refit the regulator in the reverse order, ensuring that the plastic spacer on the end of the unit is replaced.

BOSCH ALTERNATOR . . . . [3]

The Bosch alternator fitted to the Escort is similar in construction to the Lucas unit. However, there are certain detail differences between the types. See the Lucas Alternator section for details of the precautions and charging circuit test. See the TECHNICAL DATA for details of output and regulated voltage.

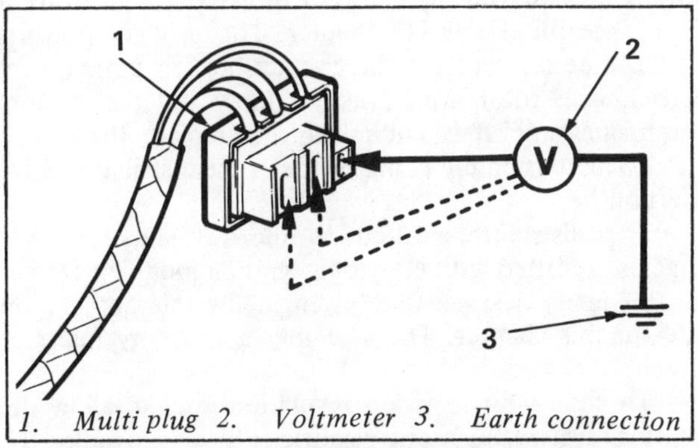

Fig. E:1 Lucas alternator wiring loom check

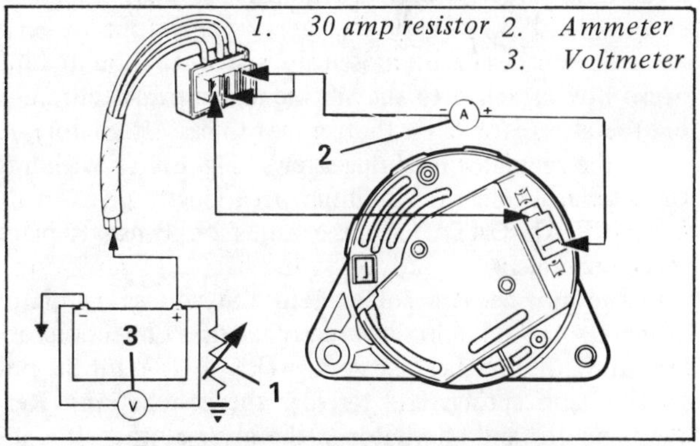

Fig. E:2 Lucas alternator output check

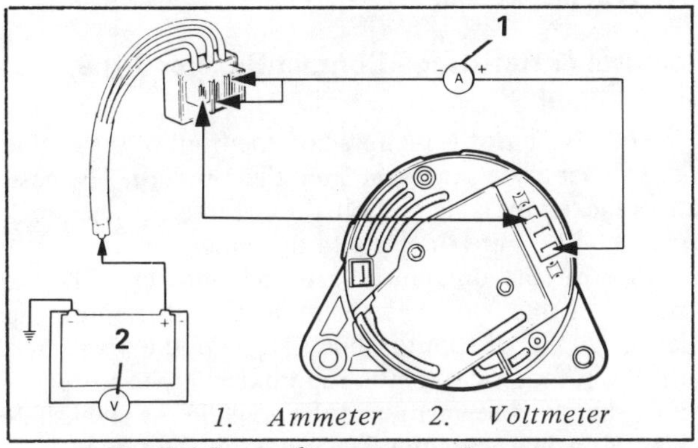

Fig. E:3 Lucas alternator regulator test

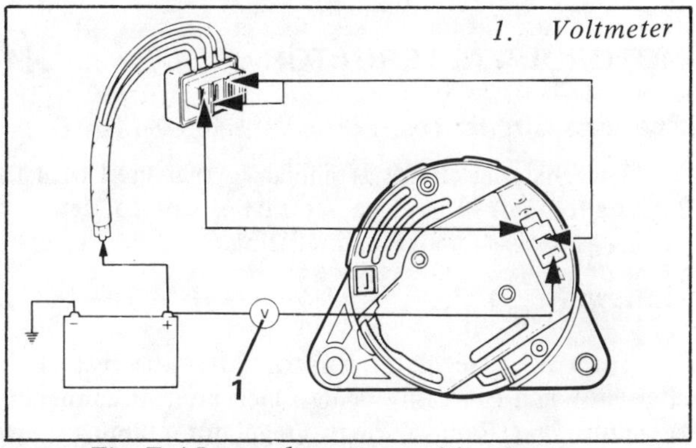

Fig. E:4 Lucas alternator positive drop test

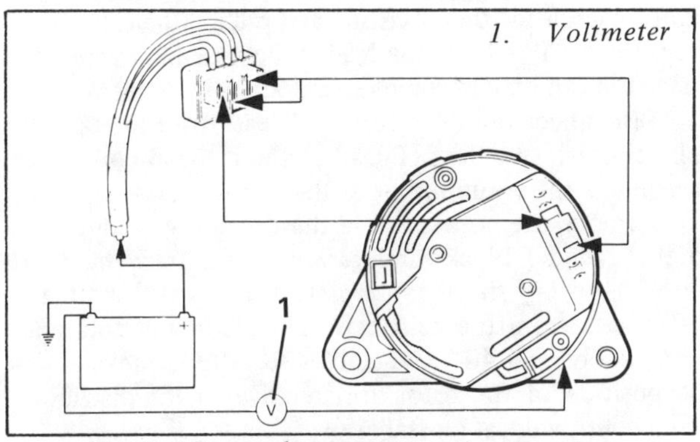

Fig. E:5 Lucas alternator negative drop test

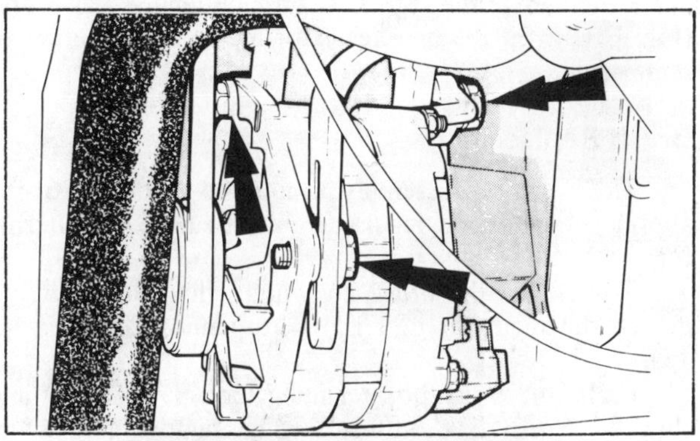
Fig. E:6 Alternator mounting bolts seen from above (arrowed)

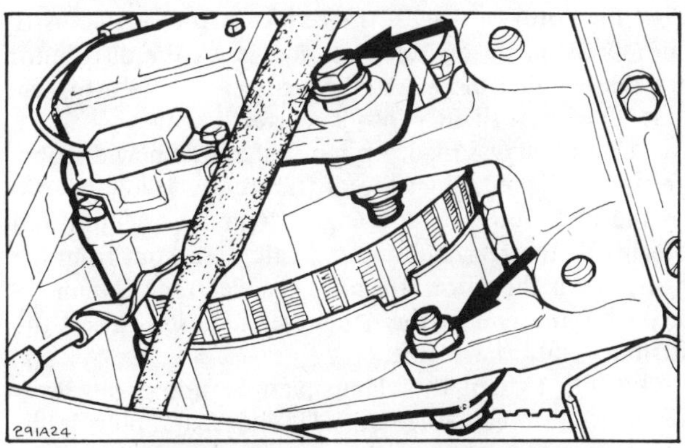

Fig. E:7 Alternator mounting bolts seen from below (arrowed)

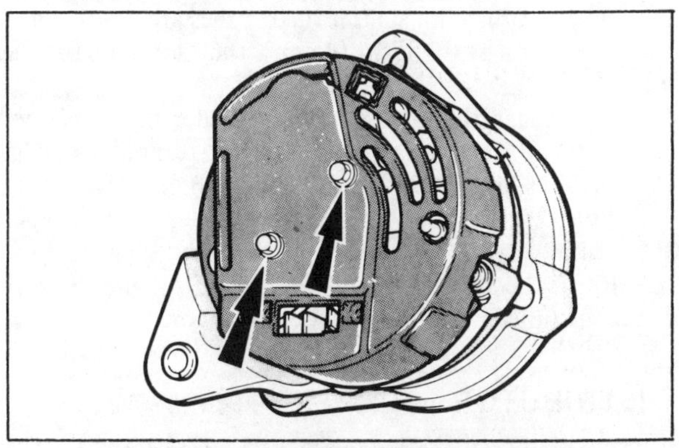
Fig. E:8 Lucas alternator rear cover bolts

Brush Replacement & Voltage Regulator

The brushes on the Bosch alternator are mounted in a brush box attached to the voltage regulator. After removing the alternator from the car (see Lucas Alternator), remove the regulator retaining screws (Fig. E:12). Withdraw the regulator. Measure the brush free length, given in the TECHNICAL DATA. See the notes on Brush Replacement previously.

The brushes are soldered to the voltage regulator. Using a small soldering iron unsolder the connections and withdraw the brushes and springs (Fig. E:13). Fit the new brushes and springs and resolder the connections. Refit the regulator and alternator in the reverse order. If a new voltage regulator is to be fitted, new brushes should be fitted.

MOTOROLA ALTERNATOR .... [4]

Charging Circuit Test

The charging circuit is similar to that used with the Lucas alternator. Therefore, see that section for details of testing as well as removal and refitting.

Voltage Regulator

The voltage regulator is mounted at the rear of the alternator and can easily be detached after disconnecting the multi-plug. Remove the two regulator retaining screws (Fig. E:15), disconnect the two wires and remove the regulator.

If defective, the regulator must be renewed as a unit (Fig. E:18). Fit the new regulator in the reverse order of removal.

Brush Replacement

The brush box assembly is mounted at the rear of the alternator under the regulator unit. Remove the alternator.

To remove the brush box, undo the two screws to detach the regulator unit and wire connections (Fig. E: 15).

Undo the brush box retaining screw (Fig. E:16) and turn the brush box outwards when removing to avoid damaging the protruding brushes (Fig. E:17).

When new, the brushes protrude 0.4 in (10 mm) beyond the brush housing. If the brushes are worn near to the wear limit of 0.157 in (4 mm), they should be replaced.

To replace the brushes, first unsolder the brush connections and remove the brushes with springs. Install the new brushes and springs in the brush holder, making sure that both brushes slide freely before soldering the leads to the terminals.

Refit the brush box to the alternator and refit the regulator unit in the reverse order of removal.

DISTRIBUTOR........................ [5]

The distributor fitted to 1.1 OHV and CVH engine versions is of the contact breaker ignition type. The assembly may be made by Lucas or Bosch, for identification details see the TUNE-UP chapter. The only components that can be replaced are the contact breaker points, condenser, cap, rotor arm, drive assembly and the vacuum diaphragm unit. It is not possible to replace the centrifugal weight assembly components or the distributor drive shaft bushes.

The distributors fitted to the 1.3 and 1.6 CVH engines are fitted with electronic ignition modules. Details of this ignition system are given in the relevant section later in this chapter. The assembly is made by Lucas or Bosch.

To cope with emission regulations and to allow the ignition system to work as efficiently as possible, the vacuum pipes to the distributor are fitted with devices to control the intake depression. See the 'Vacuum Line' Devices in this section.

Removal & Refitting - Contact Breaker Type, OHV Engine

The distributor is mounted on the front of the engine and is driven by a skew gear from the camshaft. The position is locked by a clamp bolt through the base clamp. A separate bolt clamps the plate to the block.

Remove the distributor cap and note or mark the position of the No. 1 segment in the distributor cap relative to the distributor body. Rotate the crankshaft until the rotor arm is pointing approximately to the position of the No. 1 segment, and the notch in the crankshaft pulley is in line with the TDC mark on the timing scale (Fig. E:21).

Disconnect the plug leads by grasping the moulded cap of each lead, then twisting and pulling the cap off the plug. Do NOT pull on the lead, otherwise the connection inside the cap may become separated.

Disconnect the distributor LT lead from the ignition coil terminal and the HT lead at the ignition coil tower. Disconnect the vacuum pipe at the distributor.

Remove the bolt securing the distributor clamp plate to the cylinder block, and carefully withdraw the distributor, allowing the rotor arm to turn as the shaft gear dis-engages from the camshaft. The shaft will rotate approximately 35 - 40° anti-clockwise. After removal, mark the position of the rotor arm relative to the distributor body as this will ease installation.

Before refitting, ensure that the engine is at TDC position with No. 1 cylinder on its compression stroke. Align the rotor arm with the scribed mark on the distributor body made on removal, and insert the distributor into the cylinder block. As the gears mesh, the rotor will rotate clockwise slightly into its correct position.

If no mark was made on removal, or if a new distributor is being fitted, align the distributor as follows: Position the distributor on the engine with the vacuum unit pointing to the rear and approximately 35° out from the engine. Turn the rotor arm until the electrode is immediately over the end of the condenser cannister. Insert the distributor into the engine.

Fit and tighten the clamp plate securing bolt. Refit the distributor cap, the LT, HT and spark plug leads. Ensure that the plug leads are refitted in their correct respective positions. Reconnect the vacuum line.

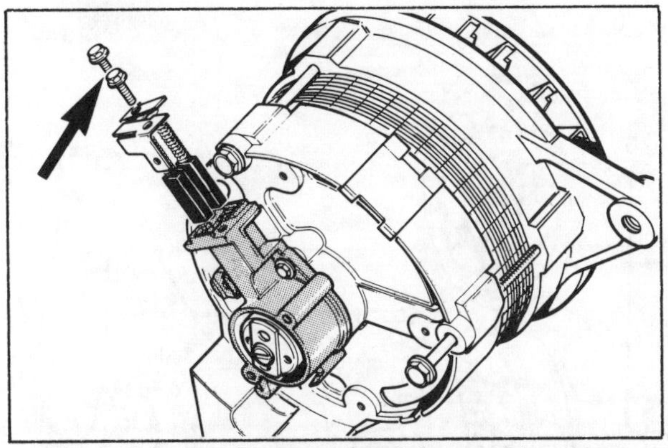

Fig. E:9 Lucas alternator brush box and brushes

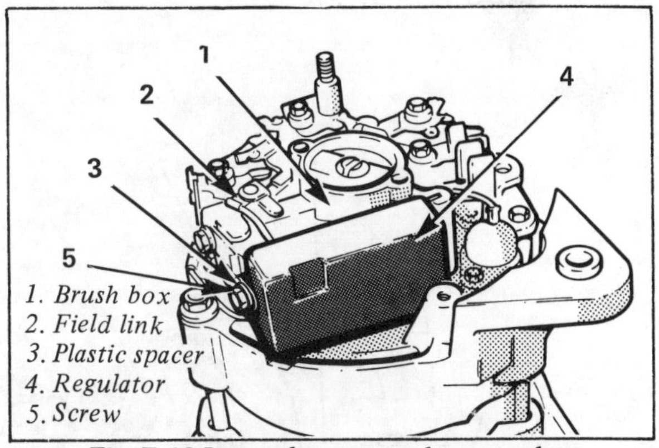

Fig. E:10 Lucas alternator voltage regulator

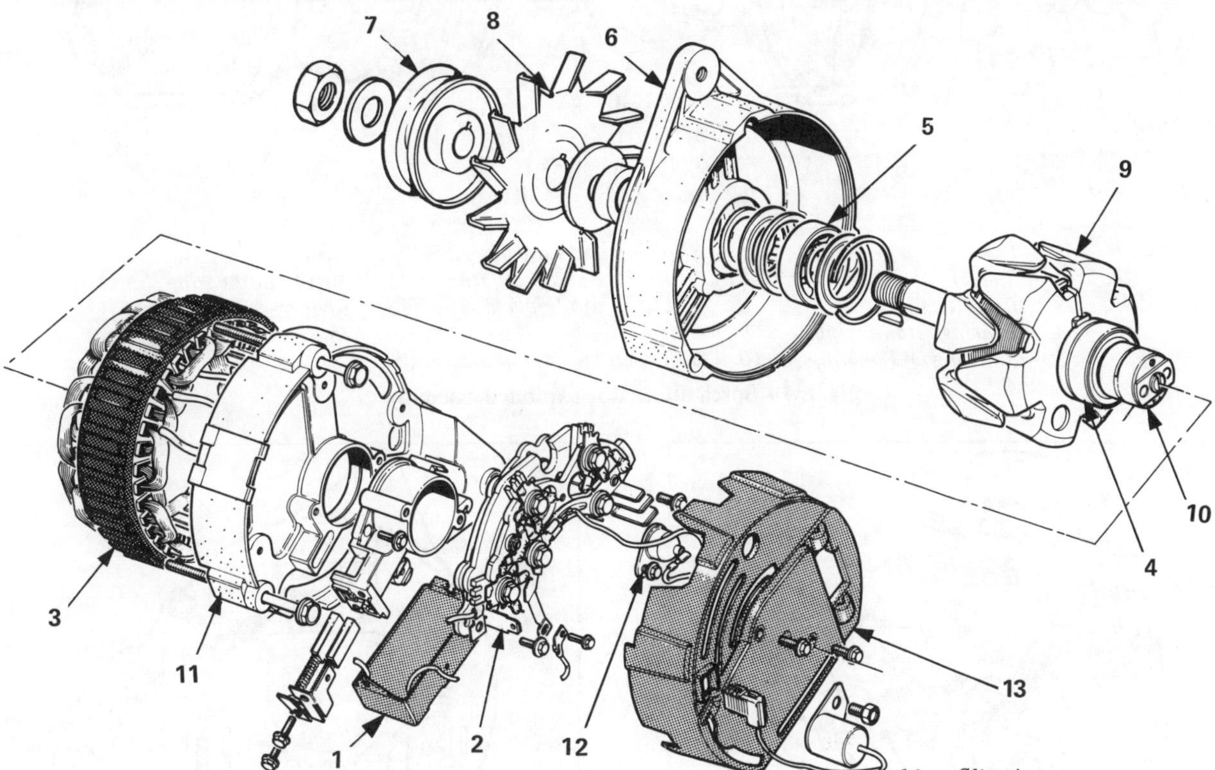

1. *Voltage regulator*
2. *Rectifier diode pack*
3. *Stator*
4. *Slip ring end bearing*
5. *Drive end bearing*
6. *Drive end housing*
7. *Pulley*
8. *Fan*
9. *Rotor*
10. *Slip ring*
11. *Slip ring end housing*
12. *Surge diode*
13. *End housing*

Fig. E:11 Lucas alternator exploded assembly

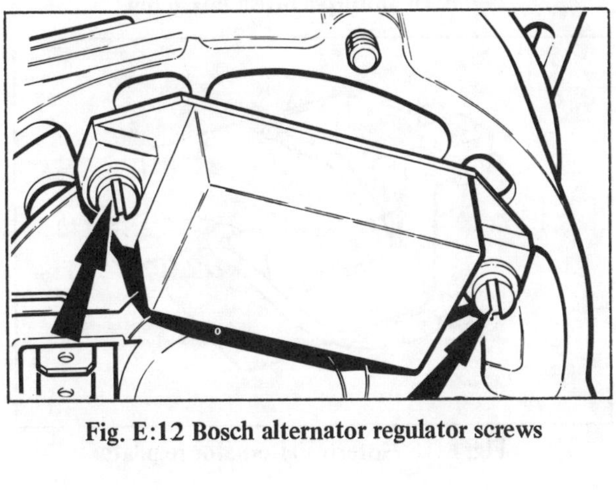

Fig. E:12 Bosch alternator regulator screws

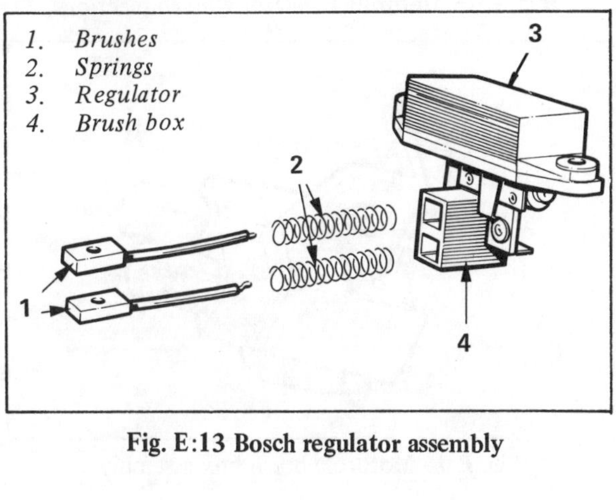

Fig. E:13 Bosch regulator assembly

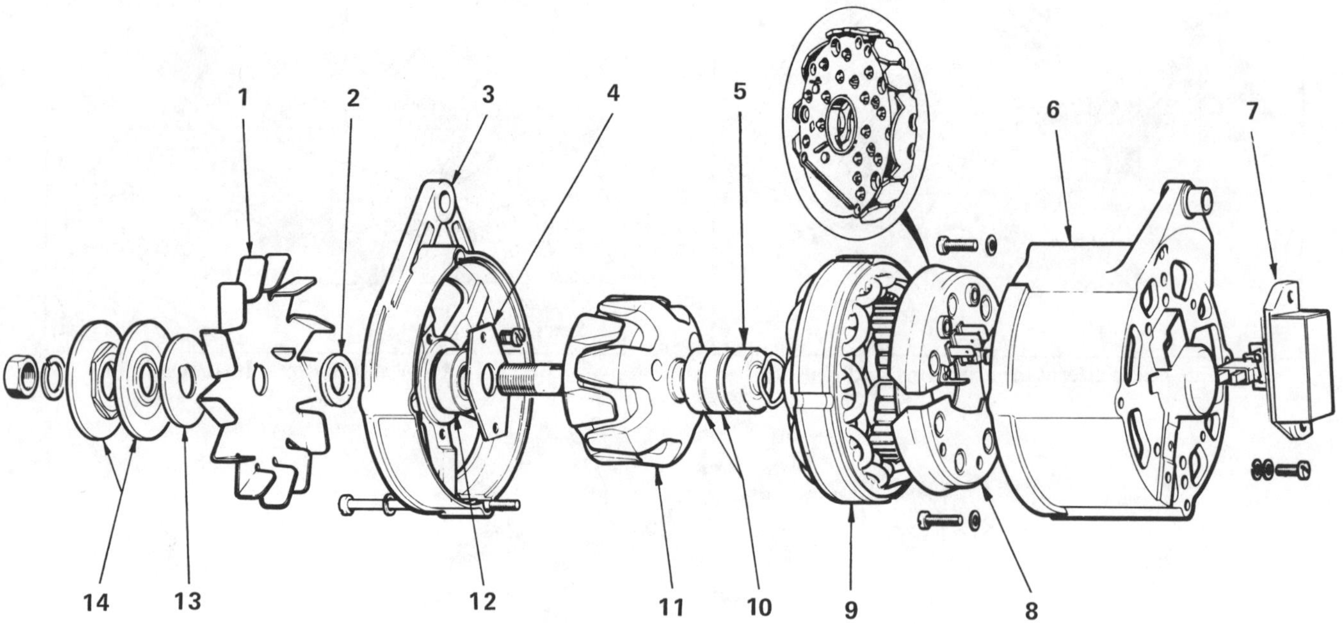

1. *Fan*
2. *Spacer*
3. *Drive end housing*
4. *Bearing retaining plate*
5. *Slip ring end bearing*
6. *Slip ring end housing*
7. *Brush box and regulator*
8. *Rectifier diode pack*
9. *Stator*
10. *Slip rings*
11. *Rotor*
12. *Drive end bearing*
13. *Spacer*
14. *Pulley*

Fig. E:14 Bosch alternator exploded assembly

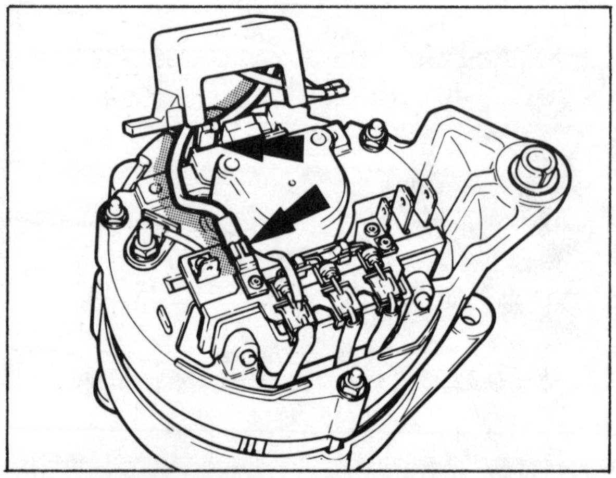

Fig. E:15 Motorola regulator wire connections

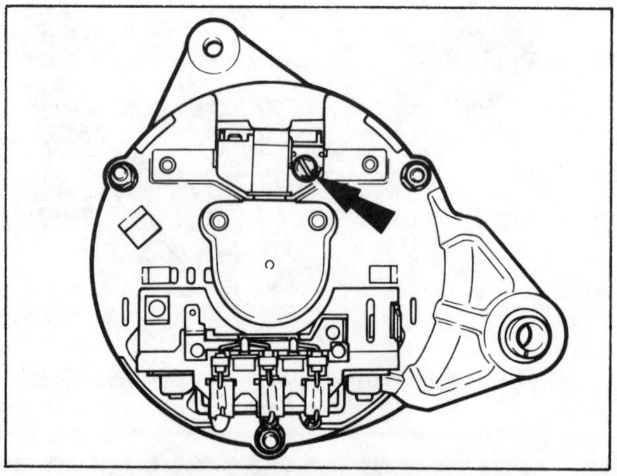

Fig. E:16 Motorola brush box screw

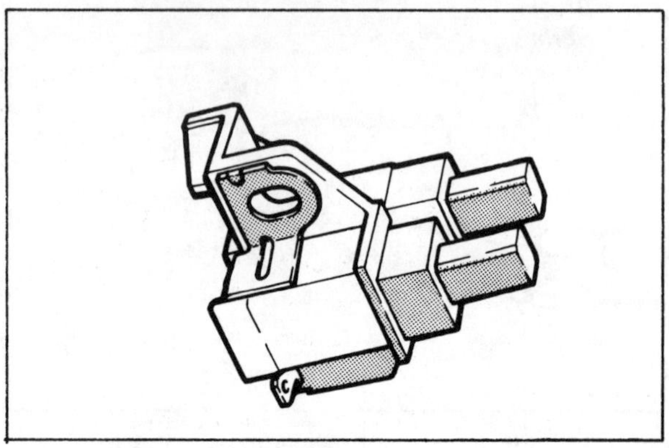

Fig. E:17 Motorola brush box assembly

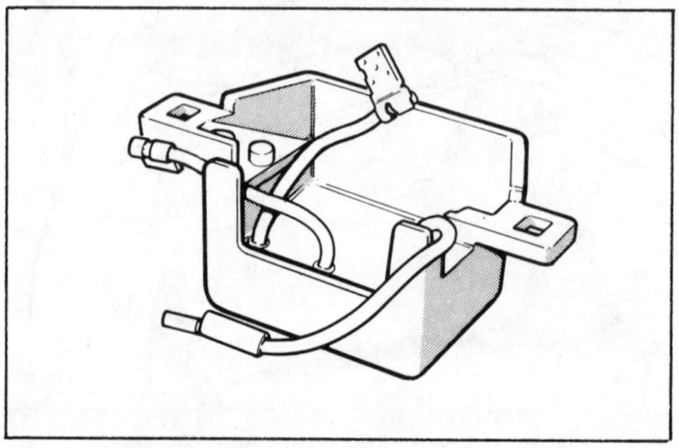

Fig. E:18 Motorola alternator regulator

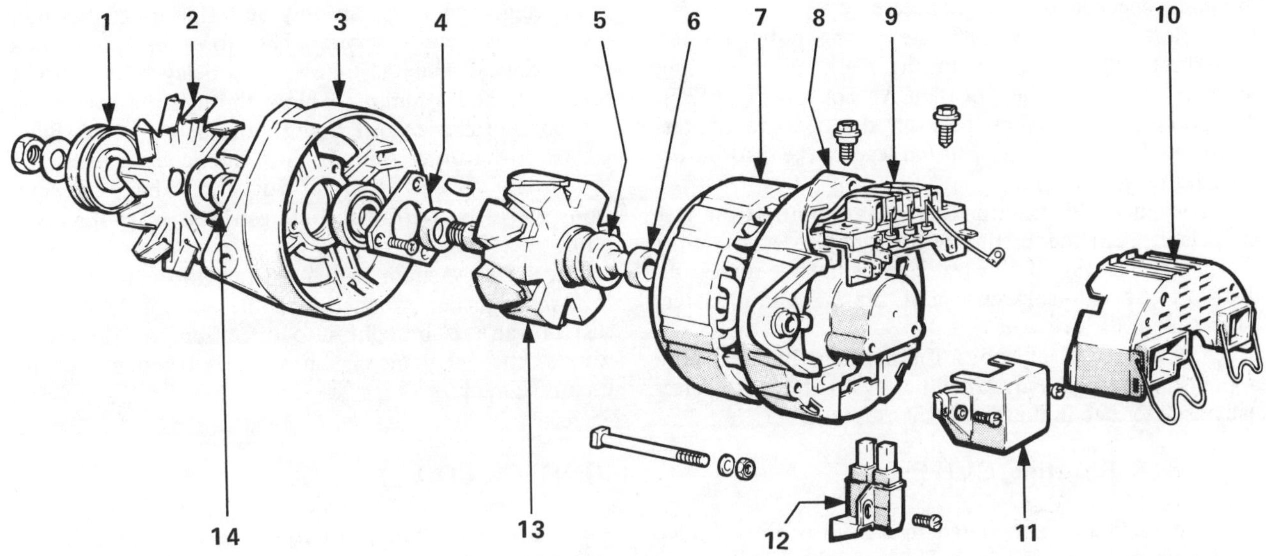

1. *Pulley*
2. *Fan*
3. *Drive end housing*
4. *Bearing retaining plate*
5. *Slip ring*
6. *Slip ring end bearing*
7. *Stator*
8. *Slip ring end housing*
9. *Rectifier diode pack*
10. *End cover*
11. *Regulator*
12. *Brush box*
13. *Rotor*
14. *Spacer*

Fig. E:19 Motorola alternator exploded assembly

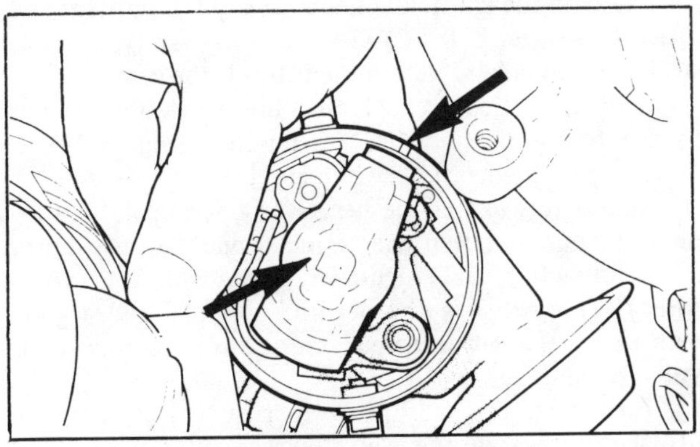

Fig. E:20 OHV engine distributor rotor at TDC, no. 1 cylinder

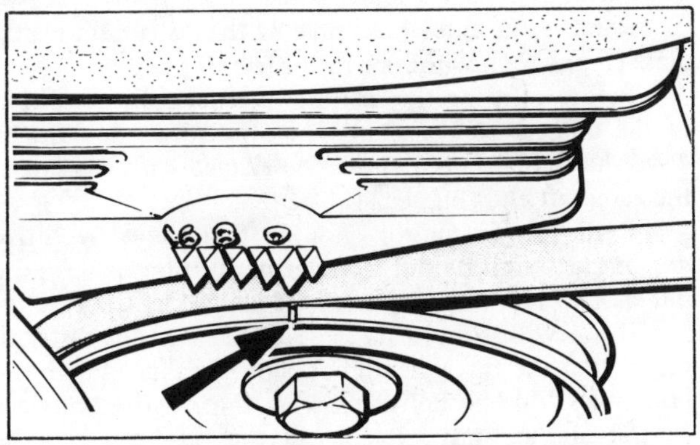

Fig. E:21 CVH engine timing marks at TDC

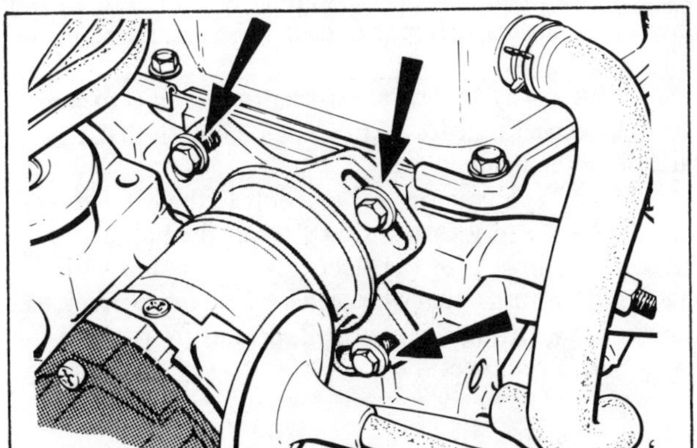

Fig. E:22 CVH engine distributor clampbolts

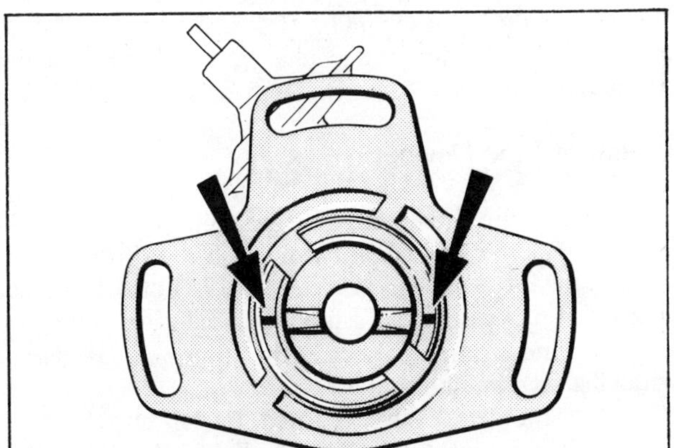

Fig. E:23 CVH engine distributor drive dog

After refitting the distributor, the ignition timing should be checked and reset as necessary.

1. Rotate the engine until the timing notch on the crankshaft pulley aligns with the mark on the timing scale as specified in the TECHNICAL DATA (Fig. E:20).
2. Connect a test lamp between the negative LT terminal on the ignition coil and the lead to the distributor. Switch the ignition on.
3. Slacken the distributor shaft clamp bolt. If the test lamp is lit, turn the distributor body anti-clockwise until the lamp goes out. If the test lamp is not lit, rotate the distributor body clockwise until the lamp illuminates. Then turn anti-clockwise.
4. Tighten the clamp bolt. The ignition timing is now approximately set, and should be set accurately dynamically as described in the TUNE-UP chapter.

Removal & Refitting, CVH Engine

The distributor is mounted on the side of the cylinder head, driven by the camshaft. The assembly fitted to contact breaker ignition cars is similar in appearance to the electronic ignition fitted to larger engine cars, but a different procedure is necessary when refitting.

1. Disconnect the battery earth lead. Rotate the engine to TDC (Fig. E:21). Disconnect the distributor low tension lead from the ignition coil. Pull the spark plug leads from the distributor cap. Disconnect the vacuum tube from the vacuum capsult. Note the punch marks on the distributor body and cylinder head.
2. Remove the three bolts holding the distributor clamp plate (Fig. E:22). Withdraw the distributor body.
3. Before refitting the distributor, check the condition of the oil seal on the shaft. The offset drive dog on the drive shaft (Fig. E:23), should be aligned with the slot in the camshaft end (Fig. E:24).
4. Refit the distributor ensuring that the offset drive dog engages with its slot. Rotate the distributor until the punch marks on the body and head align (Fig. E:25).
5. If a new distributor is to be fitted, align the distributor so that the three clamp bolt slots are aligned with the bolt holes. Fit the bolts, checking that all the bolts are central in their slots. Reconnect the HT leads and the vacuum pipe.
6. Connect the LT lead to the ignition coil. Reconnect the battery earth lead.
7. Check the ignition timing as described in the TUNE-UP chapter. Tighten the clamp bolts to the specified torque.

Vacuum Line Devices

All cars are fitted with a fuel/vacuum separator in the vacuum pipe between distributor and carburettor. Its purpose is to stop fuel from damaging the vacuum diaphragm. If the valve is suspect, it can be replaced after pulling the vacuum lines from either end. The filter body is marked showing vacuum line connection.

A spark delay valve is fitted to 1.1 and 1.3 CVH engine high compression cars (Fig. E:26). A spark sustain valve is fitted to the XR 3 variant (Fig. E:27). The delay valve is to aid combustion and lower emission levels. The sustain valve is to aid throttle response under part load when cold. The valve can only be tested using a vacuum pump, gauge and reservoir. Therefore, an approximate check can be made as follows. If a delay valve is fitted, remove it and attempt to blow through the valve from the carburettor side. Air flow should be restricted. Blowing from the other side air flow should be unrestricted. If a sustain valve is fitted, air flow should be unrestricted from the carburettor side but restricted from the distributor side.

A ported vacuum switch is fitted to the intake manifold water jacket. When the engine is cold, the delay or sustain valve is brought into operation. As the engine warms, the delay or sustain valve is taken out of the vacuum circuit.

IGNITION COIL. . . . . [6]

The ignition coil is mounted on the right hand side of the engine bay (seen from the front of the car) on the inner wing panel (Fig. E:28). Note that a coil used on contact breaker type ignition is not suitable for electronic ignition. If replacement is necessary, it is most important that the correct replacement is obtained. A blue labelled coil is fitted to contact breaker ignition cars. A red labelled coil is fitted to electronic ignition cars.

Ignition Coil Test

Faults in the ignition coil are usually intermittent and difficult to trace. If the coil is completely faulty, a simple check can be made to determine if this is the case.

1. With a test lamp check that low tension current is flowing to the coil positive (+) terminal (Fig. E:29). If none, suspect the ignition switch and lead.
2. Repeat this test at the negative (-) terminal. First try the test lamp between the terminal and the distributor lead (contact breaker ignition only, with the points closed). Test between the negative terminal and a good earth point. If there is no current at this point, the coil is probably faulty.
3. These tests can only prove that the coil is operating sufficiently to start the car. However, after the car has been running for a while and the coil has warmed up, problems can sometimes develop. A constant misfire at high revs and whilst under load, assuming that the rest of the ignition circuit is in order, can usually be attributed to the coil.
4. The best way to test an intermittent coil is to substitute the suspect unit for a known good coil so that a comparison can be made.

The coil is located in a single bolt fixing bracket. Disconnect the battery earth lead. Slacken the bolt and slide the coil out. Exchange the electrical connections with the new unit, ensuring that the leads go to the correct terminals. When fitting the new coil, do not overtighten the fixing bolt. Reconnect the battery.

Ballast Resistor

A ballast resistor is fitted to the ignition circuit of contact breaker ignition cars to aid starting. The ignition

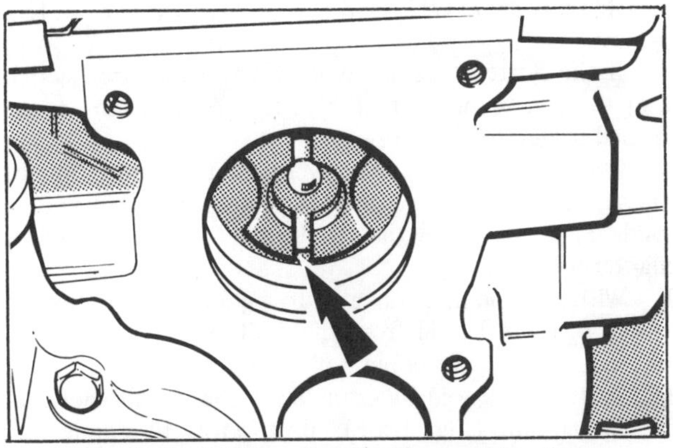
Fig. E:24 CVH engine distributor drive slot

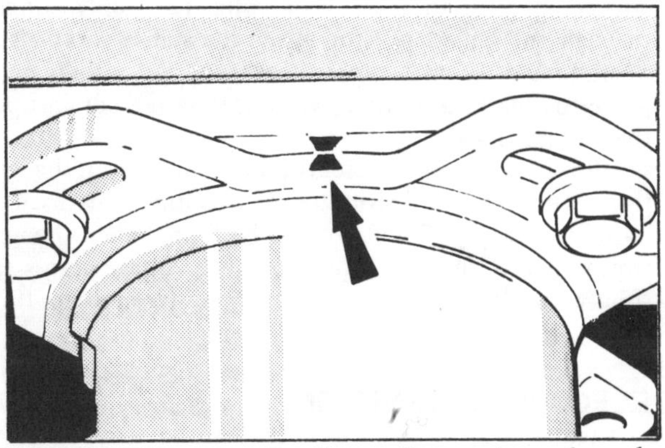
Fig. E:25 CVH engine distributor alignment punch marks

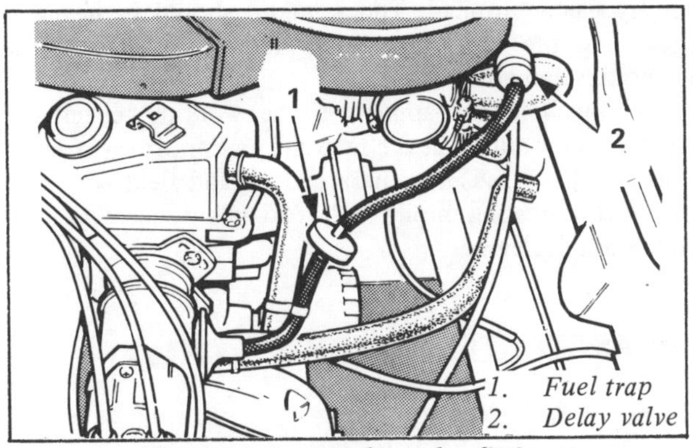

Fig. E:26 Spark delay valve fitting

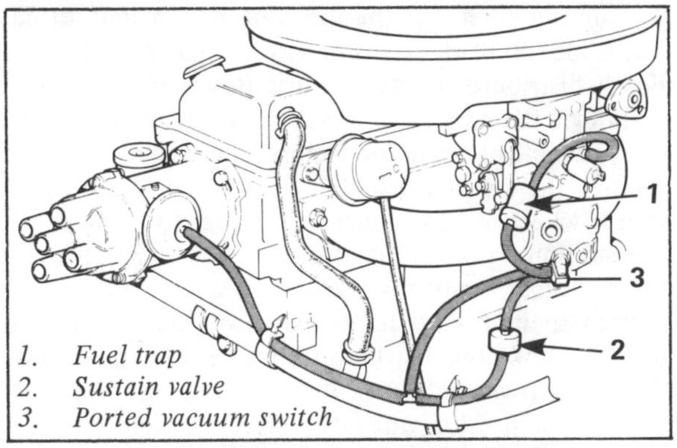

Fig. E:27 Spark sustain valve fitting

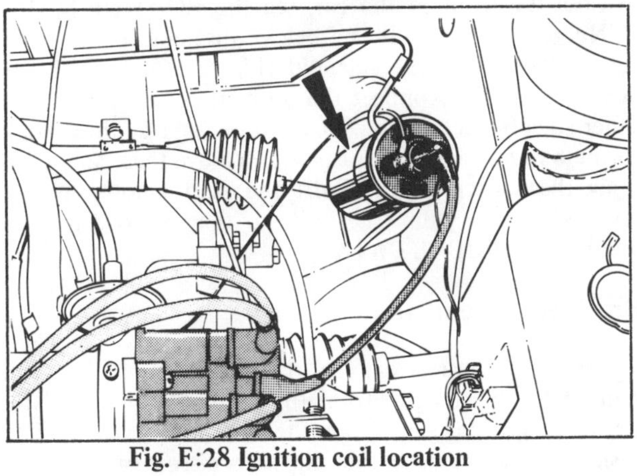
Fig. E:28 Ignition coil location

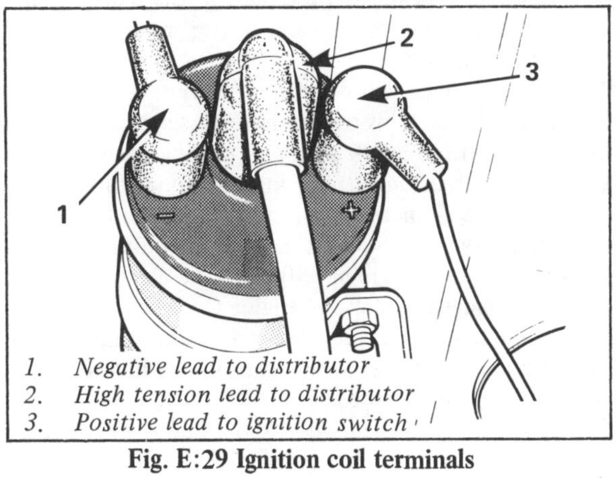

Fig. E:29 Ignition coil terminals

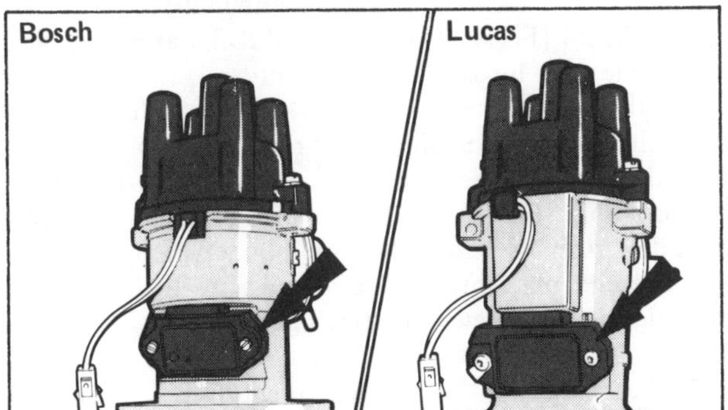

Fig. E:30 Electronic ignition distributor, ignition module arrowed

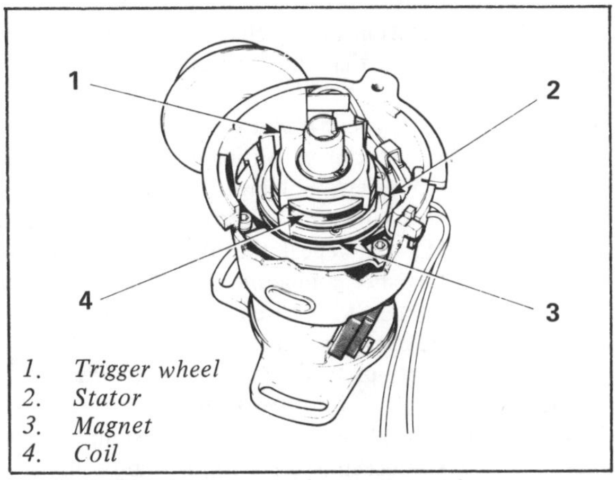

Fig. E:31 Bosch electronic ignition

coil is designed to work at 7 volts, its supply coming through the ballast resistor. When the starter motor is in operation, a 12 volt supply feeds the coil, producing a powerful spark, at a time when the starter's drain upon the battery is high.

The resistor on the Escort is in the wire from the ignition switch to the distributor. If a voltmeter is used to check the supply, 7 volts should be found when the ignition switch is in the on position. The voltage across the supply should be 12 volts when the starter motor is turning.

ELECTRONIC IGNITION ....[7]

The electronic ignition circuit is controlled by a sealed unit mounted on the distributor. The unit fitted to Lucas distributors is manufactured by AC Delco, the Bosch distributor is fitted with a Bosch unit (Fig. E:30). Although the two systems are different in construction, the principle of operation is similar. The camshaft drives a shaft in the distributor body. On the end of the shaft is a trigger wheel. As a cylinder approaches its firing point on the ignition stroke, a metal lobe or trigger passes in front of a magnet, inducing current in a coil winding. The current so induced is fed to the electronic unit mounted on the side of the distributor body. The module controls the strength and duration of the spark (Figs. E:31 & 32). The ballast resistor wire of contact breaker ignition systems is replaced by the circuitry in the module.

If there appears to be a fault in the ignition circuit, after checking the battery and starter motor connections proceed as follows.

1. Test the high tension current output from one of the spark plug leads. This can be done by smashing the insulation from an old spark plug and cutting away the centre electrode. Fit the electrode into the plug cap. Hold the cap no more than 0.2 in (5 mm) from the engine block with insulated pliers. This is most important - the gap must be no more than specified or the electronic module may be damaged. Crank the engine (Fig. E:33).
2. If no spark is made, disconnect the plug lead from the distributor and fit to the ignition coil high voltage terminal. Crank the engine again.
3. If no spark is made, check the coil connections. Repeat the high voltage test.
4. Check the distributor static timing, as described in the Distributor section of this chapter. Remove the distributor cap and check that the drive shaft rotates when the engine is cranked.
5. If at this point there is still no spark, check the connections to the distributor ignition module (Fig. E:34). Remove the two screws holding the module, remove it and examine the connectors. If tarnished, clean with fine glass paper. Refit and repeat the high voltage test. Note that a special heatsink compound (part no. 81SF-12103-AA) should be lightly smeared over the back of the module to ensure a good earth connection.
6. If the connections to the electronic module are clean and the system will still not work, the electronic module is probably at fault. If possible, test the system further by replacing the suspect module with a module known to work.

STARTER MOTOR TEST ....[8]

If the starter does not work when the ignition switch is in the starter position, the following tests can be performed to determine the cause.

1. Check that the battery is in an adequate state of charge, and that the battery terminals are clean and in good condition. See the ROUTINE MAINTENANCE chapter for details.
2. With a test lamp connected to a good earth, check the supply to the solenoid from the ignition switch. An assistant may be needed to operate the switch.
3. Test the solenoid by connecting the circuit shown in the illustration (Fig. E:35). Both push fit connectors should be removed from the solenoid. Disconnect the battery earth terminal. The solenoid should be heard to click, and the test bulb should illuminate. If not, the solenoid is defective.
4. If the solenoid appears to be working, check the brushes as described in the relevant section. If these are in good condition, suspect the armature and field windings. Any further testing should be left to an auto electrician or your Ford dealer.

LUCAS STARTER MOTOR ....[9]

Removal & Refitting

Disconnect the battery earth lead. It will be necessary to raise the front of the car, see the ROUTINE MAINTENANCE chapter. From below disconnect the main starter cable from the solenoid (Fig. E:36). The cable is held by a nut. Disconnect the push fit connectors on the other solenoid terminals. The starter is held by three bolts to the bell housing adaptor plate. Remove these and withdraw the starter motor. Refit the starter motor in the reverse order. Tighten the bolts to the specified torque, see the TECHNICAL DATA.

Brush Replacement

Remove the starter from the engine as detailed then remove the plastic cap from the commutator end plate. Remove the star clip from the end of the shaft. The easiest way to remove the clip is to distort or break off the prongs of the clip with a small chisel until the clip can be withdrawn (Fig. E:37). Discard the clip after removal. A new one must be used on reassembly.

Withdraw the commutator end plate sufficiently to allow the two field brushes to be removed from their holders, then remove the end plate complete with the brush box assembly (Fig. E:38). If new brush springs are necessary, the brush box assembly should be replaced.

The end plate brushes come ready to fit in place of the originals. Unscrew the terminal nuts and fit the new set.

When renewing the field winding brushes, cut the old brush leads away from the terminal post then use a file or hacksaw to make a groove in the head of the terminal deep enough to accommodate the new brush leads. Ensure that the different lengths of lead are fitted the correct way round, then solder the new leads in position.

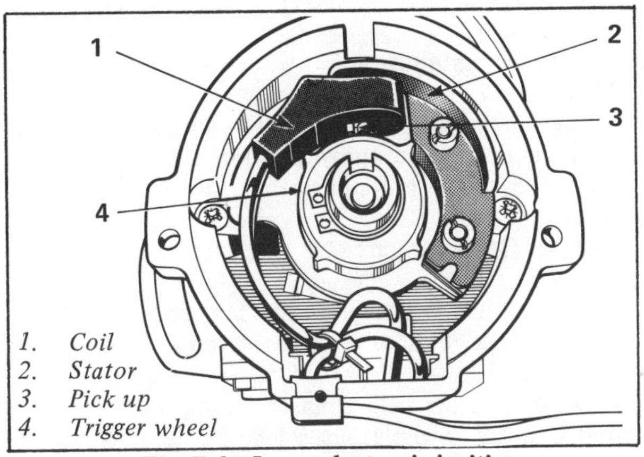

Fig. E:32 Lucas electronic ignition

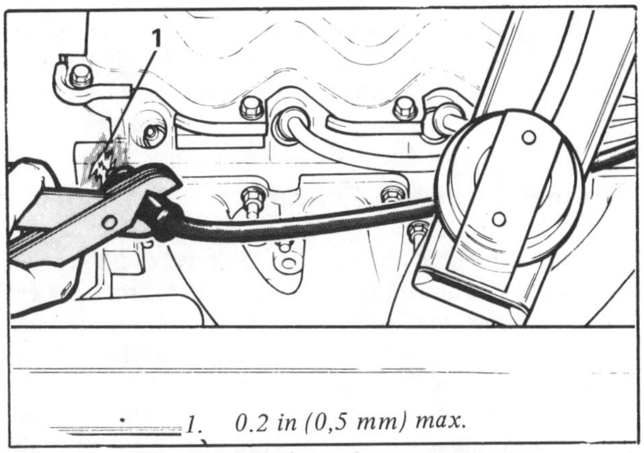

Fig. E:33 High tension output test

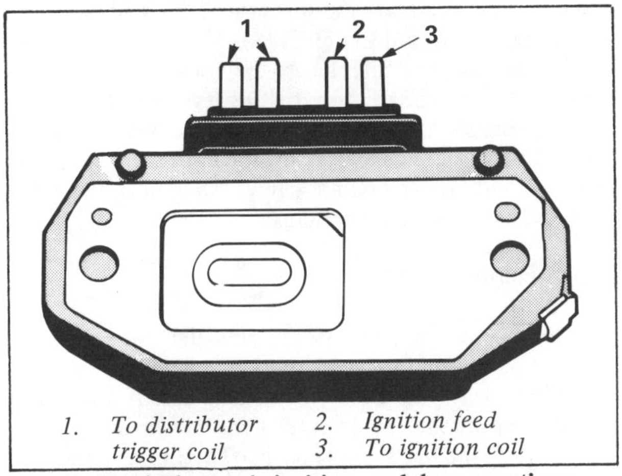

Fig. E:34 Electronic ignition module connections

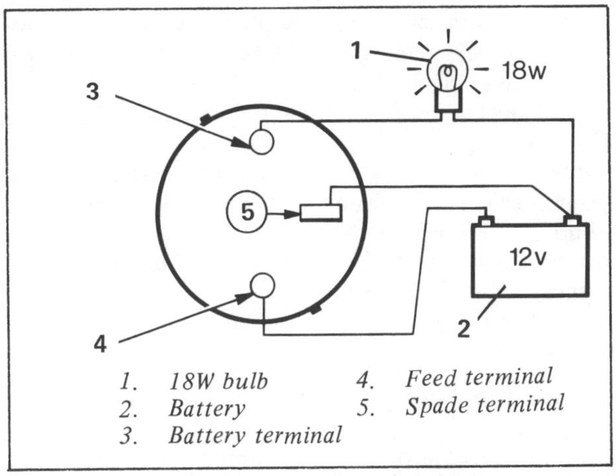

Fig. E:35 Starter motor solenoid test

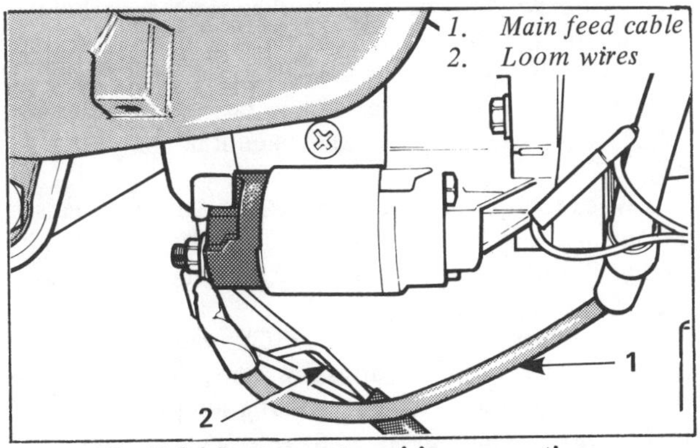

Fig. E:36 Starter motor wiring connections

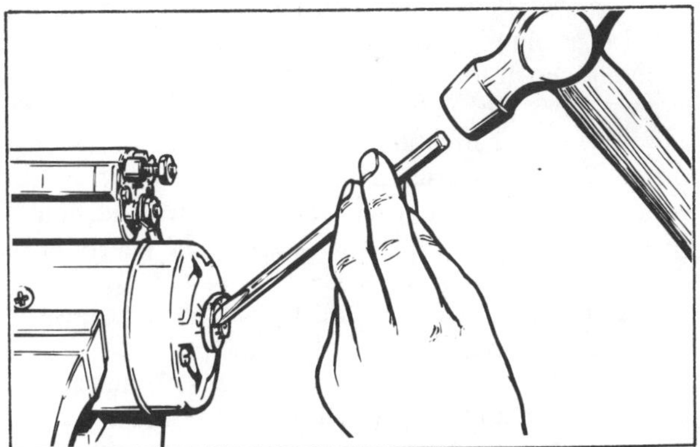

Fig. E:37 Lucas starter starclip removal

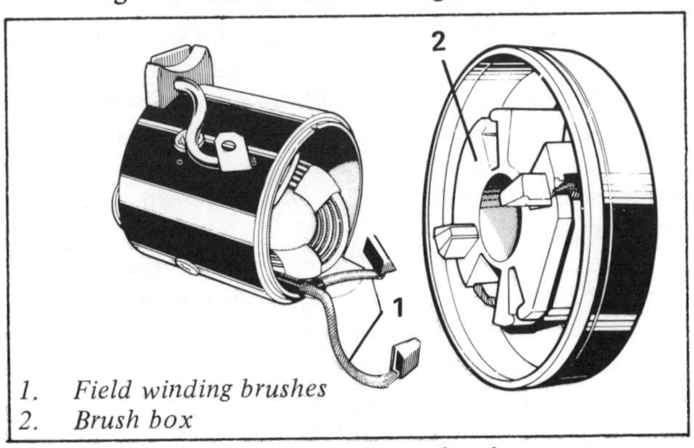

Fig. E:38 Lucas starter brushes

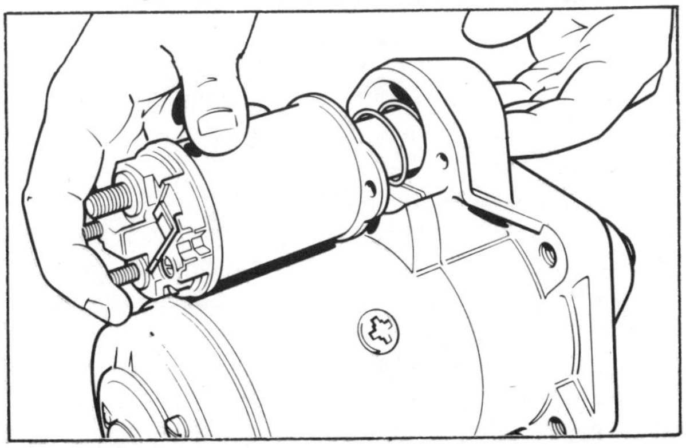

Fig. E:39 Lucas starter solenoid removal

After assembling the end plate on the starter yoke, fit the new star clip on the end of the armature shaft and press it fully home to eliminate any endfloat in the armature. Refit the plastic cap over the star clip and shaft end.

Solenoid Removal & Refitting

The solenoid is mounted on the drive end housing of the starter motor and is easily removed for replacement.

Disconnect the field connecting link from the lower terminal on the starter motor by unscrewing the retaining nuts and washers. Remove the two bolts securing the solenoid to the drive end housing and draw the solenoid body and spring away from the housing (Fig. E:39). Unhook the solenoid armature from the actuating lever inside the housing by moving the plunger upwards and away from the lever. Remove the plunger together with the rubber seal and brass insert.

To refit the solenoid, hook the solenoid plunger on to the drive engagement lever inside the drive end housing, ensuring that the brass insert and rubber seal are correctly positioned. Fit the solenoid body and spring over the armature and secure in position with the two retaining bolts. Reconnect the field link to the solenoid terminal (Fig. E: 40).

Drive Pinion Assembly

Replacement of the drive pinion assembly involves dismantling most of the starter motor.

Remove the solenoid and commutator end plate as detailed previously. At the drive end bracket, remove the retaining clip from the drive engagement lever pivot pin, and withdraw the pin from the housing. Separate the drive end bracket and armature assembly from the starter yoke, then withdraw the armature and pinion assembly together with the engagement lever from the drive end housing (Fig. E:41). The lever fork engages in the groove in the drive assembly clutch.

A suitably sized tube will be needed to remove the drive pinion assembly from the armature shaft (Fig. E: 42). Secure the armature in a vice with soft jaws and, using a piece of tube, tap the thrust collar down the shaft to expose the jump ring. Extract the jump ring from its groove in the armature shaft and slide the thrust collar and drive pinion assembly up off the shaft (Fig. E:43).

Slide the new drive pinion assembly and thrust collar on to the armature shaft. Fit the jump ring into the groove in the shaft, then pull the thrust collar up over the ring, using a two-legged puller.

Engage the actuating lever fork into the locating lugs on the drive pinion assembly and insert the armature assembly and lever into the drive end housing. Assemble the starter yoke to the drive end housing, ensuring that the rubber sealing block is in position at the pinion actuating lever. Locate the lever pivot pin through the drive end bracket and lever and secure with the retaining clip.

Assemble the commutator end plate and the solenoid as detailed under the appropriate headings previously.

BOSCH STARTER MOTOR . . . . [10]

Removal & Installation

The Bosch starter motor fitted to the Escort is similar in construction to the Lucas unit. For details of removal and refitting see the previous section.

Solenoid Replacement

The solenoid is secured to the drive end housing. The solenoid can be replaced after the starter has been removed.

Disconnect the field connecting link from the lower terminal on the solenoid. The solenoid is secured by two screws. Withdraw the solenoid housing. Unhook the solenoid armature from the operating lever and withdraw it with the return spring (Fig. E:45).

To refit the solenoid, locate the hook on the solenoid plunger over the end of the actuating lever inside the drive housing. Ensure the armature return spring is in position when fitting the solenoid body. Secure the solenoid in position with the retaining screws, then reconnect the field lead to the solenoid terminal.

Brush Replacement

Remove the two screws securing the protective end cap to the commutator end bracket and detach the cap and rubber seal (Fig. E:46). Wipe the grease from the end of the armature shaft, then extract the 'C' clip and remove the shims from the shaft (Fig. E:47). Unscrew the two nuts and washers and detach the commutator end housing. Withdraw the two field brushes from their holders on the brush plate, then remove the two earthed brushes by prising the springs clear then sliding the brushes from their housings. The brushes should be withdrawn before attempting to remove the brush mounting plate (Fig. E: 48).

When renewing the brushes, cut the old brush leads at a point mid-way between their base and the brush. Solder the new brush leads to the original connections, ensuring a good soldered joint.

To reassemble the brush gear, fit the brush holder plate over the end of the armature shaft, aligning the cut-outs in the plate with either the fixing studs or the 'loops' in the field windings, as applicable. Position the four brushes in their respective holders and retain with the brush springs. Position the spring ends on the brushes and check that the brushes seat correctly on the commutator.

Refit the commutator end housing, ensuring that the rubber insulator engages correctly in the cut-out in the end housing. Secure the end housing with the two bolts and washers. Fit the shims and 'C' clip on the end of the armature shaft.

There should be no endfloat on the armature when the 'C' clip is in place. If necessary, take up any play by altering the thickness of the shim pack under the 'C' clip.

Smear a quantity of lithium-based grease on the end of the armature shaft, then refit the protective end cover with its sealing ring and secure with the two retaining screws.

Fig. E:40 Solenoid connecting link

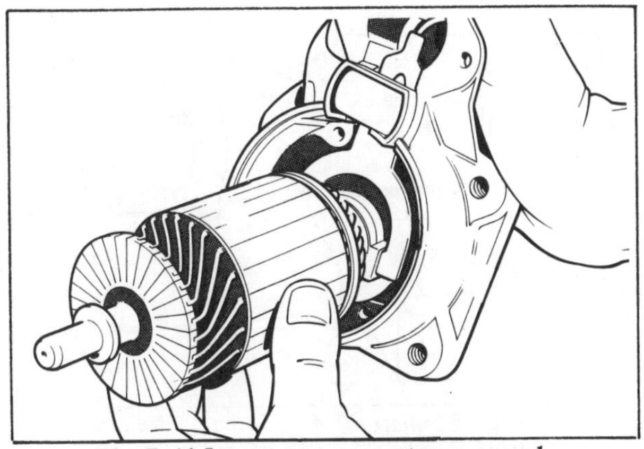

Fig. E:41 Lucas starter armature removal

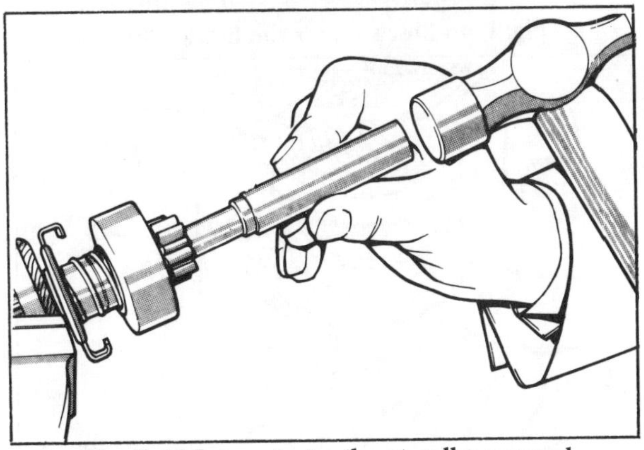

Fig. E:42 Lucas starter thrust collar removal

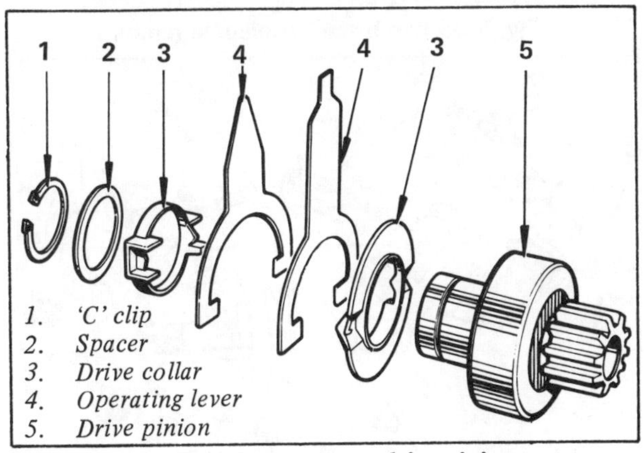

Fig. E:43 Lucas starter drive pinion

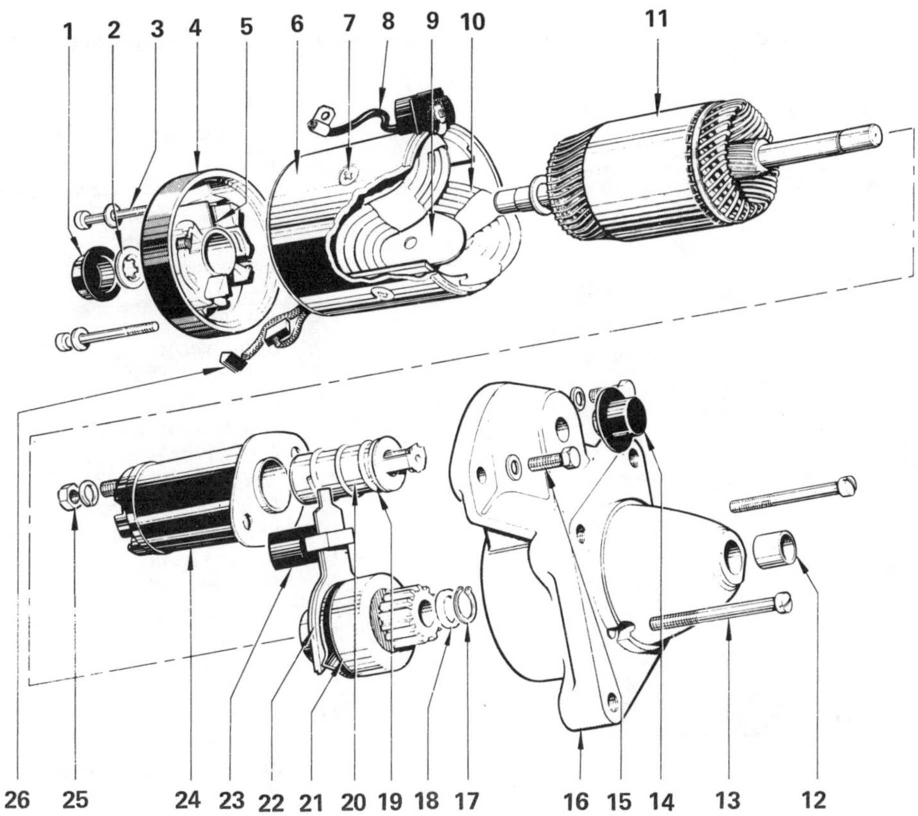

1. *Dust cover*
2. *Star clip*
3. *Commutator end plate retaining bolt*
4. *Commutator end plate*
5. *Brush housing*
6. *Main casing*
7. *Pole screw*
8. *Connector link*
9. *Pole screw*
10. *Field coils*
11. *Armature*
12. *Bearing bush*
13. *Housing retaining screws*
14. *Dust cover*
15. *Solenoid retaining bolt*
16. *Drive end housing*
17. *'C' clip*
18. *Spacer*
19. *Return spring*
20. *Solenoid armature*
21. *Drive assembly*
22. *Engagement lever*
23. *Pivot*
24. *Solenoid body*
25. *Terminal nut and washer*
26. *Brushes*

Fig. E:44 Lucas starter motor exploded view

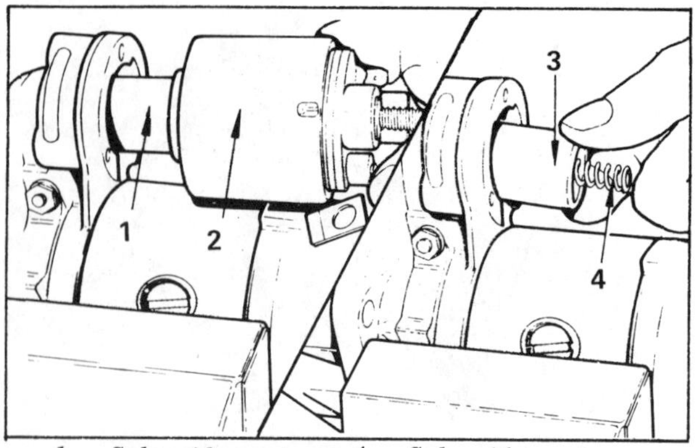

1. *Solenoid armature*
2. *Solenoid yoke*
3. *Solenoid armature*
4. *Solenoid armature return spring*

Fig. E:45 Bosch starter solenoid removal

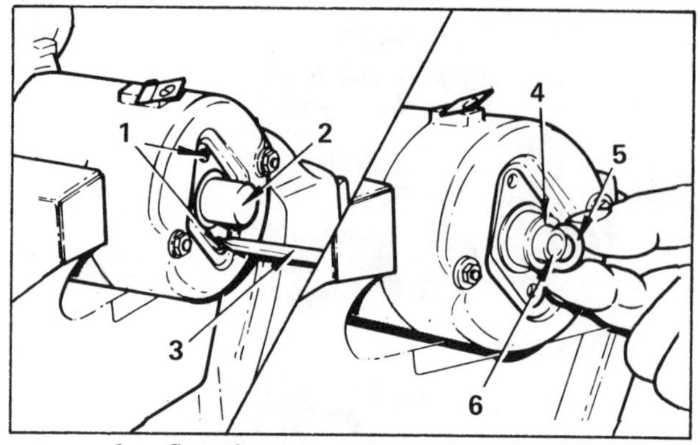

1. *Securing screws*
2. *Housing cap*
3. *Screwdriver*
4. *Shims*
5. *'C' clip*
6. *Armature shaft*

Fig. E:46 Bosch starter end bracket fixings

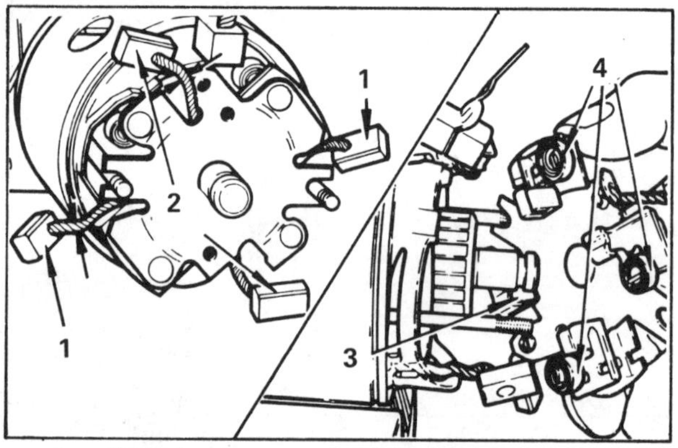

1. *Field brushes*
2. *Terminal brushes*
3. *Brush plate*
4. *Brush retaining springs*

Fig. E:47 Bosch starter brush plate removal

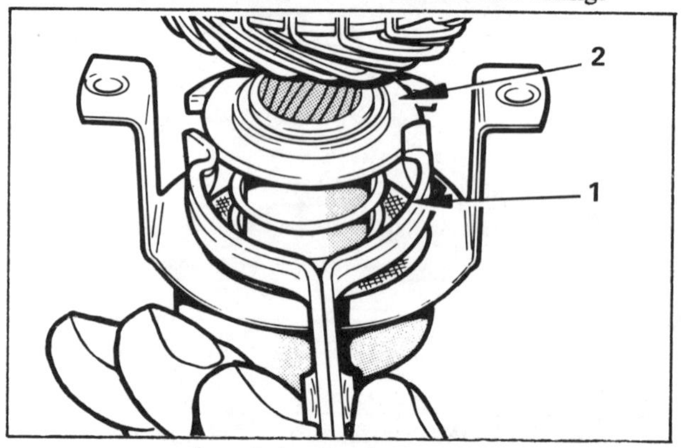

1. *Actuating lever*
2. *Lever locating flange*

Fig. E:48 Bosch starter actuating lever location

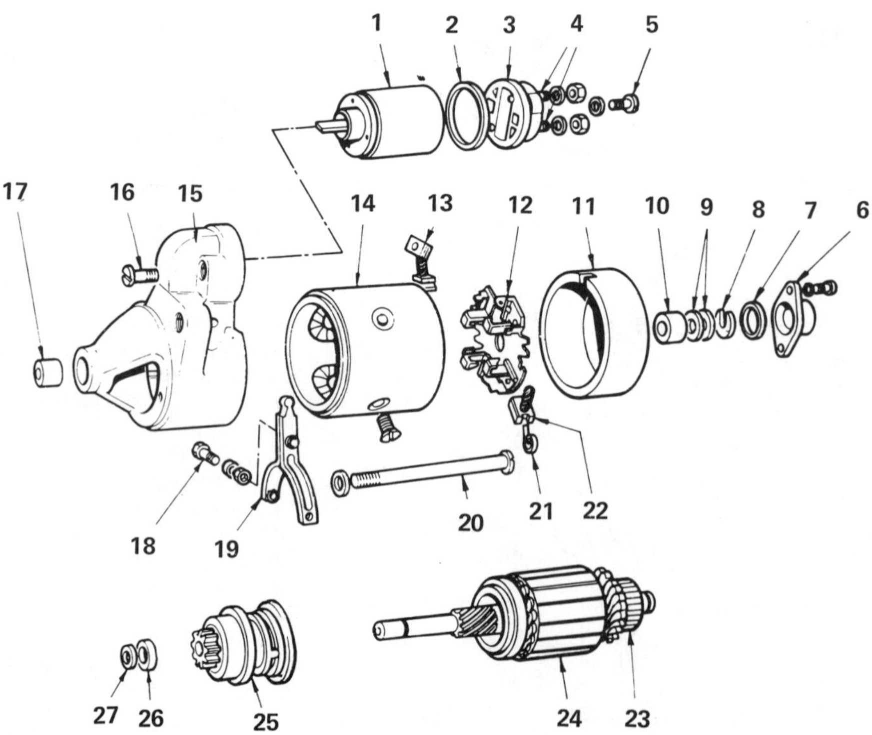

1. *Solenoid body*
2. *Gasket*
3. *Switches contacts and cover*
4. *Terminals (main)*
5. *Retaining screw*
6. *End cover*
7. *Seal*
8. *'C' clip*
9. *Shim washer*
10. *Bearing bush*
11. *Commutator end housing*
12. *Brushbox assembly*
13. *Connector link*
14. *Main casing (yoke)*
15. *Drive end housing*
16. *Solenoid retaining screw*
17. *Bearing bush*
18. *Pivot screw*
19. *Actuating lever*
20. *Through bolt*
21. *Brush spring*
22. *Brush*
23. *Commutator*
24. *Armature*
25. *Drive pinion and roller clutch assembly*
26. *Bearing bush*
27. *Thrust washer*

Fig. E:49 Bosch starter exploded assembly

Fig. E:50 Nippon starter brush plate

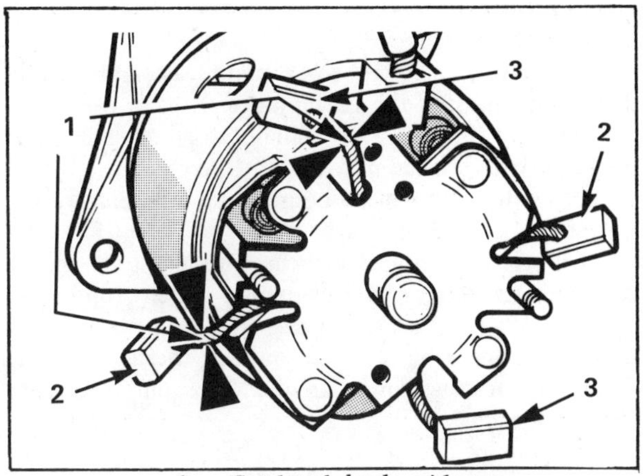

1. *Cut brush leads mid way*
2. *Field brushes*
3. *Terminal brushes*

Fig. E:51 Nippon starter brush replacement

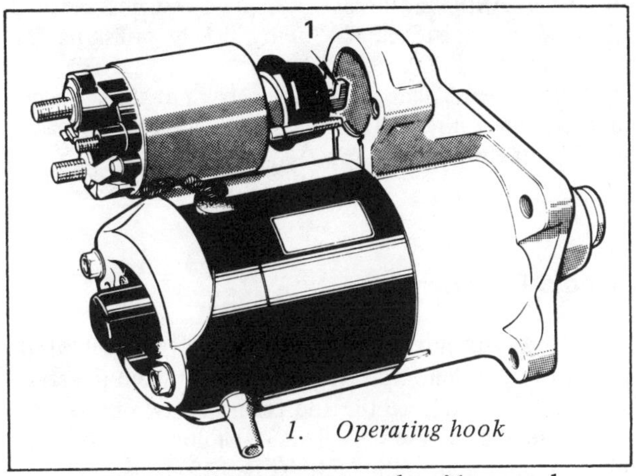

Fig. E:52 Nippon starter solenoid removal

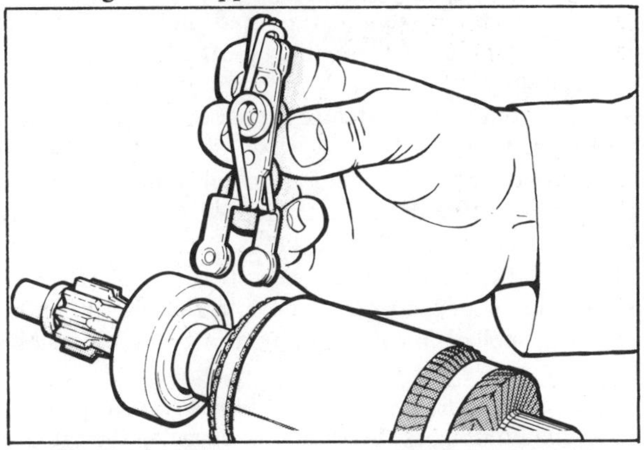

Fig. E:53 Nippon starter actuating lever removal

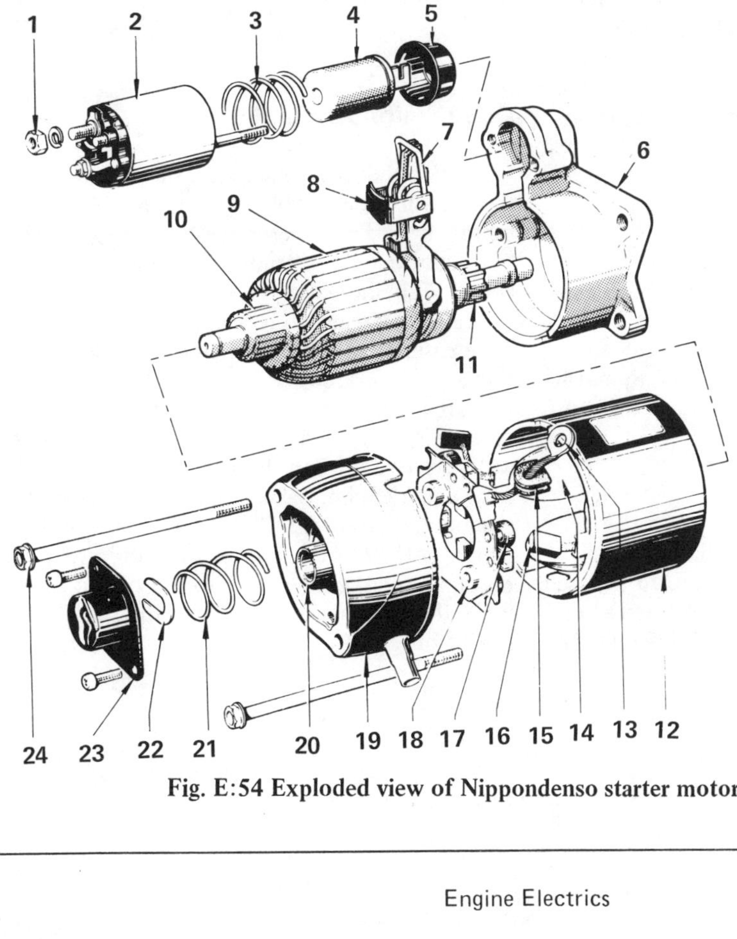

1. *Solenoid terminal nut*
2. *Solenoid body*
3. *Solenoid return spring*
4. *Solenoid armature*
5. *Seal*
6. *Drive end housing*
7. *Actuating lever*
8. *Pivot assembly*
9. *Armature*
10. *Commutator*
11. *Drive pinion and roller clutch assembly*
12. *Main casing*
13. *Connecting link solenoid to starter*
14. *Pole shoe*
15. *Seal*
16. *Brush*
17. *Brush spring*
18. *Brush box assembly*
19. *Commutator end housing*
20. *Bush*
21. *Spring*
22. *'C' - clip*
23. *End cover*
24. *Through bolt*

Fig. E:54 Exploded view of Nippondenso starter motor components

Drive Pinion

Replacement of the drive pinion assembly involves dismantling most of the starter motor.

Remove the solenoid, commutator end housing and brush plate as detailed previously.

Separate the drive end housing and armature from the yoke by tapping them apart.

On starter motors which have fixing studs, unscrew the studs to release the drive pinion clutch stop bracket.

Withdraw the armature assembly from the drive end housing at the same time uncoupling the actuating arm from the locating flange on the drive pinion assembly (Fig. E:48).

A suitably dimensioned tube will be required to remove the drive pinion assembly from the armature shaft.

Secure the armature in a vice with soft jaws and, using the piece of tube, tap the thrust collar down the shaft to expose the jump ring. Extract the jump ring from its groove in the armature shaft and slide the thrust collar and drive pinion assembly up off the shaft.

The drive pinion and clutch assembly are serviced as a complete unit and no attempt should be made to repair or even separate them.

Slide the new drive pinion assembly and thrust collar onto the armature shaft. On starters with a separate clutch stop bracket, ensure the bracket is in place on the armature shaft before fitting the new drive pinion assembly. Fit the jump ring into the groove in the shaft, then pull the thrust collar up over the ring, using a two-legged puller.

Insert the armature assembly into position in the drive end housing, at the same time engaging the actuating arm fork onto the drive pinion flange (Fig. E:48).

On starters which use fixing studs in place of through-bolts, align the clutch retaining bracket with the drive end housing and secure in position with the two fixing studs.

Ensure that the rubber sealing block is in position at the actuating lever in the drive end housing, then assemble the starter yoke to the drive end housing and tap it home.

Assemble the brush plate, commutator end housing and solenoid to the starter yoke as detailed previously.

NIPPON DENSO STARTER MOTOR. . . . . . [11]

The Nippon Denso starter motor fitted to some models, is similar to the Lucas and Bosch units. Removal and refitting is similar to that described in the Lucas Starter Motor section.

Brush Replacement

Remove the two cross head screws holding the protective cap on the end plate. Remove the cap to expose the 'C' washer. Tap the washer from the shaft and withdraw with the spring. Remove the two bolts holding the end plate. Tap the end plate from the main casing (Fig. E:50).

Free the brushes from their housings. If the brush length is less than 0.4 in (10 mm) the brushes should be replaced.

Inspect the brush contact face on the side of the commutator for signs of oil contamination; if present, the brushes should be renewed. If the commutator surface is blackened or dirty, clean it with a petrol-moistened cloth. If there is evidence of burning, this can be cleaned off with very fine glass paper. On no account must emery cloth or sand paper be used.

To replace the brushes, cut the original brush leads at a point midway between the base and the original soldered connection (Fig. E:51). The new brush leads should be soldered to the old stubs. A good joint must be made. Fit the brushes into the end plate. Check that the brushes can move freely. Sticking brushes can be freed and cleaned with a petrol moistened rag, or by lightly polishing the sides with a fine file.

Refit the end plate, securing bolts, spring, 'C' washer and then smear the end of the shaft with lithium based grease. Refit the bearing cap.

Solenoid Replacement

Remove the nut and washer securing the field winding lead to the solenoid, then free the cable from the stud. The solenoid is held to the end casing by two bolts. Remove these and release the solenoid. Unhook the solenoid pull rod from the operating lever (Fig. E:52).

The solenoid assembly is refitted in the reverse order.

Drive Pinion Assembly

When the drive pinion is worn it can be replaced, although this does involve dismantling most of the starter motor.

Remove the solenoid and end plate as described above. Tap the casing free and withdraw with the field windings. The armature can now be removed from the drive casing and the operating lever pulled from the armature (Fig. E:53).

To release the drive pinion from the armature, a piece of tube slightly wider than the armature shaft thickness will be required. Tap the thrust collar down the shaft to expose the 'C' clip. Take care not to damage the armature assembly. Remove the 'C' clip from the retaining groove by tapping off with a chisel, or by using circlip pliers. Slide the clip, thrust collar and the drive pinion from the shaft. Refit the pinion and reassemble the starter motor in the reverse order.

ENGINE ELECTRICS

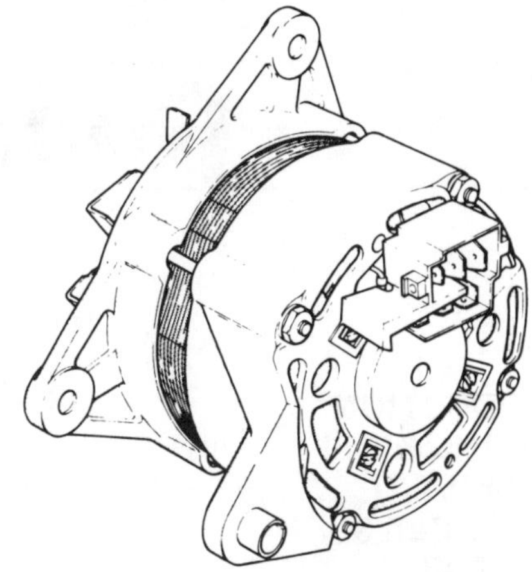

Trouble Shooter

FAULT	CAUSE	CURE
STARTER		
Starter doesn't turn (lights dim).	1. Battery flat or worn. 2. Bad connection in battery circuit	1. Charge or fit new battery. 2. Check all feed and earth connections.
Starter doesn't turn (lights stay bright).	1. Faulty ignition switch. 2. Broken starter circuit.	1. Check switch. 2. Check starter circuit.
Solenoid switch chatters.	1. Flat battery.	1. Charge or replace battery.
Starter just spins.	1. Bendix gear sticking.	1. Remove starter and clean or replace Bendix gear.
CHARGING CIRCUIT		
Low or no charge rate	1. Broken or slipping drive belt. 2. Poor connections on or faulty alternator.	1. Fit new belt. 2. Check and replace alternator.
IGNITION SYSTEM		
Engine Fails to Start (Contact Breaker Points Closed)		
Low or no voltage at coil positive terminal (ign on)	1. Open or short circuit in coil circuit.	1. Proceed to next test.
Battery voltage at coil lead (disconnected)	1. Faulty coil.	1. Replace coil.
No voltage at coil lead (disconnected from coil).	1. Open circuit or short in wiring.	1. Reconnect or replace wiring.
Battery voltage at coil	1. Open circuit primary coil winding. 2. Short circuit condenser. 3. Contact breaker lines earthed.	1. Check coil primary windings. 2. Replace condenser. 3. Replace or insulate lead.
High voltage reading (over 0.2V max) between coil negative and earth (contact points closed)	1. Poor C.B. earth. 2. Dirty contact breaker points.	1. Clean earth connections. 2. Clean or renew CB points.
Poor or no spark between coil HT, lead and earth when contacts flicked open	1. Faulty HT lead. 2. Faulty condenser.	1. Replace HT lead. 2. Check or replace condenser.
Poor or no spark between coil HT lead and earth when contacts flicked open (with known good condenser).	1. Coil secondary winding faulty.	1. Replace coil.

Cooling System

INTRODUCTION ....[1]

The cooling system on the Escort is of the conventional semi-pressurised type. On CVH engine cars, water is circulated around the cylinder block and head by a pump on the right hand side of the engine and driven by the camshaft drive belt. Coolant flow from the engine to the raidator is controlled by a wax element type thermostat mounted in a housing on the end of the cylinder head. The 1.1 litre models are fitted with a simple coolant overflow tank, with a pressure cap on the thermostat housing (Fig. F:1). 1.3 and 1.6 litre models are fitted with a 'Degas' tank, and pressure cap, connected by a pipe from the thermostat housing (Fig. F:2). Excess coolant pressure when hot, is vented into the tank and the displaced air is bled off through the pressure cap. As the engine cools, the displaced coolant is siphoned back into the engine. Hot water for the heater matrix is supplied by a pipe connection at the thermostat housing, coolant is returned via the water pump.

The cooling system fitted to the OHV engine cars is similar in operation to the CVH engine models (Fig. F:3), except that the water pump is driven by the alternator drive belt. Hot water for the heater matrix is supplied from the engine via the carburettor automatic choke.

All models are fitted with an electric cooling fan, although only the 1.3 and 1.6 engine models are fitted with a thermostatic operating switch. The temperature gauge and sensor switch are described in the GENERAL ELECTRICS chapter.

All of the cooling system components are quite easy to get at for service or replacement although the heater matrix unit inside the car does require some extra dismantling of the trim. Before topping up the system, take special note of the section on antifreeze in the ROUTINE MAINTENANCE chapter.

It is possible to overhaul the OHV engine water pump, a kit is available containing the necessary bearing and impellor. However, as a press will be needed for the overhaul, it will probably be easier and quicker to fit a replacement unit, which will be guaranteed. Overhaul of the water pump is not possible on CVH engines.

No special tools are required for servicing. A selection of metric spanners, screwdrivers and pliers should be sufficient for most dismantling. Note that a torque wrench should be used if possible when tightening threads in alloy components. Tightening torque figures are given in the TECHNICAL DATA at the end of the book.

DRAINING & REFILLING ....[2]

1. Before attempting to drain the cooling system when the engine is warm, release any accumulated pressure in the system through the thermostat housing cap on 1.1 litre cars, or the degassing tank cap on 1.3 and 1.6 litre cars. Rotate the cap through 90 degrees only.

NOTE: Considerable pressure can build up in the cooling system when hot causing steam and scalding water to be ejected when releasing the cap. To prevent accidental injury, muffle the cap with a heavy cloth and unscrew it slowly to release the pressure in the system. When satisfied that all pressure in the system has been released, remove the cap fully.

2. Place a suitable container beneath the cylinder block drain plug (Fig. F:4). Remove the plug and allow the block to drain.
3. Loosen the radiator bottom hose clamp. Pull the hose from the radiator connection and allow the system to drain (Fig. F:5).
4. Refit the cylinder block drain plug. Refit the radiator bottom hose, after checking its condition.
5. Refill the cooling system with a water/anti-freeze mix. See ROUTINE MAINTENANCE chapter for details of the recommended type. Fill through the thermostat housing until slowly, allowing any air to bubble through. Refit the thermostat housing cap. Top up degassing or overflow tank to maximum mark.
6. Run the engine for a while, checking connections, hoses and the drain plug for leakage. Check coolant level frequently as air will bleed into the tank. Top up as necessary.

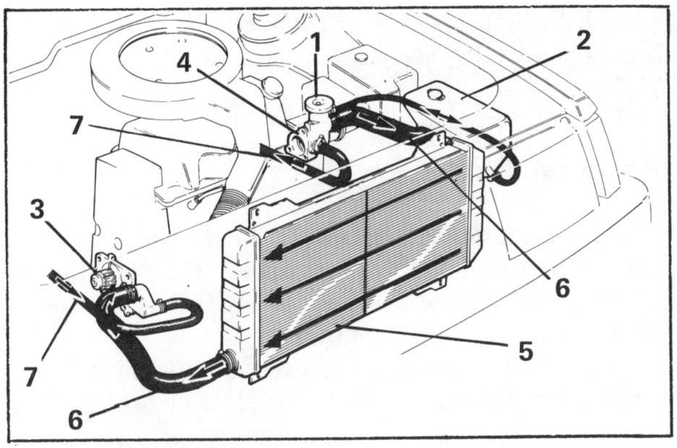

1. *Filler cap*
2. *Overflow tank*
3. *Water pump*
4. *Thermostat housing*
5. *Radiator*
6. *Radiator hoses*
7. *Heater hoses*

Fig. F:1 CVH 1.1 engine cooling system

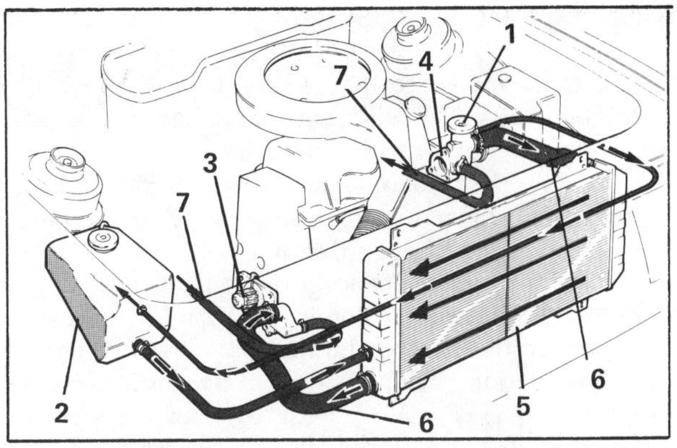

1. *Filler cap*
2. *Degas tank*
3. *Water pump*
4. *Thermostat housing*
5. *Radiator*
6. *Radiator hoses*
7. *Heater hoses*

Fig. F:2 CVH 1.3 and 1.6 engine cooling system

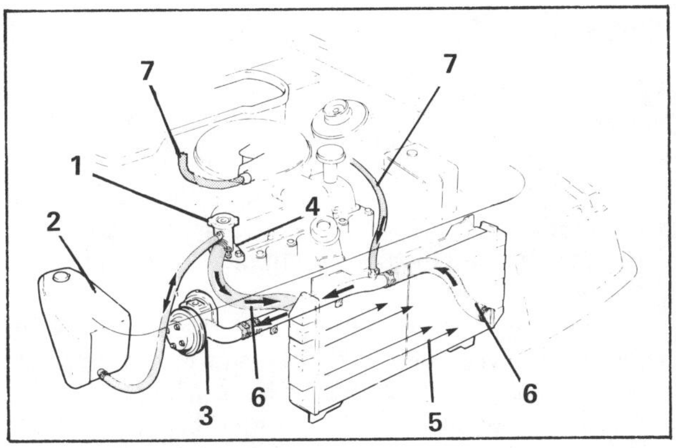

1. *Filler cap*
2. *Overflow tank*
3. *Water pump*
4. *Thermostat housing*
5. *Radiator*
6. *Radiator hoses*
7. *Heater hoses*

Fig. F:3 OHV engine cooling system

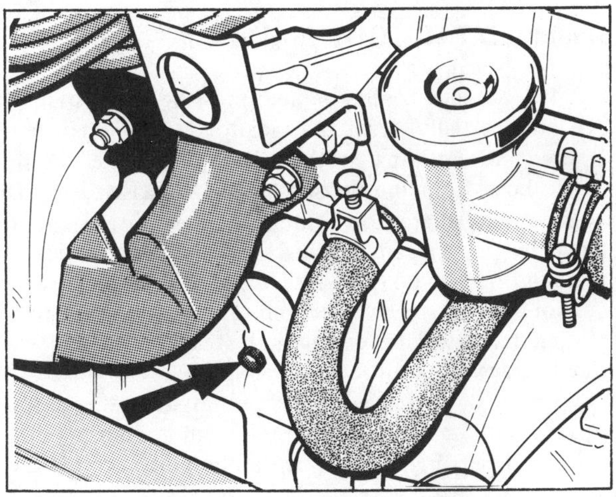

Fig. F:4 Cylinder block drain plug

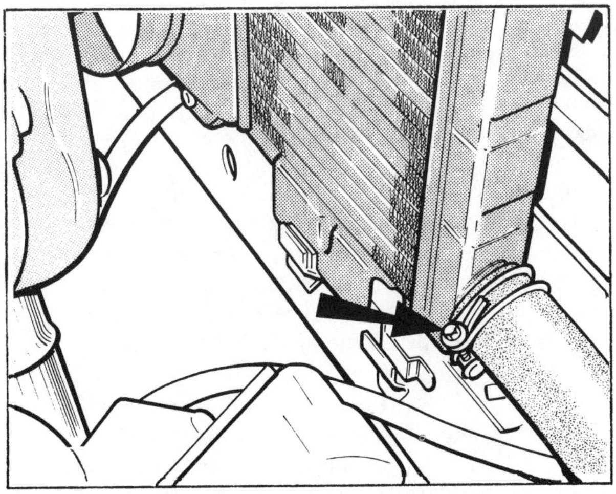

Fig. F:5 Radiator bottom hose

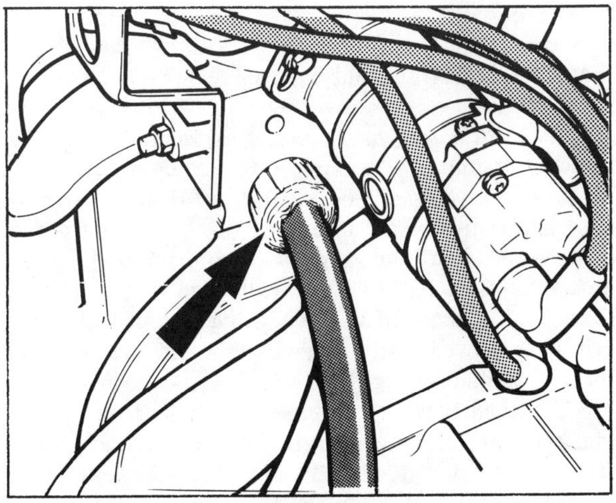

Fig. F:6 Flushing engine with high pressure hose

Flushing the Cooling System

It is recommended that the cooling system is flushed every two years. This will involve back flushing the radiator and cylinder block with a hose.

1. Remove the radiator as described later in this chapter. Turn upside down, fit the water hose into the bottom hose location and flush until clear.
2. Place a container beneath the radiator bottom hose. Remove the thermostat housing, see the Thermostat section later in this chapter. Fit the hose into the thermostat housing location packing with rags or plastic to make a water tight seal (Fig. F:6). Flush the cooling system until clear. Remove the hose and allow to drain after removing the cylinder block drain plug.
3. Refit the drain plug, radiator and then refill as described above.

THERMOSTAT........................[3]

Remove & Refit, OHV Engine

Release any cooling system pressure by undoing the cap and partially drain the radiator as described in the Draining & Refilling section. Disconnect the overflow bottle hose from the thermostat housing (Fig. F:7). Disconnect the radiator top hose clamp, then free the hose. Remove the thermostat housing from the cylinder head by undoing the two bolts. If the housing will not come away, do not attempt to lever it from the cylinder head or the mating faces will be damaged. Instead, gently tap from side to side with a soft headed hammer.

Clean the mating surfaces to remove any traces of old gasket. Withdraw the thermostat from its housing in the cylinder head (Fig. F:8). Test the thermostat or replace it as necessary.

Place the thermostat into the housing, then position a new gasket on the mating face. Refit the housing and tighten the bolts to the specified torque - see TECHNICAL DATA. After reconnecting the thermostat housing hoses, refill the cooling system as detailed previously.

Remove & Refit, CVH Engine

After releasing any cooling system pressure and partially draining the cooling system as detailed previously, detach the degassing or overflow bottle hose, the heater hose and radiator top hose from the thermostat housing (Fig. F:9). Undo the three bolts and remove the thermostat housing from the cylinder head (Fig. F:10). The thermostat is retained in the housing by a wire clip which can be removed with thin nosed pliers. Withdraw the thermostat and sealing ring.

Clean all traces of old gasket material from the mating faces. Fit a new sealing ring before fitting the thermostat and securing it with the clip. Fit a new gasket and refit the housing. Tighten the bolts to the specified tightening torque, see TECHNICAL DATA. Reconnect the hoses to the housing and refill the cooling system as detailed previously.

Thermostat Test

The thermostat is a relatively inexpensive item, and should be replaced if thought suspect. The operation of the thermostat can be checked, see TECHNICAL DATA for details of opening temperature.

1. Suspend the thermostat in a suitable container full of water so that it does not touch the sides or bottom (Fig. F:11)
2. Gradually heat the water, checking the temperature frequently with an accurate thermometer.
3. If the thermostat does not open at the correct temperature, refuses to open fully, or does not open at all, replace it with a new unit.

HOSES & CLIPS........................[4]

Checking

The cooling system hoses should be checked periodically - particularly before the beginning of winter - to ensure that they are in good condition and that the clips are secure.

Examine each hose carefully for cracks; particularly on the bends, separation of the layers, swelling or excessive softness of the rubber. Also inspect them for damage, especially chafing due to contact with other components.

Deterioration of the hose rubber is best detected by pinching the hose and looking to see if the surface shows signs of cracking. You can often tell by the feel if a hose is deteriorating. It will be soft and soggy (Fig. F:12).

Looking for leaks is best carried out with the system full of water, it makes identifying the location of leaks a lot easier. The bright colour of anti-freeze will leave stains, and rust particles in suspension will adhere to a surface once the water evaporates. If the cooling system is usually topped up with hard water, the deposits will be white.

Hose Replacement

Changing a hose can sometimes present problems. Access to the hoses on the Escort is quite good, as the lower hoses can be reached from above. Before attempting to remove a hose, ensure that the cooling system is depressurized and that the coolant level is below the relevant hose connection. See the 'Draining & Refilling', section. Removal and refitting will be easier if the engine has cooled.

The clips fitted to the hoses may be of several types (Fig. F:13). Large hoses are fitted with wire clips, which have a tendency to cut through hoses if overtightened. These joints should be regularly checked for damage, and replaced if necessary. Replacement clips should, if pos-

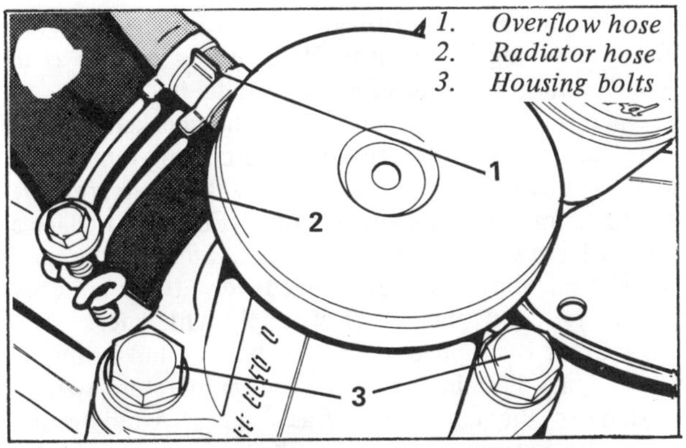

Fig. F:7 OHV engine thermostat housing

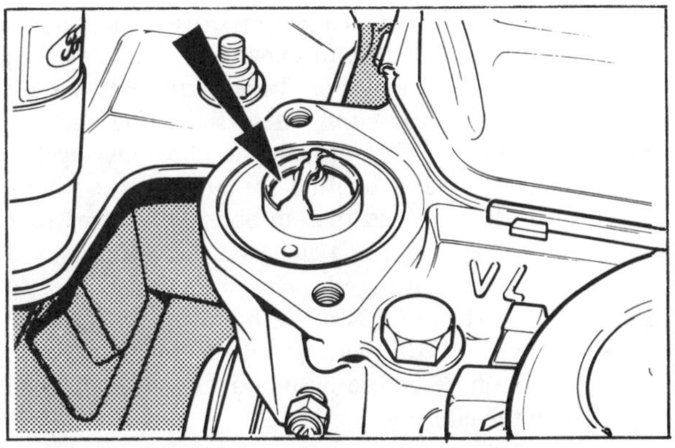

Fig. F:8 OHV engine thermostat in cylinder head

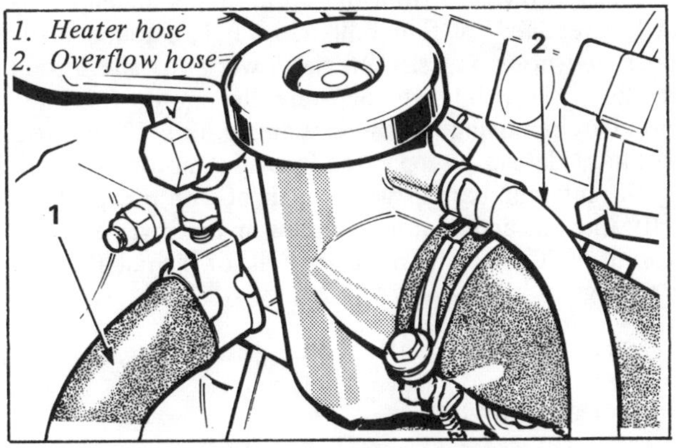

Fig. F:9 CVH engine thermostat housing

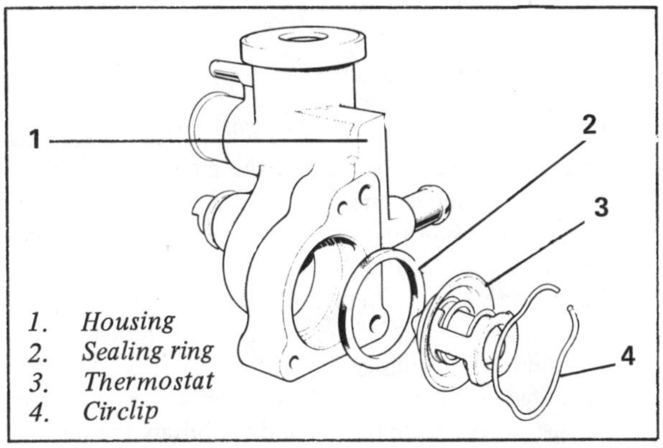

Fig. F:10 CVH engine thermostat assembly

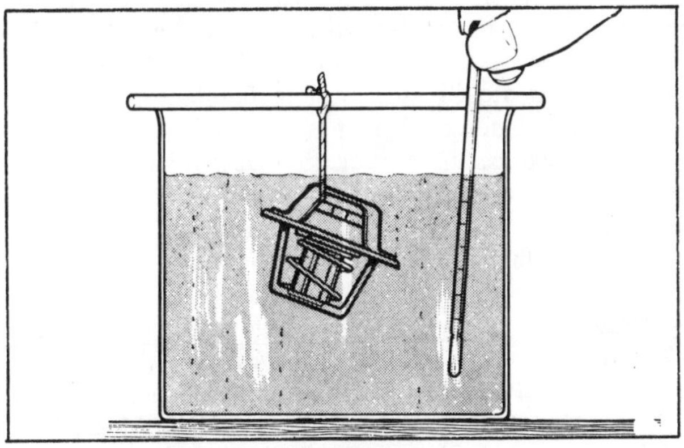

Fig. F:11 Thermostat test in hot water

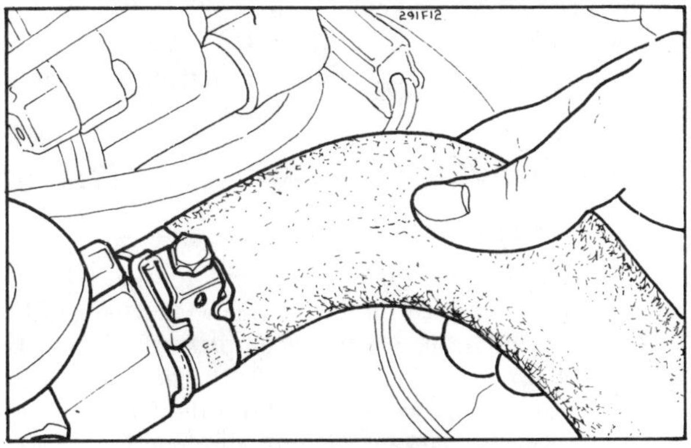

Fig. F:12 Test hose condition by feel

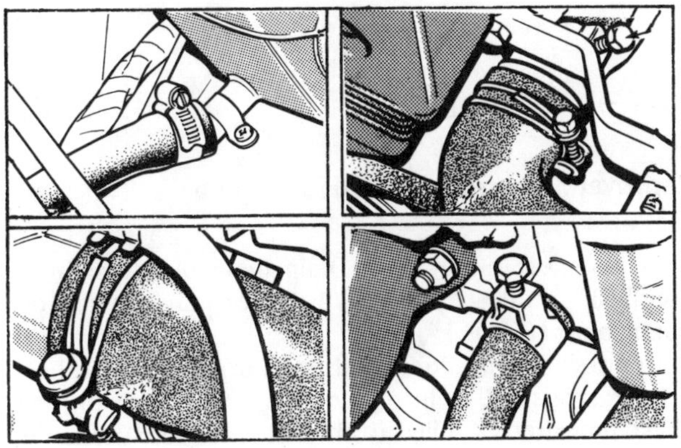

Fig. F:13 Hose clip types

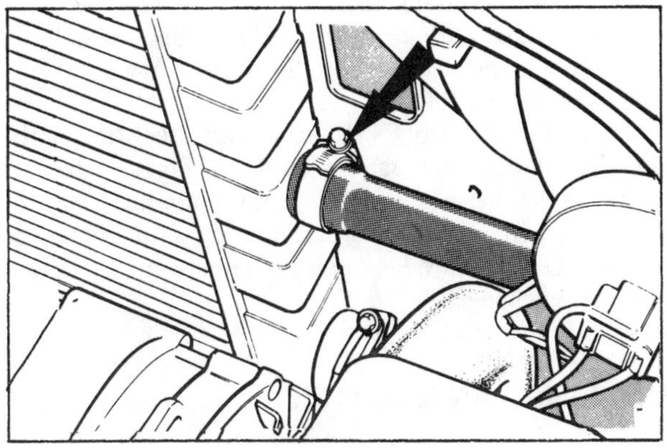

Fig. F:14 Radiator degas tank hose

sible be of the flat bladed worm drive type, such as Jubilee clips. The radiator to expansion tank hose clips are all of this type. Small bore hoses may be fitted with spring wire clamp type clips. These should be slackened and refitted using pliers, pushing the clip tabs together.

There is no real substitute for brute force when it comes to shifting a hose that stubbornly adheres to its spigot - waggling, wrenching and twisting will usually do the trick. To ease fitting the new hose, clean the stub smooth with a strip of emery cloth and smear the inside of the hose with a little grease or petroleum jelly. Position the clip on the hose before pushing it on to the stub and then tighten the clip.

RADIATOR .... [5]

Removal & Refitting

1. Disconnect the battery earth lead. Drain the cooling system as described at the start of the chapter.
2. If fitted, remove the hose to the expansion tank, mounted on the side of the radiator (Fig. F:14). Remove the radiator top hose (Fig. F:15).
3. Detach the wiring plug from the rear of the fan motor (Fig. F:17). Unclip the wiring loom from the fan shroud.
4. The radiator assembly is retained by two bolts at the top, and held by lugs at the bottom (Fig. F:16). Remove the bolts, pull the radiator up and withdraw from the car.
5. Before refitting the radiator, check the condition and position of the lower rubber mountings. Refit the radiator in the reverse order, tightening the bolts to the specified torque, given in the TECHNICAL DATA. After refitting the hoses, refill as described at the start of the chapter.

COOLANT OVERFLOW TANK........... [6]

The coolant overflow tank fitted to the 1.1 OHV engine cars is mounted on the off side wing panel in front of the strut mounting. It is an un-pressurized tank fed from the thermostat housing (Fig. F:3). The tank is secured by a clamp and two screws above the feed pipe. Remove the screws, slacken the hose clip, remove the hose and withdraw the tank. Refit in the reverse order.

The overflow tank fitted to 1.1 CVH engined cars is similar to that fitted to the OHV engine cars, but is mounted on the near side of the engine bay (Fig. F:1). Removal and refitting is similar.

The 1.3 and 1.6 CVH engine cars are fitted with a pressurized degassing tank (Fig. F.2). Pressure caps are fitted to the thermostat housing and to the tank. An additional pipe from the thermostat housing should be disconnected before removal. Note that the pressure caps should never be exchanged.

COOLING FAN & SWITCH .... [7]

The radiator cooling fan is mounted on a plastic shroud on the radiator casing. The fan can be removed without removing the radiator. The motor is serviced as an assembly, and it is not possible to overhaul it.

Operation of the cooling fan can be checked. On 1.1 engine cars, turn on the ignition. The cooling fan should operate immediately. On 1.3 and 1.6 engine cars, remove the plug from the temperature switch (Fig. F:17). Place a jump lead between the two terminals. Turn on the ignition. The fan should operate immediately. If the fan does not operate, check the electrical supply with a test lamp. If no current is supplied, check the battery and wiring. If the test lamp illuminates, check the temperature switch.

Remove the fan assembly as follows. After disconnecting the battery earth lead, pull the fan motor multiplug from the rear of the fan (Fig. F:18).

Release the lead from the shroud retaining clip. The cowl is retained by four bolts (Fig. F:16). Remove the upper two bolts and slacken the lower two. The shroud can then be withdrawn. Separate the motor from the shroud by removing the three retaining nuts. If required, the fan blade can be separated from the motor by releasing the retaining circlip with a pair of thin nosed pliers. Refit the fan and motor assembly in the reverse order. Do not overtighten the retaining bolts or the radiator will be damaged.

Temperature Switch

A temperature switch is fitted to 1.3 and 1.6 engine versions only (Fig. F:17). The switch is mounted on the thermostat housing next to the water outlet pipe. The switch can be tested as follows. A faulty thermostatic switch is indicated if the cooling fan fails to operate although the engine coolant is at boiling-point. To check a suspect switch, disconnect the wiring harness from the thermostatic switch, bridge the two terminals in the wiring harness and switch on the ignition. If the fan operates, the fan motor circuit is in good order and the switch is faulty.

Drain the cooling system to below the level of the switch, see the start of the chapter. The switch can then be unscrewed. Fit a new switch using a new gasket. Tighten to the specified torque, see the TECHNICAL DATA. Refill and reconnect the supply lead.

WATER PUMP .... [8]

Removal & Installation, OHV Engine

Disconnect the battery. Drain the cooling system as described previously. Loosen the three pump pulley retaining bolts (Fig. F:19).

This should be done whilst the drive belt is still attached as this will help prevent the pulley from turning whilst pressure is applied to the bolts. If difficulty is experienced, a second ring spanner or socket can be used to

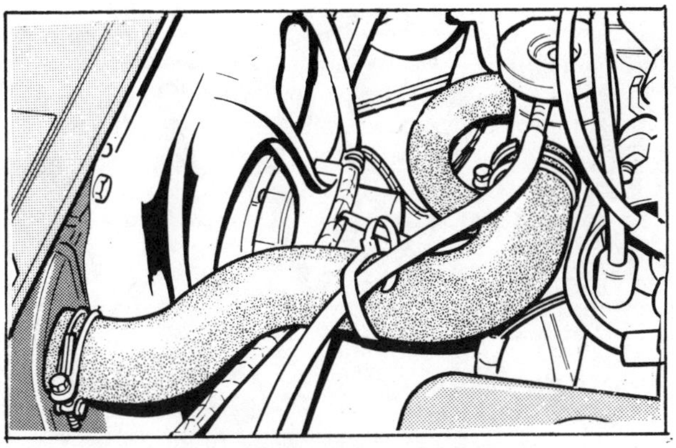

Fig. F:15 Radiator top hose

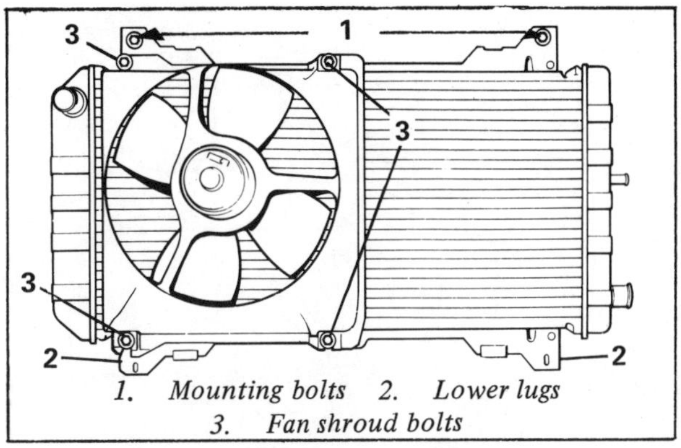

Fig. F:16 Radiator mounting points

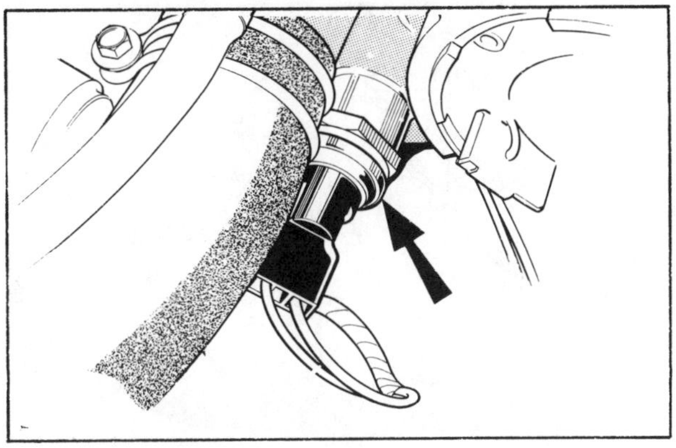

Fig. F:17 Cooling fan radiator temperature switch

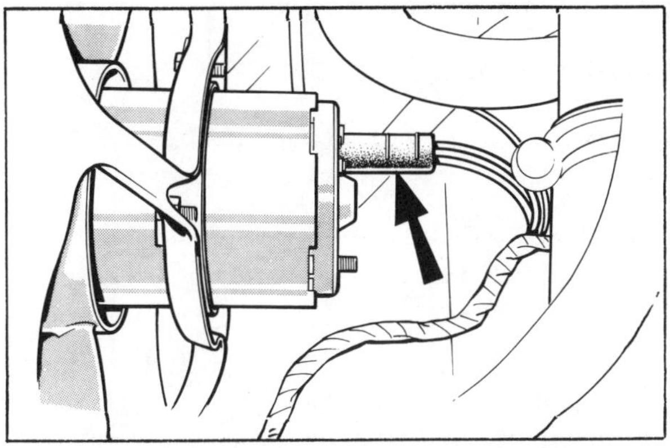

Fig. F:18 Fan motor multiplug (arrowed)

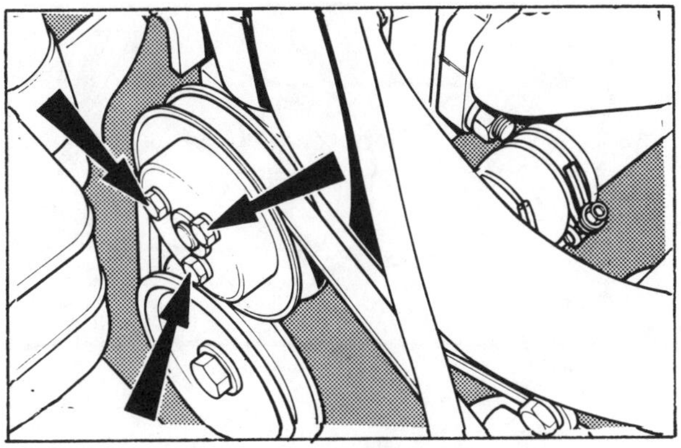

Fig. F:19 OHV engine water pump pulley

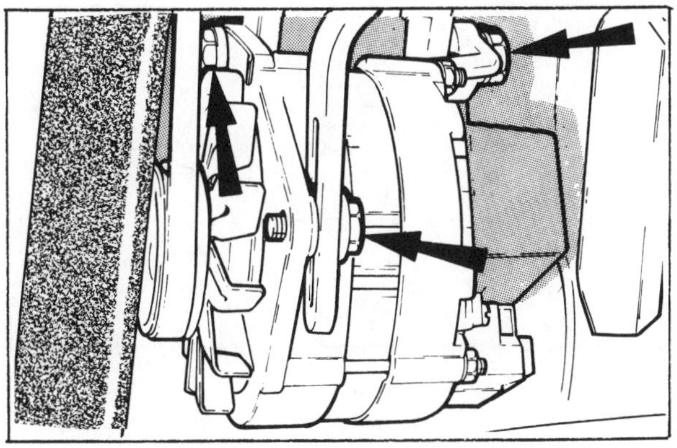

Fig. F:20 Alternator mounting bolts

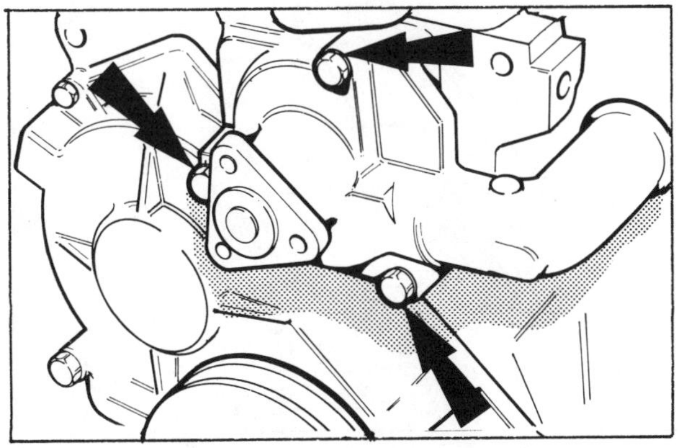

Fig. F:21 Water pump retaining bolts (arrowed)

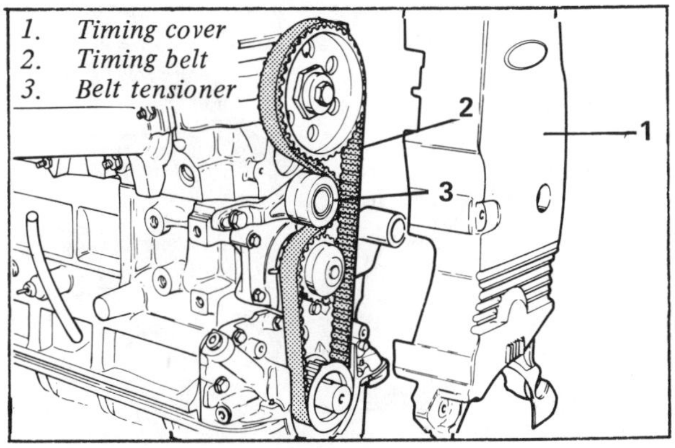

Fig. F:22 CVH engine timing belt removal

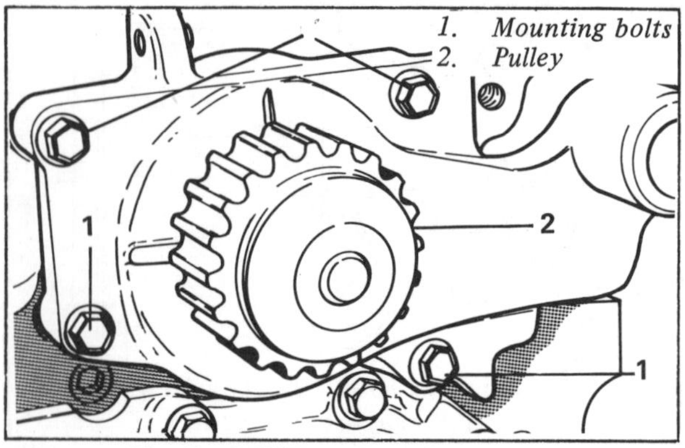

Fig. F:23 CVH engine water pump

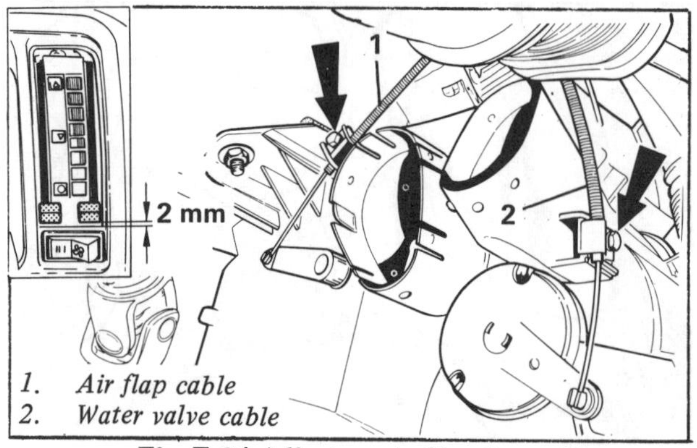

Fig. F:24 Adjustment of heating cables

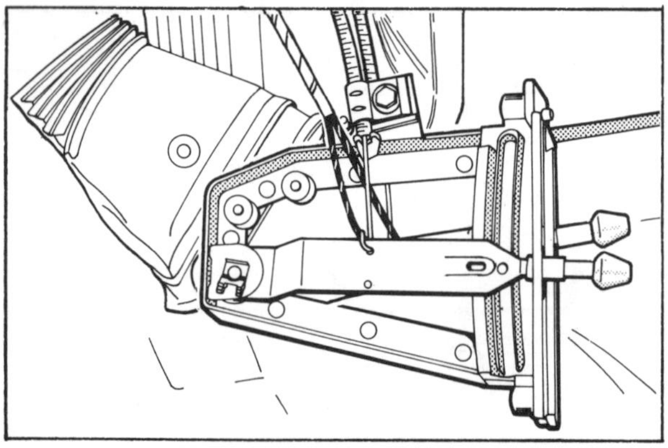
Fig F:25 Heater control lever unit

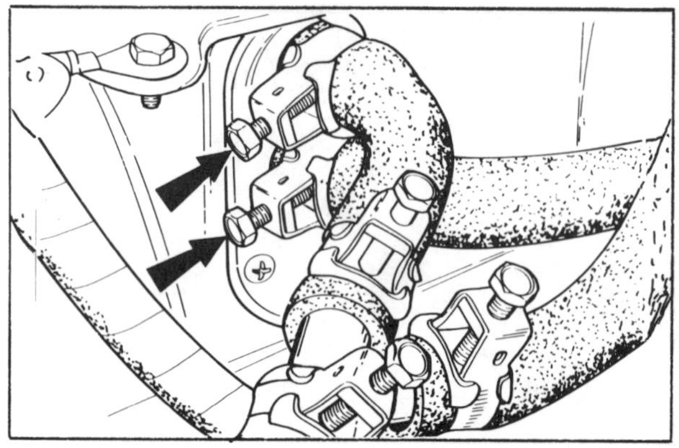
Fig. F:26 Heater hose pipe connections

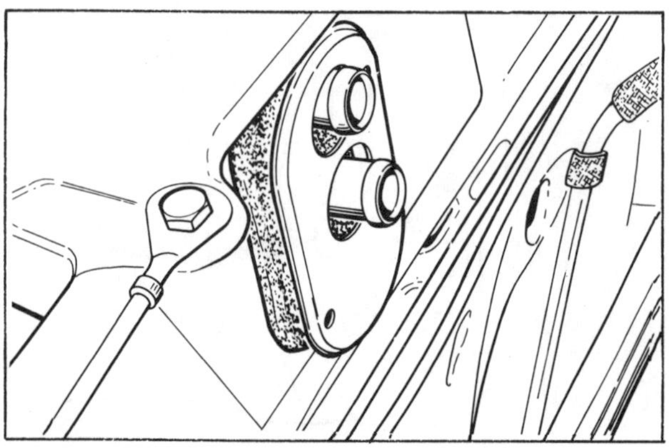
Fig. F:27 Heater pipe retaining plate

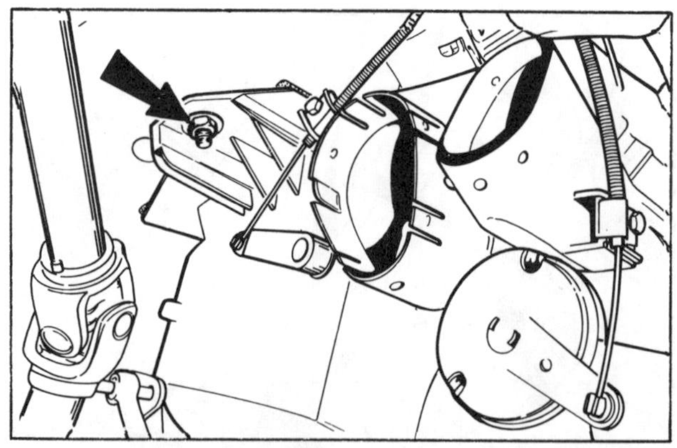
Fig. F:28 Heater housing retaining nut, right hand side shown

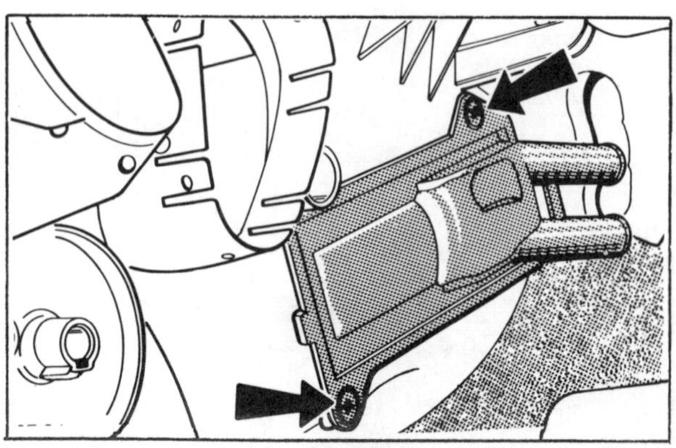
Fig. F:29 Heater matrix retaining plate screws

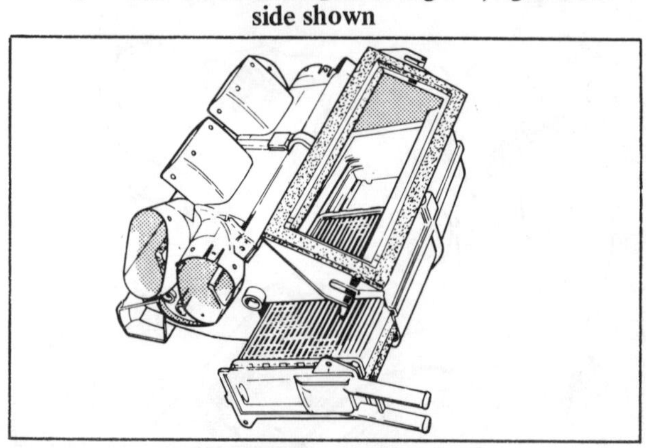
Fig. F:30 Heater matrix withdrawal

hole one bolt head whilst turning another. This will prevent the pulley turning. When two bolts have been loosened, a heavy screwdriver can be placed across them and used to hold the pulley whilst loosening the third bolt.

Loosen the alternator mounting bolts followed by the bolt securing the adjusting arm to the cylinder block. Swing the alternator towards the block sufficiently to be able to remove the drive belt (Fig. F:20).

Loosen the clip securing the radiator hose to the pump and remove the hose. Remove the pump pulley retaining bolts and lift away the pulley. Remove the three pump retaining bolts and lift away the pump (Fig. F:21).

Remove all traces of old gasket material from the cylinder block and pump mating surfaces.

Push the radiator hose on to the water pump, making sure the retaining clip is already over the hose, and using a new gasket refit the pump to the block. Refit the pulley. Refit the drive belt, adjusting the tension as described in the ROUTINE MAINTENANCE chapter.

Tighten the water pump pulley retaining bolts to the torque specified in the TECHNICAL DATA.

Refit the radiator hose and tighten the retaining clips. Refill the cooling system, reconnect the battery and run the engine to check for leaks. Top up the system as necessary.

Removal & Installation, CVH Engine

The water pump fitted to the CVH engine is mounted behind and driven by, the camshaft drive belt. The drive belt must be removed first before the pump can be removed - see CVH ENGINE chapter for details of drive belt removal.

1. Drain the cooling system as detailed earlier in this chapter.
2. Slacken the bottom hose and heater clamps, then pull both hoses from the water pump connections.
3. Remove the timing belt tensioner, held by two bolts (Fig. F:22).
4. The water pump is held by four bolts. Remove these and withdraw the pump (Fig. F:23) from the engine.
5. Check the condition of the pump impellor, shaft bearings and replace the pump if suspect.
6. Remove all traces of gasket from the mating surfaces. Refit the pump in the reverse order of removal using a new gasket. Tighten the bolts to the specified torque, see TECHNICAL DATA.
7. Refit the timing belt as described in the CVH ENGINE chapter. Refill the cooling system as described previously in the 'Draining & Refilling', section.

HEATER MATRIX & CONTROLS. . . . . [9]

Heater Control Cables - Replacement

1. The heater control panel is connected to the heater unit and ventilation ducts by Bowden cables. To replace the cables, proceed as follows.
2. Remove the fascia lower trim panel on the right hand side, held by two clips and two metal tags.
3. Free the air hoses from the right-hand heater duct connection.
4. Position the hose over the right-hand Bowden cable, clear to the right, then unscrew the clamp bolts securing each outer cable and disconnect the inner cables from the levers (Fig. F:24).
5. Pull the control lever knobs from the levers, then press down on the top of the heater control plate until the lower edge is pushed out. The plate can then be withdrawn.
6. Undo the screws holding the lever control unit to the fascia and withdraw the assembly, pulling the cables free. Disconnect the light holder (Fig. F:25).
7. Refit the cables and lever control assembly in the reverse order of removal. Adjust the control cables by positioning the outer cable in the clamp so that there is a 2 mm gap between the lever and the lower lever stop (Fig. F:24). Tighten the cable clamp bolts.

Heater Matrix - Replacement

The heater matrix can be removed leaving the majority of the heater assembly attached.

1. Disconnect the battery earth strap, then free the heater hoses from the matrix pipes in the engine compartment bulkhead (Fig. F:26).
2. Position the hose ends upwards to avoid excessive coolant leakage from the engine.
3. Blow air into the upper heater matrix pipe to force any residual water out of the matrix.
4. Remove the crosshead screws holding the pipe plate at the bulkhead, then withdraw the plate (Fig. F:27).
5. Detach the fascia lower trim panels on each side, held by two metal tags and two clips each. Pull the air hoses from the heater casing, and move the right-hand vent nozzle hose to the rear.
6. Remove the nuts holding the heater casing to the bulkhead (Fig. F:28). Pull the casing back until the matrix pipes are free of the bulkhead.
7. Remove the two crosshead screws holding the heater matrix (Fig. F:29) then pull the matrix from the housing (Fig. F:30). The heater matrix is refitted in the reverse order of removal. Top-up the cooling system as detailed previously in this chapter.

COOLING
Trouble Shooter

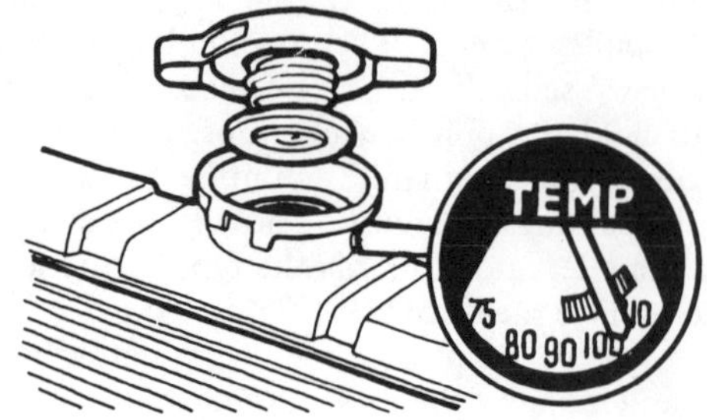

FAULT	CAUSE	CURE
Loss of coolant	1. Leaks. 2. Damaged radiator or heater matrix. 3. Damaged cylinder head gasket. 4. Cracked cylinder block. 5. Cracked cylinder head. 6. Loose cylinder head bolts.	1. Locate and repair. 2. Repair or replace radiator or matrix. 3. Replace gasket. Check engine oil and refill as necessary. 4. Replace cylinder block. Check engine oil in crank-case for mixing with water. 5. Replace cylinder head. 6. Tighten cylinder head bolts.
Poor circulation	1. Restriction in system 2. Insufficient coolant. 3. Inoperative water pump. 4. Loose fan belt. 5. Thermostat not opening.	1. Check hoses for kinking. Clear the system of rust and sludge. 2. Replenish. 3. Replace water pump. 4. Adjust belt. 5. Replace thermostat.
Corrosion	1. Insufficient anti-freeze. 2. Infrequent flushing and changing of coolant.	1. Add anti-freeze. 2. Flush thoroughly at least every two years.
Overheating	1. Inoperative thermostat. 2. Radiator fins obstructed. 3. Incorrect engine tune. 4. Inoperative water pump. 5. Loose fan belt. 6. Restricted radiator. 7. Inaccurate temperature gauge. 8. Faulty sender unit.	1. Replace thermostat. 2. Clean fins. 3. Tune engine. 4. Replace. 5. Adjust tension. 6. Flush radiator. 7. Replace temperature gauge. 8. Replace sender unit.
Overcooling	1. Inoperative thermostat. 2. Faulty temperature gauge. 3. Faulty sender unit.	1. Replace thermostat. 2. Check gauge. 3. Check sender unit.

Fuel System

INTRODUCTION . . . . [1]

The Escort is equipped with a conventional fuel system consisting of the fuel tank, fuel pump and carburettor. A Ford variable venturi (VV) single-choke carburettor is fitted to all models except the XR3, which is fitted with a twin-choke Weber unit. All models are fitted with an automatic choke mechanism, water heated on the Ford unit and electrically operated on the Weber carburettor. The carburettors are designed to run as economically as possible, with enrichment devices to cope with load conditions.

The fuel pump is mechanically driven from the camshaft, on OHV engine versions through a lever, CVH engines by a pushrod. Details are given for a fuel pump pressure text, however, it is not possible to overhaul the sealed diaphragm units, and if the pump is found to be faulty, a replacement should be fitted.

The fuel tank is mounted in the rear of the car below the load carrying area. The filler cap is mounted in the right-hand rear wing. A ventilation pipe in the top of the tank runs to the front, while an anti-splash back pipe runs up along the filler neck.

The carburettor is probably the most important component of the engine in terms of performance and economy. It is accurately calibrated to suit the individual requirements of the engine to which it is fitted and should not be interfered with unnecessarily. It is therefore recommended that only the adjustments covered in this chapter be attempted. Any further work should preferably be left to a Ford dealer or carburettor specialist.

Overhaul kits which contain all the necessary replacement parts are generally available for most carburettors, but it is recommended that work of this nature should be undertaken only by someone with previous experience. A sensible alternative, especially in the case of a carburettor which has done considerable mileage, would be to exchange the carburettor for a new or reconditioned unit, with the added advantage of getting a warranty with the new unit.

When working on any part of the fuel system, ALWAYS disconnect the battery earth strap first. Keep all components as clean as possible and clean the area around any components or connections before removing them. Use new gaskets, sealing rings, hoses, clips, etc., where required.

No special tools are necessary for the operations described in this chapter, apart from a selection of metric spanners, various length screwdrivers, a pair of cutters for clamp type fuel line clips, and a suitable drift for removal of the sender unit. A small pressure tester will be useful to measure the fuel pump output pressure. These testers are available from most accessory shops. Note that the adjustment procedures for the idle speed and mixture are described in the TUNE-UP chapter.

FUEL PUMP . . . . [2]

The fuel pump fitted to the OHV engine Escort is lever operated from the camshaft (Fig. G:1). The CVH engine unit is operated by a pushrod from the camshaft. Neither type can be overhauled if the pump diaphragm is damaged.

Pump Output Test

The fuel pump operating efficiency can be checked by measuring the fuel output pressure. A pressure tester will be required, see the 'Introduction'.

Fuel pump pressure is measured with the engine idling. After the engine has idled for a time, disconnect the outlet pipe at the carburettor end and connect the tester to the pump (Fig. G:2). See the 'Fuel Line' section. With the engine idling, note the pressure. Slightly rev the engine and note the pressure again. The specified output pressure is given in the TECHNICAL DATA at the end of the manual. If after cleaning the pressure is below specification, the pump should be replaced.

Fuel Pump Cleaning

Although the fuel pump diaphragm is sealed, the filter assembly can be cleaned (Fig. G:3). Remove the crosshead screw holding the cover and detach the cover. Clean the filter and housing in petrol, and check that the seal is not damaged. If suspect, it should be replaced. Refit the cover with the notch locating in the groove in the pump housing. Do not overtighten the securing screw.

Removal & Refitting

The removal and refitting of the fuel pump is similar for both engine types. After disconnecting the battery, disconnect the outlet and inlet hoses at the pump. See the 'Fuel Lines' section. The pump is held by two bolts. Remove these and withdraw the pump assembly (Figs. G:4 & 5).

Clean the pump and block mating faces. Position a new gasket and fit the pump and bolts. Tighten to the specified tightening torques, see TECHNICAL DATA. Reconnect the hoses. After reconnecting the battery check that the fuel pump delivers petrol to the carburettor.

If signs of an oil leak are visible around the base of the fuel pump, remove the pump and push rod from the cylinder head. Clean the mating faces and refit the pump without using a gasket. Tighten the nuts to a torque of 1.5 to 3 lb ft (2 to 4 Nm). Measure the gap between pump and head. If greater than 0.010 in (0,25 mm) the fuel pump should be replaced.

Remove the fuel pump and refit the push rod. Refit the fuel pump using a new gasket. Tighten the securing nuts to the specified torque, see TECHNICAL DATA.

CARBURETTOR CONTROLS............[3]

Throttle Cable - Replacement

Remove the trim mat below the driver's side fascia panel (Fig. G:6). Disconnect the pedal from the shaft by levering the shaft forward. Free the cable from the bulkhead by pushing the grommet into the engine bay. Remove the air cleaner assembly, see Carburettor. Free the cable end from the carburettor throttle shaft lever (Fig. G:7). Disconnect the cable from the bracket. If a new cable is to be fitted, the retaining clip can be snapped. If the cable is to be refitted, free the clip by levering the cable end of the clip outwards (Fig. G:8). Press the retaining latches inwards using a screwdriver and withdraw the clip. The cable can then be withdrawn.

Fit the cable in the reverse order of removal. Adjust the cable tension by turning the cable adjuster nut (Fig. G:7). Hold the throttle pedal in the wide open position, jamming it in place with a block of wood. Wind back the adjusting nut until the throttle shaft lever is in the fully open position. Release the pedal. Check that the throttle is wide open at the bottom of the pedal stroke. Refit the air cleaner assembly.

Accelerator Pedal

The accelerator pedal is mounted on a bracket on the bulkhead. The assembly can be removed after moving aside the insulating trim mats. Disconnect the throttle shaft from the pedal by pushing the shaft forward (Fig. G: 6). Remove the two bolts holding the pedal box to the bulkhead. The assembly can then be withdrawn.

Refit the pedal box in the opposite order. Check the cable tension as described above.

FORD VV CARBURETTOR ....[4]

The carburettor fitted to all models apart from the XR3 is a single choke variable venturi (VV) unit (Fig. G: 9). The carburettor idle speed and mixture should be adjusted as described in the TUNE-UP chapter. The adjustments in this chapter should normally be necessary only if the carburettor is under repair or has been overhauled.

Removal & Refitting

Disconnect the battery earth strap. Remove the air cleaner assembly. After removing the screws from the top of the cleaner remove the emission valve from the lower face. Pull the cold air tube from the lower pipe if fitted. The assembly can now be withdrawn.

Remove and refit the cooling system pressure cap to depressurize the system, see the COOLING SYSTEM chapter for details. Disconnect the hoses to the automatic choke cover. (Fig. G:10). Position the hoses pointing upwards to minimise coolant leakage. Disconnect the push fit connector to the anti-dieseling valve. Pull the vacuum pipe from the carburettor. Disconnect the throttle linkage, see 'Carburettor Controls'. Disconnect the fuel feed pipe. See the 'Fuel Lines' section. The carburettor is held by two nuts on studs. Remove the nuts and washers. Withdraw the carburettor by pulling upwards (Fig. G:11).

Clean the carburettor and intake manifold mating faces. Refit the carburettor in the reverse order of removal. Tighten the nuts to the specified torque. Top up the cooling system if much coolant has been lost. Set the idle speed and check the mixture as described in the TUNE-UP chapter.

Automatic Choke Settings

The following adjustments should only be required if the setting has been inadvertently altered or when the carburettor has been overhauled.

Prepare the carburettor for adjustment. Remove the air cleaner. Detach bi-metal spring housing complete with coolant hoses and position to one side. Remove plug from the access hole (Fig. G:12).

Choke Gauging

This adjustment controls the air/fuel mixture ratio when choke is in operation.

1. Turn choke operating lever until drilling in central shaft lines up with hole in carburettor housing.
2. Insert shank of drill into choke gauging hole, see TECHNICAL DATA for specified size, ensuring drill fully locates into choke central shaft (Fig. G:13).
3. Loosen nut and turn choke lever clockwise up to its stop, then tighten nut.
4. Remove drill, check and adjust choke plate pull down as described below.

Choke Pull Down

This adjustment ensures that the vacuum pull down system, which controls fast idle, is operating correctly.

1. Bend back tab on pull down lever with a pair of pliers

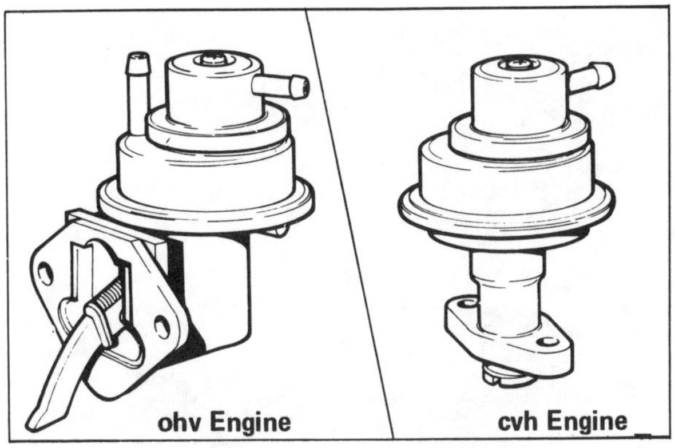

Fig. G:1 Fuel pumps fitted

Fig. G:2 Fuel pump output pressure test

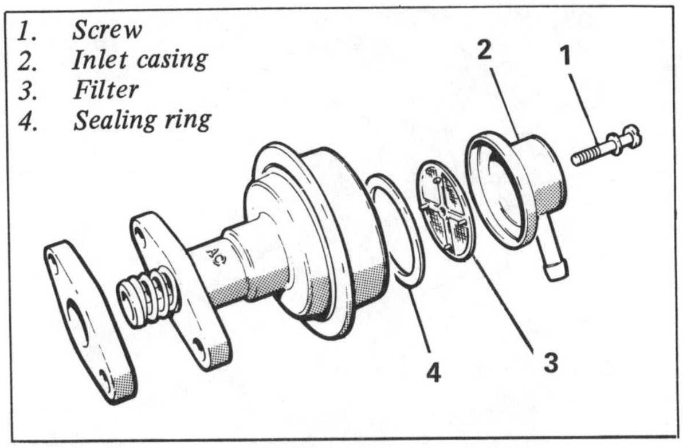

Fig. G:3 Fuel pump filter assembly (CVH engine shown)

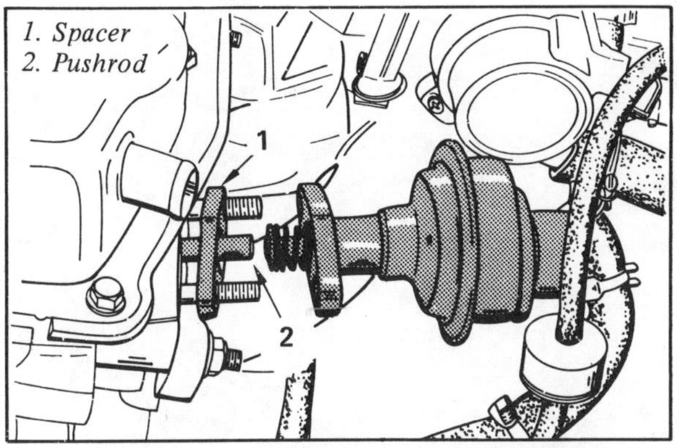

Fig. G:4 Removing fuel pump on CVH engine

Fig. G:5 Fuel pump removal, OHV engine

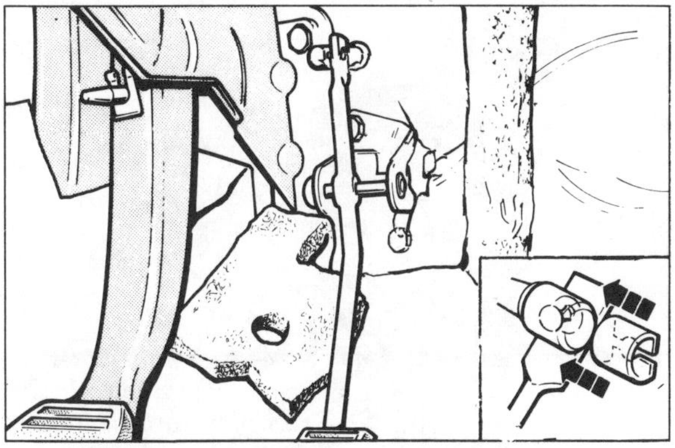

Fig. G:6 Throttle cable to pedal assembly

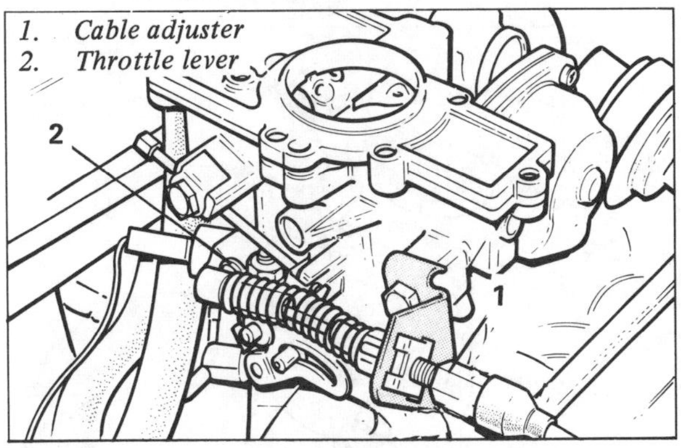

Fig. G:7 Throttle cable to carburettor

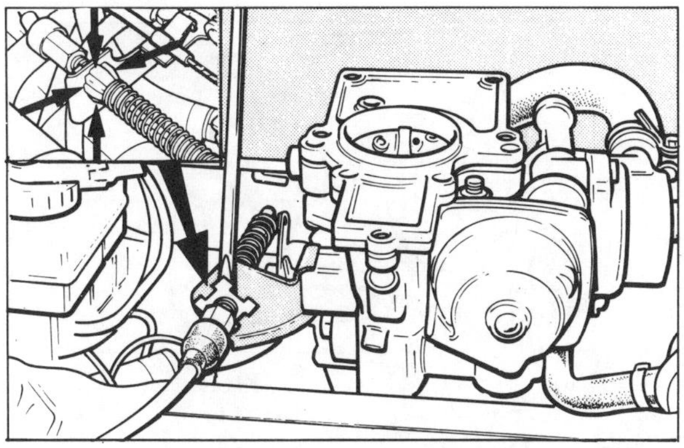

Fig. G:8 Throttle cable removal

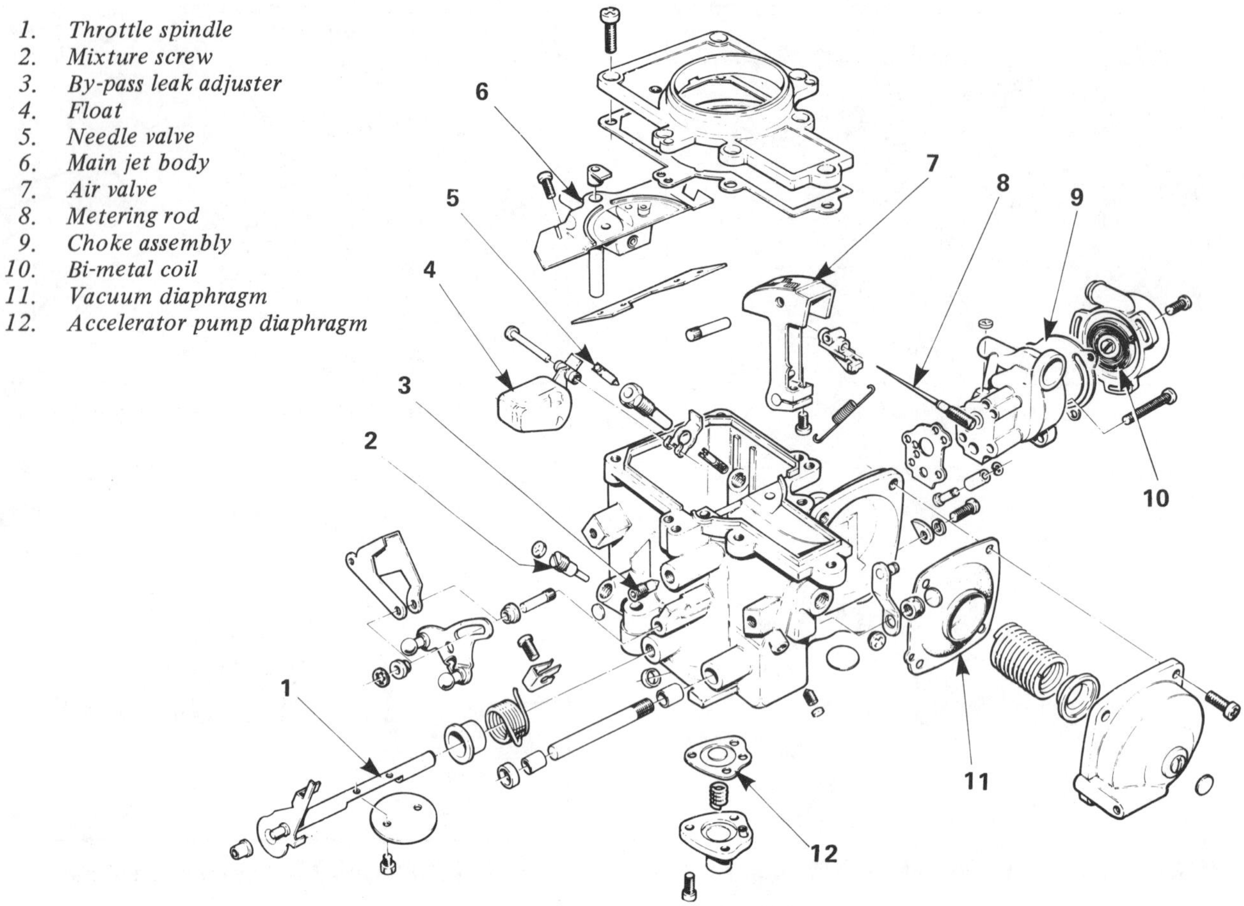

Fig. G:9 Ford VV carburettor exploded view

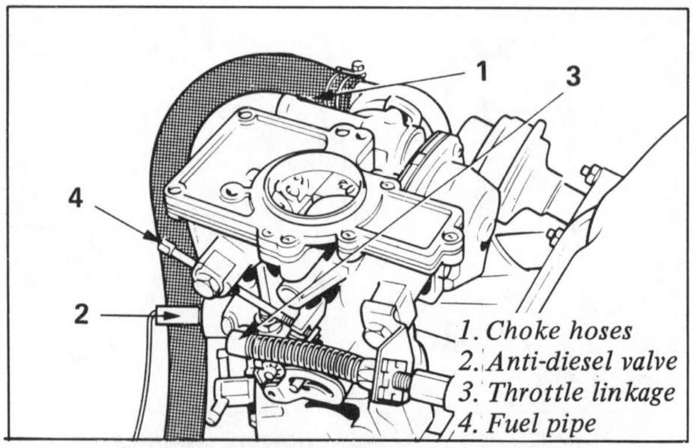

Fig. G:10 Ford carburettor connections

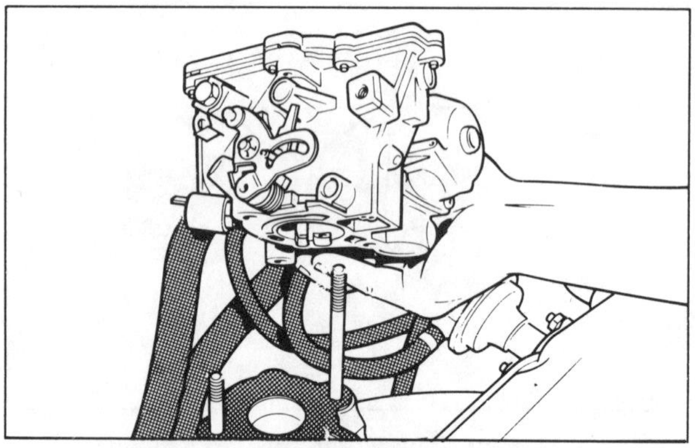

Fig. G:11 Ford carburettor removal

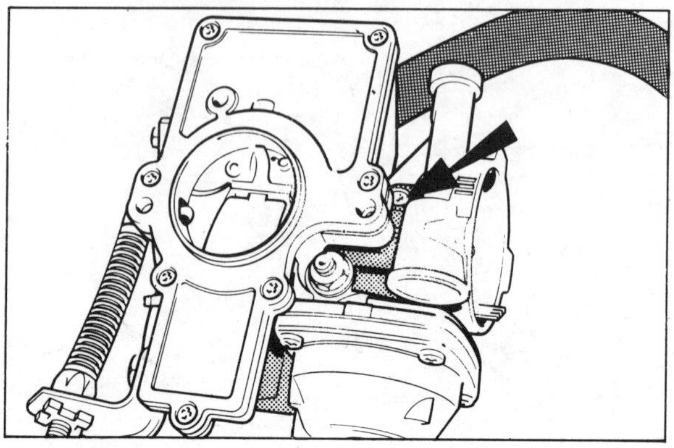

Fig. G:12 Ford carburettor gauging hole plug

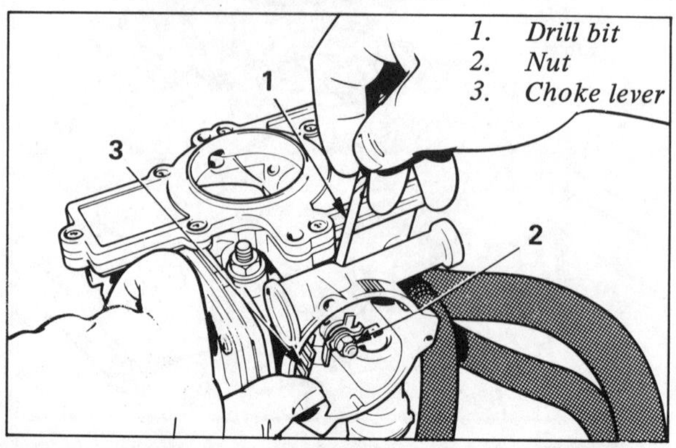

Fig. G:13 Ford carburettor choke gauging adjustment

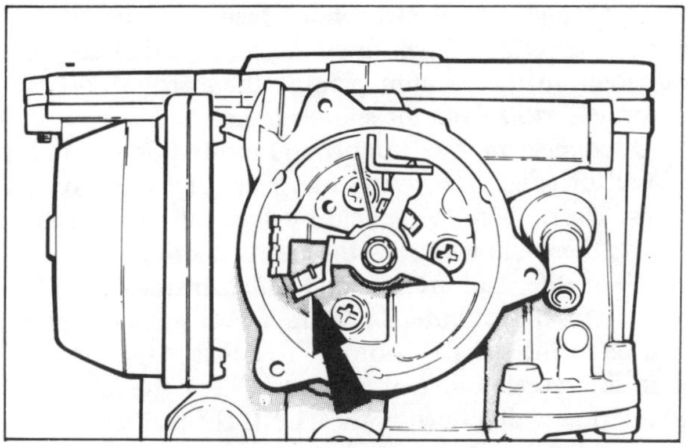

Fig. G:14 Ford carburettor pulldown lever

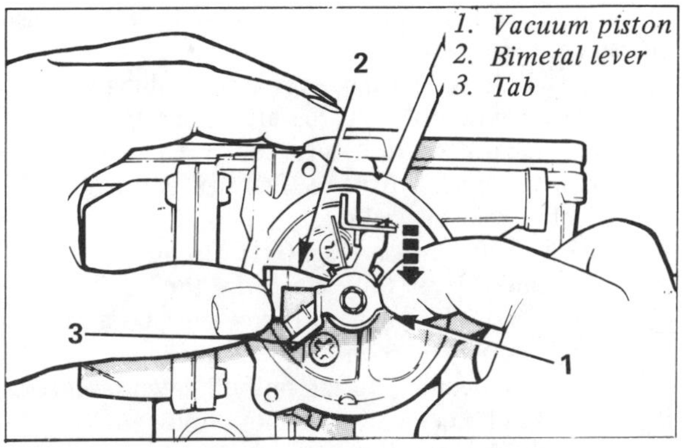

Fig. G:15 Ford carburettor choke plate pulldown

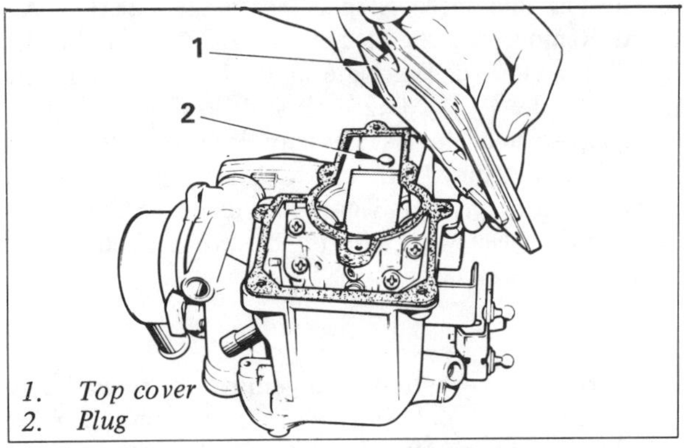

Fig. G:16 Ford carburettor top cover removal

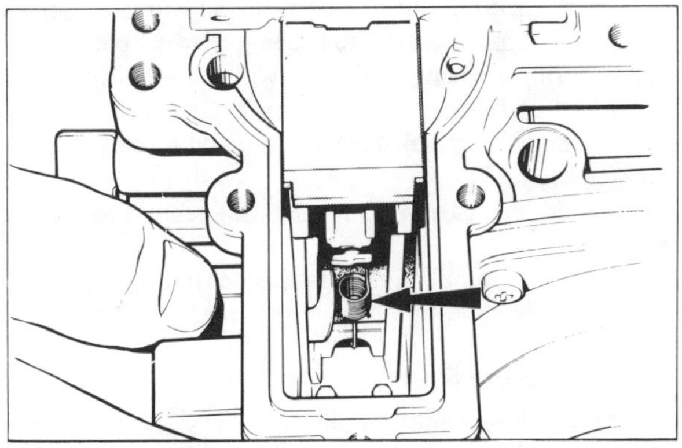

Fig. G:17 Ford carburettor metering rod bias spring

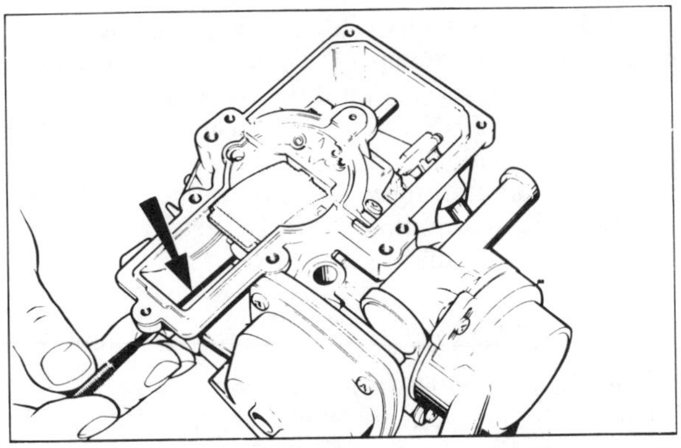

Fig. G:18 Ford carburettor metering rod removal

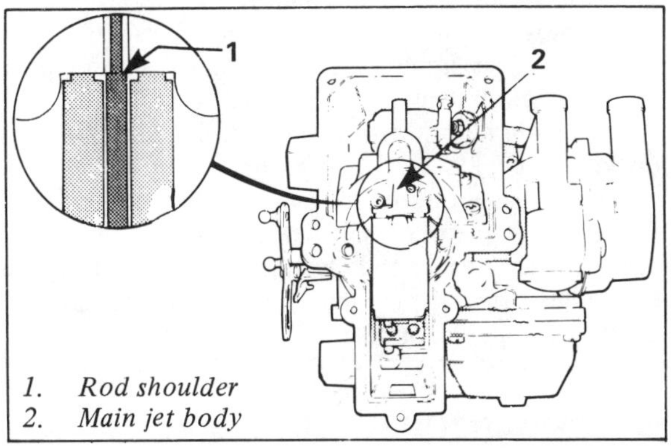

Fig. G:19 Ford carburettor metering rod adjustment

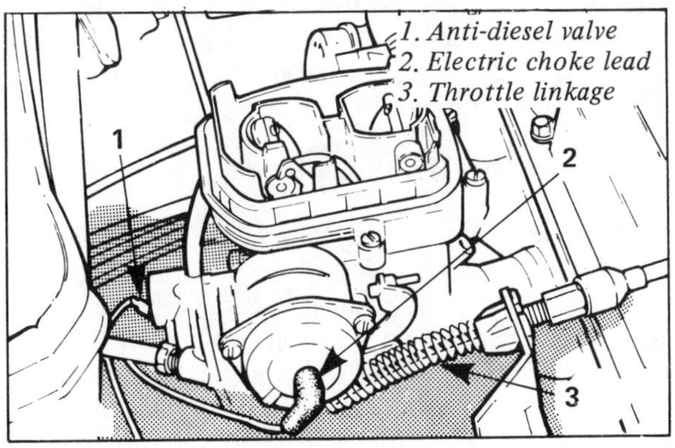

Fig. G:20 Weber carburettor controls

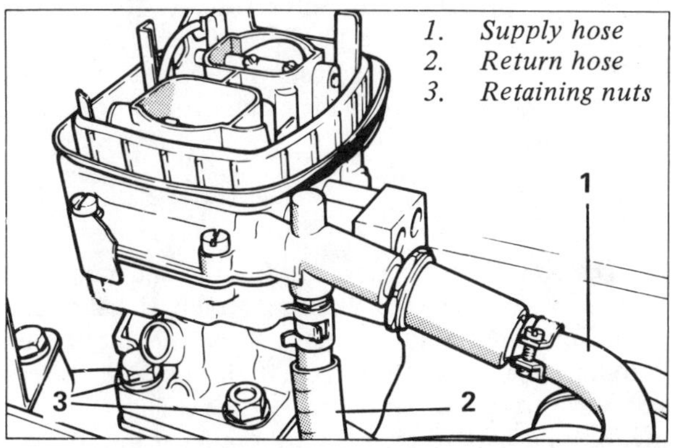

Fig. G:21 Weber carburettor fuel lines

to position shown (Fig. G:14) so that vacuum piston movement is not restricted.

2. Turn choke operating lever until drilling in central shaft lines up with hole in carburettor housing.

3. Insert shank of drill into choke gauging hole, see TECHNICAL DATA for specified size, ensure drill fully locates into choke central shaft.

4. Push vacuum piston (Fig. G:15) to bottom of travel and hold choke lever fully clockwise for 1,1 OHV, anti-clockwise for CVH engines. There must be a clearance between pull down lever and bi-metal choke lever.

5. Adjust by bending up tab on pull down lever so that it just touches bi-metal lever. Remove drill.

Choke Housing Positioning

After carrying out the previous adjustments fit new tamperproof plug to choke gauging access hole. Refit bi-metal spring housing complete with gasket. Ensure that outer end of bi-metal spring is correctly engaged in cental slot of choke lever. Position housing and loosely fit retaining screws.

To aid installation, lower securing screw should be fitted first.

Line up notch on cover with central index mark on choke housing and tighten screws. Finally refit air cleaner.

Main Metering Rod

This adjustment should be necessary only if the metering rod has been replaced or if a poor cold starting condition is apparent.

1. Disconnect battery and remove air cleaner.
2. Remove carburettor top cover and plug (Fig. G:16).
3. Unclip bias spring (Fig. G:17) and then unscrew and carefully withdraw main metering rod (Fig. G:18).
4. Screw in new metering rod until shoulder on rod lines up with vertical face on main jet body (Fig. G:19).
5. Refit bias spring and fit new plug to access hole. Refit carburettor top cover, using a new gasket.
6. Refit air cleaner and reconnect battery. Finally check and adjust idle speed and CO level as necessary, see TUNE-UP chapter for details.

WEBER CARBURETTOR . . . . [5]

The Escort XR3 model is fitted with a twin choke Weber downdraught carburettor (Fig. G:20). The idle speed and mixture should be adjusted as described in the TUNE-UP chapter. The adjustments in this chapter are normally only necessary if the carburettor is under repair or overhaul.

Removal & Refitting

Disconnect the battery earth strap and remove the air cleaner assembly. The cleaner assembly is held to the engine by three bolts. Remove these and lift the box. Withdraw the spout from the hot air box. Pull the emission valve from the bottom half of the casing. Disconnect the vacuum supply pipe from the inlet manifold. Withdraw the assembly.

Disconnect the electric choke feed wire pushfit connector (Fig. G:20). Then disconnect the anti-diesel valve lead. Remove the vacuum pipe to the distributor. Lever the throttle cable from the spindle lever.

Disconnect the fuel supply and return hoses from the carburettor (Fig. G:21). See the 'Fuel Lines' section for details. The carburettor is held by four nuts and washers. Remove these and withdraw the carburettor upwards.

Clean the carburettor and inlet manifold mating faces. Refit the carburettor using a new gasket. Tighten the nuts to the specified torque. Reconnect the carburettor in the reverse order of removal. Check the idle speed and mixture as described in the TUNE-UP chapter.

Fast Idle Speed

The fast idle speed is specified in the TECHNICAL DATA. Remove the air cleaner. Partially open throttle (Fig. G:23) close choke plate and release throttle. The adjustment screw (Fig. G:24) should now be resting on the highest stage of the fast idle cam. Release choke plates and make sure they fully open. Start engine without touching throttle pedal and note fast idle speed. If necessary, correct speed with adjustment screw. Recheck fast idle speed. Refit air cleaner.

Float Level

Disconnect the battery and remove air cleaner. Disconnect fuel supply and return pipes from carburettor, see 'Fuel Lines' section. Detach carburettor upper body assembly with float. Disconnect choke electrical feed wire.

Hold upper body vertically with float down and needle valve closed without depressing spring (Fig. G:25). Measure distance from upper body face to float base. Upper body gasket must be in position. The specified figure is given in the TECHNICAL DATA. If necessary, correct by bending tag on float arm.

Refit carburettor upper body. Refit choke feed wire, fuel supply and return pipes. Refit air cleaner and reconnect battery. Check idle speed and CO level.

Automatic Choke Settings

The following adjustments ensure that the correct clearance exists between the choke plates and the air intake wall when the choke is fully applied (choke plate pull down) and partially applied (choke phasing).

Preparatory conditions for the following adjustments are: Air cleaner removed. Choke housing, bi-metal spring and internal heat shield removed. Electrical feed wire from auto. choke housing disconnected.

Choke Valve Gap (Pulldown)

Fit elastic band to choke plate lever, so that it holds choke plates closed (Fig. G:26). Open throttle, so that choke plates fully close, then release throttle. Push diaphragm operating rod up to stop with a screwdriver. Measure clearance between lower edge of primary choke plate and air intake wall, using drill of suitable diameter (Fig. G:27).

If necessary, correct adjustment with stop screw lo-

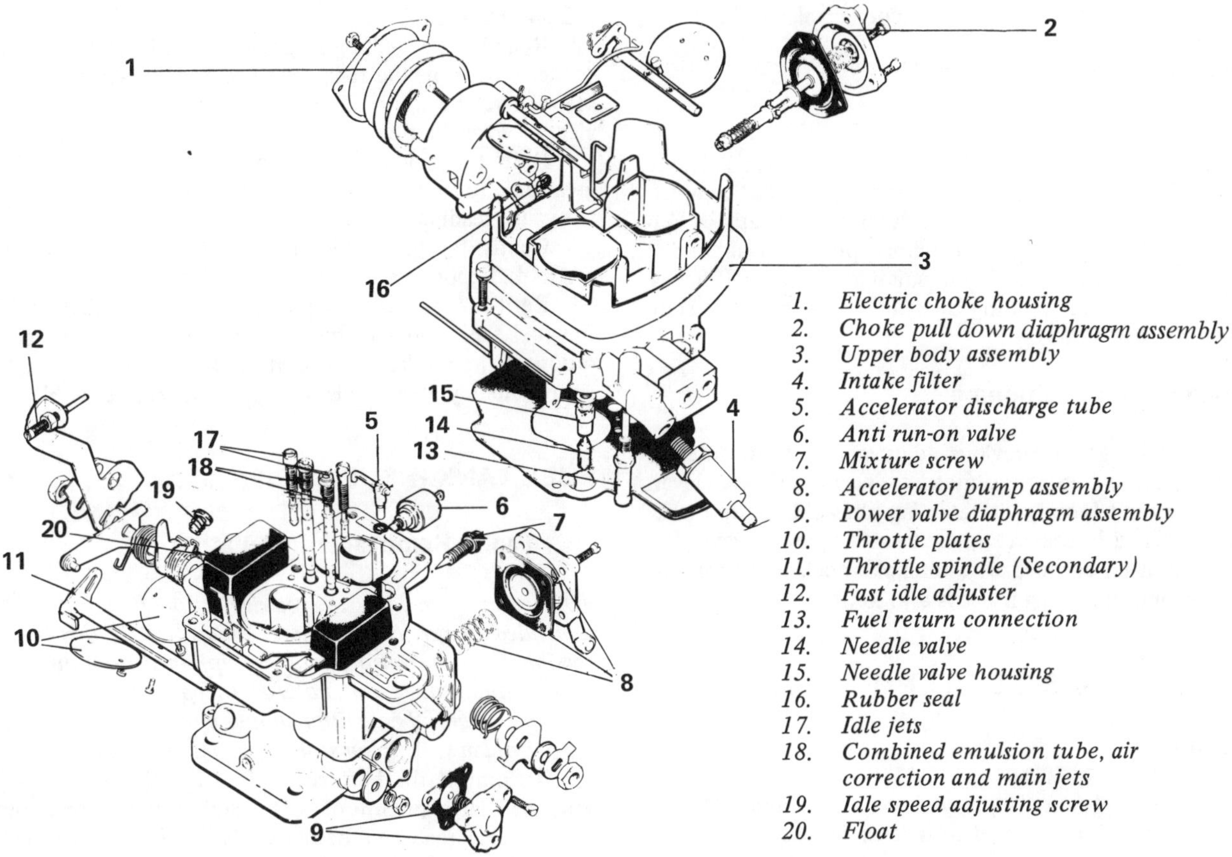

Fig. G:22 Weber twin choke carburettor exploded view

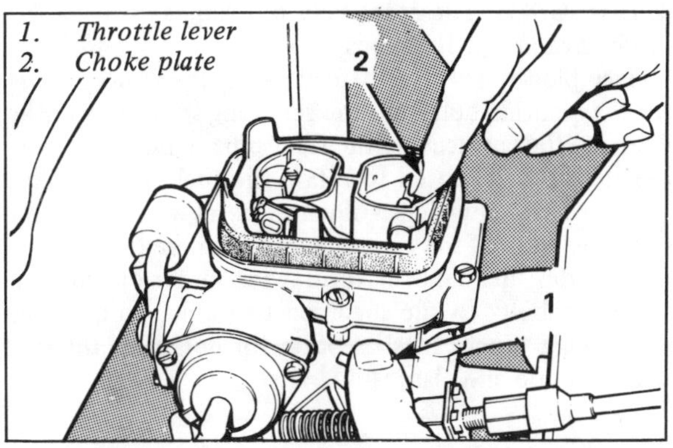

Fig. G:23 Weber carburettor fast idle check

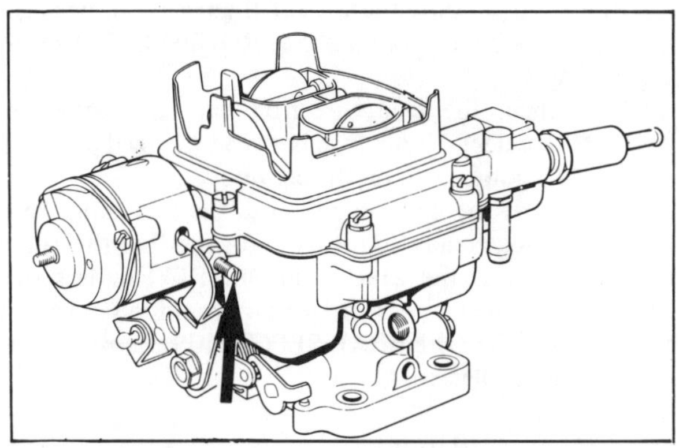

Fig. G:24 Weber carburettor fast idle screw (arrowed)

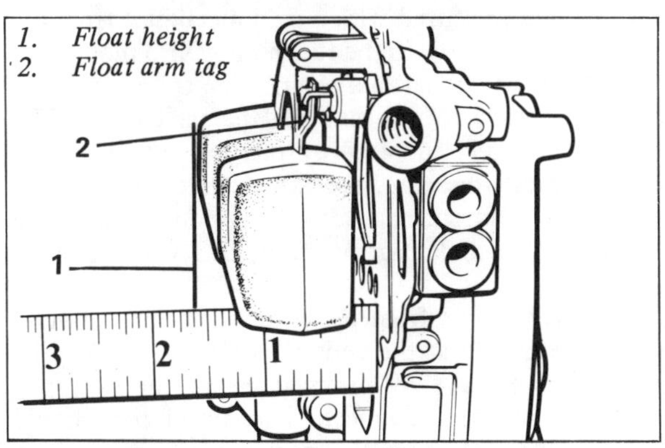

Fig. G:25 Weber carburettor float level

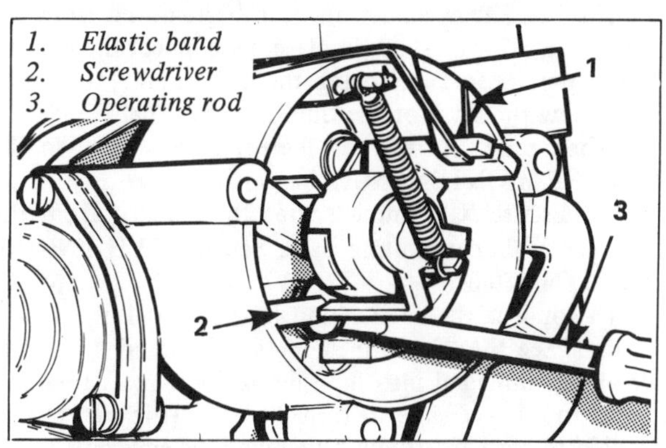

Fig. G:26 Weber carburettor choke pulldown

cated in diaphragm housing. The sealing plug must be removed first. Refit sealing plug and detach elastic band.

Choke Phasing

Hold throttle partially open and position fast idle cam as shown (Fig. G:28) then release throttle. Push choke plates down until step on cam jams against fast idle adjusting screw then release choke plates. Measure clearance between lower edge of primary choke plate and air intake wall using drill (Fig. G:27) of suitable diameter. If necessary, adjust choke phasing by bending tag on choke spindle lever (Fig. G:28).

Choke Cover Positioning

Following the previous adjustments: Refit internal heat shield and engage bi-metal spring tag on choke lever, fit cover and loosely fit retaining screws. Line up marks on cover and housing (Fig. G:29) and tighten screws. Reconnect electrical feed wire to choke housing. Check fast idle setting and adjust if necessary. Refit air cleaner.

FUEL GAUGE SENDER UNIT . . . . [6]

Testing

If the fuel gauge is not working or recording an incorrect level, it can be tested to determine whether it is the gauge itself, or the sender unit which is at fault, as follows:

Raise and support the rear of the car to allow access to the side of the petrol tank - see ROUTINE MAINTENANCE chapter.

Disconnect the feed wire from the sender unit terminal and connect a jump lead between the feed wire connector and a good earth on the car body.

Switch on the ignition and note the fuel gauge reading. If the gauge reads 'FULL', it indicates that the sender unit is faulty. If the gauge reads 'EMPTY' or less than 'FULL', then the gauge or the wiring from the sender unit is faulty. See the GENERAL ELECTRICS chapter for details of gauge replacement.

Removal & Refitting

The fuel gauge sender unit can be removed from the petrol tank with the tank in place. Disconnect the battery earth strap. Before removing the unit, drain the petrol tank below the level of the unit. See the 'Fuel Tank' section. Then raise the rear of the car as described in the ROUTINE MAINTENANCE chapter.

On Escort XR3 models, free the exhaust from the two rear rubber mounting pads. See the CVH ENGINE chapter for details.

Disconnect the sender unit lead and fuel pipes from the unit. See the 'Fuel Lines' section for details. The sender unit is threaded into the side of the tank. Although a special tool is available to turn the unit (Fig. G:31) it is possible to start it turning by tapping the edge anti-clockwise with a copper or thin, tapered hardwood drift, and a hammer, gently tap the lug on the locking ring to release it from the tank.

Remove the locking ring and carefully withdraw the sender unit from the tank aperture. It will be necessary to manoeuvre the unit around when withdrawing it to give clearance for the float arm and fuel pick-up pipe.

Refit the sender unit using a new sealing ring. Take great care to avoid bending the float arm as this will affect the gauge reading.

Gently tap the locking ring lug to secure the unit to the tank. Reconnect the fuel pipe and lead.

Refit the exhaust pipe mountings, XR3 models. Refill the fuel tank and reconnect the battery. Check the operation of the sender. Note that the fuel gauge should take approximately 30 seconds to register the correct level.

FUEL TANK & FUEL LINES. . . . . [7]

Fuel Tank - Removal & Installation

NOTE: Before tackling the removal of the fuel tank it is a good idea to run the fuel level down first. If this is not possible then the petrol should be siphoned from the tank through the sender unit hole. If the petrol is to be put back in the tank then a filter should be used when pouring the petrol back. A nylon stocking makes an ideal filter.

Disconnect the battery earth strap. If the fuel tank contains a large amount of petrol, siphon the majority out into suitable sealed containers. Disconnect the fuel lines and sender unit lead from the sender unit.

Free the filler and vent pipes from the rear end of the tank (Fig. G:33). These are held by clamps. Slacken the clamp bolts and free the pipes.

Place blocks or boxes beneath the petrol tank and remove the securing bolts and the retaining strap. Lower the tank and withdraw, leaving the filler pipe in position.

Note the bolts fitted to retain the fuel tank in place. If different from the inset illustration (Fig. G:32), new self-locking bolts of the type shown should be fitted.

Refit the fuel tank in the reverse order of removal. Tighten the bolts to the specified tightening torque. See the following 'Fuel Lines' section for details of the fuel pipe connection and clamps.

Fuel Lines

All models are fitted with a rigid metal pipe from the fuel sender and pickup to the fuel pump. XR3 models are fitted with an additional fuel return pipe from the carburettor needle valve to the tank via the sender unit. The rigid pipes are held in place by plastic clips. Whenever working beneath the car, take the time to examine the pipes and clips for damage and corrosion. If any pipe shows signs of such damage, it should be replaced as soon as possible.

To prevent pipes from being damaged by vibration, flexible hose is used at the ends of the rigid lengths. The hose may be secured by crimped clamps or by screw clamps. If the clamped connectors are fitted, they should be cut free and replaced with screw clamp or worm drive clips (Fig. G:34).

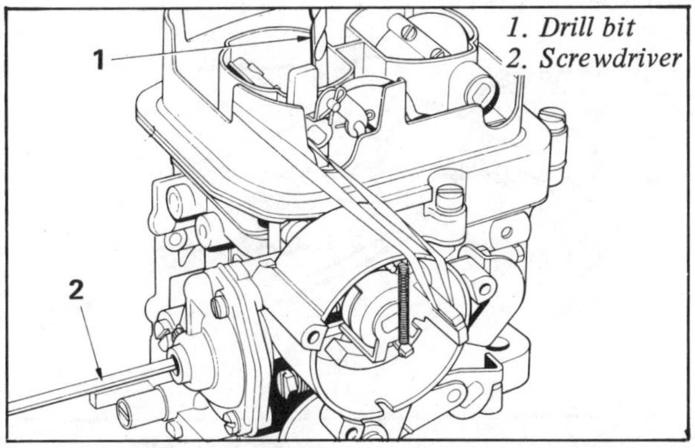

Fig. G:27 Weber carburettor choke pulldown adjustment

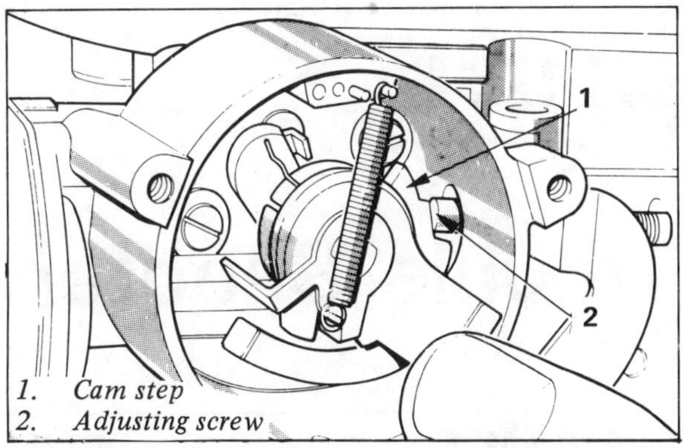

Fig. G:28 Weber carburettor choke phasing

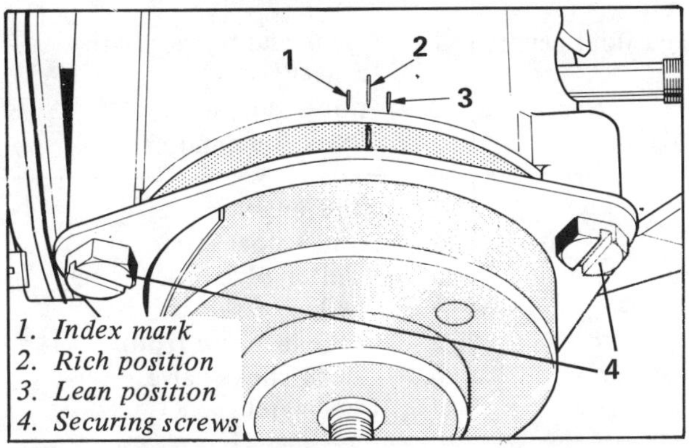

Fig. G:29 Weber carburettor choke housing alignment

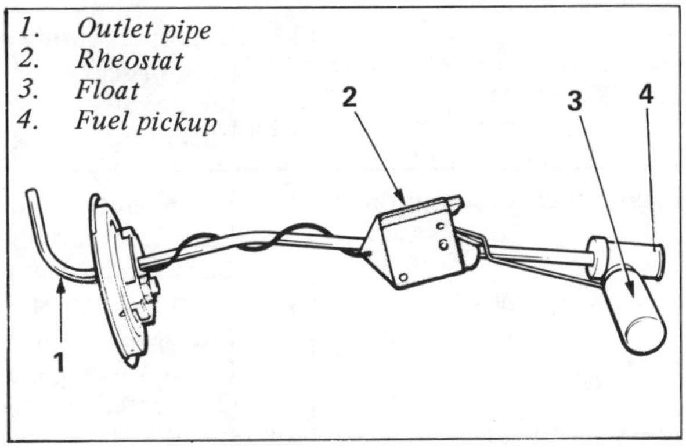

Fig. G:30 Fuel gauge sender unit

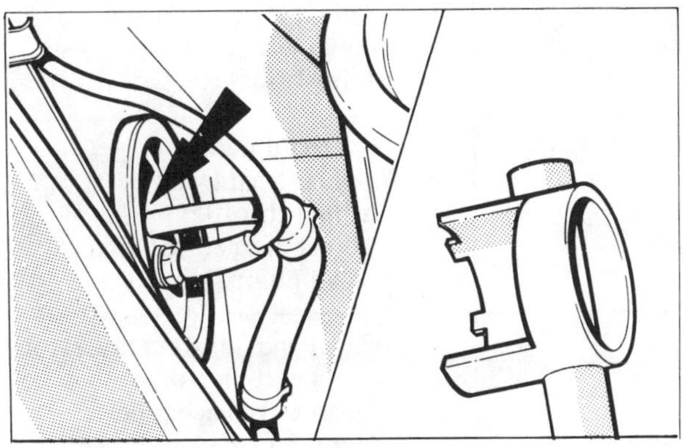

Fig. G:31 Sender unit location, showing removal tool

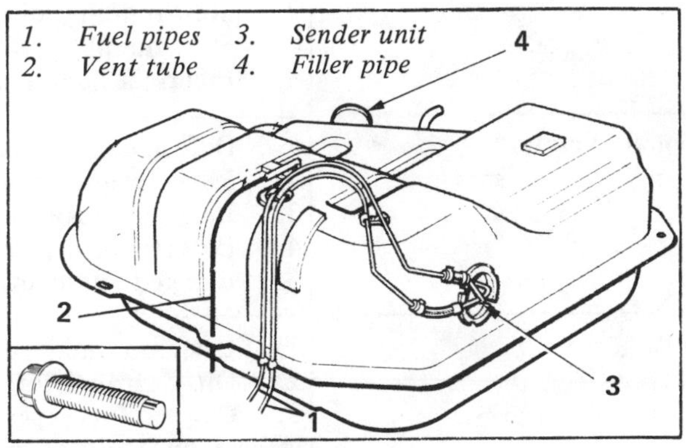

Fig. G:32 Fuel tank and sender unit

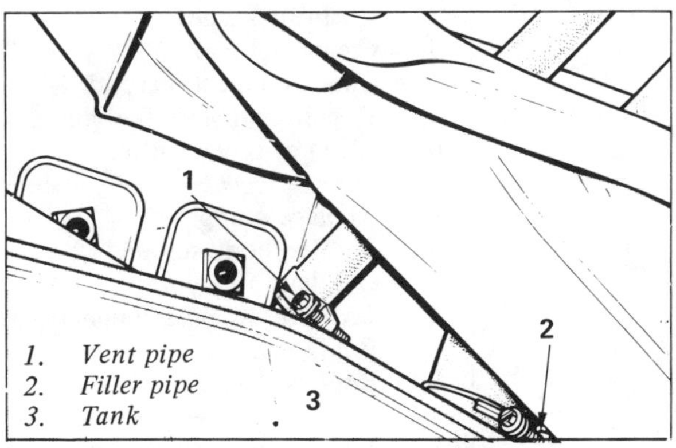

Fig. G:33 Fuel tank pipes

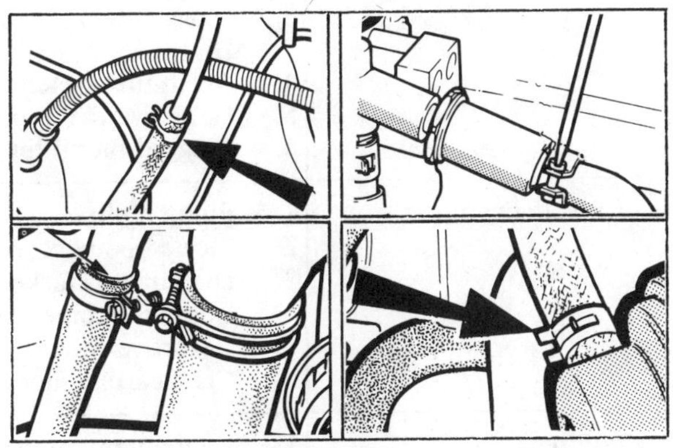

Fig. G:34 Fuel clip types

FUEL

Trouble Shooter

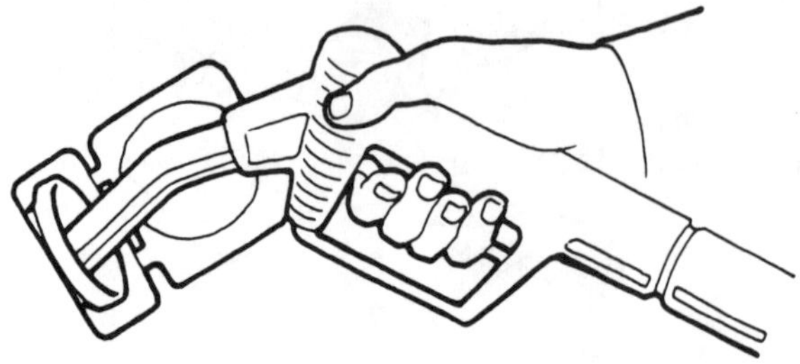

FAULT	CAUSE	CURE
Flooding	1. Improper seating or damaged float needle valve or seat 2. Incorrect float level 3. Fuel pump has excessive pressure	1. Check and replace parts as necessary. 2. Adjust float level. 3. Check fuel pump.
Excessive fuel consumption	1. Engine out of tune 2. Float level too high 3. Loose plug or jet 4. Defective gasket 5. Fuel leaks at pipes or connections 6. Choke valve operates improperly 7. Obstructed air bleed	1. Tune engine. 2. Adjust float level. 3. Tighten plug or jet. 4. Replace gaskets. 5. Trace leak and rectify. 6. Check choke valve. 7. Check and clear.
Stalling	1. Main jet obstructed 2. Incorrect throttle opening 3. Slow-running adjustment incorrect 4. Slow-running fuel jet blocked 5. Incorrect float level	1. Clean main jet. 2. Adjust throttle. 3. Adjust slow-running. 4. Clean jet. 5. Adjust float level.
Poor acceleration	1. Defective accelerator pump (if fitted) 2. Float level too low 3. Incorrect throttle opening 4. Defective accelerator linkage 5. Blocked pump jet	1. Overhaul pump. 2. Adjust float level. 3. Adjust throttle. 4. Adjust accelerator linkage. 5. Clean pump jet.
Spitting	1. Lean mixture 2. Dirty carburettor 3. Clogged fuel pipes 4. Manifold draws secondary air	1. Clean and adjust carburettor. 2. Clean carburettor. 3. Clean or replace pipes. 4. Tighten or replace gasket.
Insufficient fuel supply	1. Clogged carburettor 2. Clogged fuel pipe 3. Dirty fuel 4. Air in fuel system 5. Defective fuel pump 6. Clogged fuel filter	1. Dismantle and clean carburettor. 2. Clean fuel pipe. 3. Clean fuel tank. 4. Check connections and tighten. 5. Repair or replace fuel pump. 6. Clean or replace filter.
Loss of fuel delivery	1. Pump faulty 2. Slotted body screws loose 3. Diaphragm cracked 4. Loose fuel pipe connections 5. Defective valves 6. Cracked fuel pipes	1. Replace pump. 2. Tighten body screws. 3. Overhaul fuel pump. 4. Tighten fuel pipe connections. 5. Replace valves. 6. Replace fuel pipes.
Noisy pump	1. Loose pump mounting 2. Worn or defective rocker arm (mechanical). 3. Broken rocker arm spring (mechanical).	1. Tighten mounting bolts. 2. Replace rocker arm. 3. Replace spring.

Clutch & Gearbox

INTRODUCTION .... [1]

The clutch on all Escort models is of the conventional diaphragm spring type and is cable operated (Fig. H:1). The cable has a self-adjusting mechanism at the top of the clutch pedal to compensate for wear and therefore requires no routine maintenance.

To replace the clutch, the gearbox will have to be removed first. This can be done without removing the engine although the front of the car will have to be raised and supported high enough to enable the gearbox to be removed from under the car. Also the swivel hubs will have to be disconnected from the track control arms so that the driveshafts can be withdrawn from the gearbox.

The gearbox and final drive are in one unit, with an external gearshift rod to the remote type gear lever. The gearbox fitted to OHV engines is the same as that on CVH 1.1 and 1.3 engines, with an uprated version on all 1.6 engines. The optional 5 - speed gearbox is only fitted to 1.3 and 1.6 variants. Four speed gearboxes are known as the 'J' transmission, and 5 - speed gearboxes, the 'K' transmission.

Both the driveshaft oil seals and the gear selector rod oil seal can be replaced with the gearbox in the car but a special installer tool is essential when refitting the driveshaft seals. This tool will ensure that the seal is positioned at the correct depth inside the housing. The input shaft oil seal can only be replaced after the gearbox has been dismantled.

Overhauling the gearbox is reasonably easy for the more experienced DIY mechanic with access to a few Ford special tools, and the overhaul procedure detailed applies to 4 - speed and 5 - speed gearboxes. However, if a number of components are worn, then it may be more economical in the long run to fit a reconditioned or exchange gearbox with a guarantee.

For any work on the Escort clutch or gearbox, a comprehensive metric socket and spanner set will be needed, plus a clutch plate alignment tool (mandrel) and the special Ford tools for gearbox overhaul.

CLUTCH CABLE REPLACEMENT .... [2]

Replacing the cable is straightforward and can be done with a pair of multi-purpose pliers.

1. Remove the right-hand lower fascia insulating panel by releasing the two clips and metal tabs.
2. From inside the engine compartment, unhook the clutch inner cable from the clutch release arm (Fig. H:2) by pulling the inner cable forwards and sideways with pliers.
3. Remove the plastic clip securing the outer cable to the steering rack housing, then pull the cable through the hole.
4. From inside the car, pull the inner cable through the bulkhead hole and unhook the cable nipple from the self-adjust mechanism quadrant at the top of the clutch pedal (Fig. H:3).
5. Detach the outer cable from the bulkhead sleeve and remove the cable assembly from the engine compartment.
6. Insert the new cable through the bulkhead sleeve and attach the inner cable to the clutch pedal.
7. Align the white band on the cable with the paint spot on the steering rack housing, then refit the cable to the steering rack housing.
8. Using pliers, hook the clutch inner cable onto the clutch release lever fork.
9. Operate the clutch pedal several times to activate the clutch self-adjust mechanism.
10. Refit the fascia lower insulating panel.

CLUTCH REPLACEMENT .... [3]

Removal

1. Remove the gearbox from the engine as detailed later in this chapter.
2. Undo the six bolts securing the clutch pressure plate to the flywheel (Fig. H:4). Hold the pressure plate in position with one hand, then lever the pressure plate off the flywheel locating dowels. Remove the pressure plate and the clutch disc behind it.

Clutch Disc Inspection

Inspect the friction linings on the disc for wear, burning or contamination by oil or grease. The disc hub should be a free sliding fit on the gearbox input shaft splines without excessive side-play. If the linings are worn down to or near the rivet heads, or if any other of the above conditions are apparent, the disc must be replaced.

If oil or grease is present on the friction faces, the source must be determined and the fault rectified before fitting a new disc. This may be due to a defective gearbox input shaft oil seal or crankshaft rear oil seal.

Pressure Plate Inspection

Inspect the diaphragm fingers for any deep scoring. Inspect the contact surface of the pressure plate for cracks, signs of burning and wear. Place a flat edge across the surface of the pressure plate and check for any distortion. If necessary, replace the plate.

Flywheel Inspection

Inspect the friction surface of the flywheel - blueing or small cracks are of no particular importance, but if any deep scratches, scores, cracks or heat marks are present, the flywheel should be machined down, or preferably replaced.

Clutch Release Lever and Bearing

Both clutch release lever and bearing can be removed after the transmission has been removed. Release the bolt securing the bearing assembly to the lever shaft (Fig. H:5). Slide the shaft out of the gearbox and then remove the bearing. If the bearing shows signs of wear, damage or looseness, replace it.

Note that the release bearing bracket is located on the shaft fork by a pin (Fig. H:6).

Clutch Installation

1. First check that the clutch plate slides freely on the input shaft splines - apply a light smear of high melting point graphite grease on the splines and clutch plate hub.
2. Make sure the friction surfaces of the flywheel and pressure plate are perfectly clean and free from oil or grease.
3. Place the clutch disc on the flywheel, the flat side of the disc facing the flywheel.
4. Align the disc with a centralising tool (mandrel) and refit the clutch pressure plate (Fig. H:4).
5. Tighten the pressure plate bolts evenly to the correct torque, then remove the centralising tool.

GEARSHIFT REMOTE CONTROL . . . . [4]

Removal

1. On 4 - speed gearboxes, select 4th gear, on 5 - speed gearboxes, select reverse gear. This will ensure correct adjustment of the selector linkage when refitting.
2. Unscrew the gear lever knob and remove the rubber gaiter inside the car.
3. Raise and support the front of the car on axle stands - see ROUTINE MAINTENANCE chapter.
4. Detach the exhaust pipe from the rubber mounting at the rear and unhook the bias spring from the external gearshift rod.
5. Remove the gear lever rod connector clamp bolt and detach the gear lever rod from the gearbox selector shaft (Fig. H:7).
6. Remove the bolt securing the stabiliser rod to the gearbox. Note that there is a washer fitted between the rod and the gearbox. (Arrowed in Fig. H:8).
7. Undo the four nuts and washers securing the gear-lever housing bracket to the floorpan (Fig. H:9).
8. Remove the complete gearshift remote assembly from under the car.

Installation

Install the gear lever remote control assembly to the car in the reverse order of removal. Adjust the linkage play after installation as detailed in the CVH ENGINE chapter under 'Removal & Installation'.

Overhaul

1. Undo the four bolts and separate the gear lever housing from the stabiliser (Fig. H:12). Detach the gear lever and stabiliser from the slide block.
2. The gear lever rod can be detached from the slide block by unclipping the upper guide shell as shown in Figs. H:10 & 11.
3. To dismantle the gear lever from the housing, note the fitted position of the plastic gaiter before removing the securing 'C' clip with a screwdriver (Figs. H:10 & 11). Detach the lever and rubber spring from the housing.
4. Clean all parts and check them for wear. Replace any that are worn.
5. Smear high melting point grease at all the lever and rod contact points and reassemble the gear lever housing in the reverse order of dismantling. Make sure that the lower part of the gear lever is correctly engaged with the shift rod opening.

GEARBOX OIL SEALS - REPLACEMENT . . [5]

Gearbox Selector Shaft Oil Seal

1. Disconnect the gear lever rod from the selector shaft as detailed previously under 'Gearshift Remote Control - Removal' operation 5.
2. Pull the convoluted rubber gaiter from the selector shaft to expose the flanged end of the shaft oil seal underneath (Fig. H:13).
3. Using a screwdriver, carefully lever out the seal from the gearbox casing. Clean the seal housing and the selector shaft.

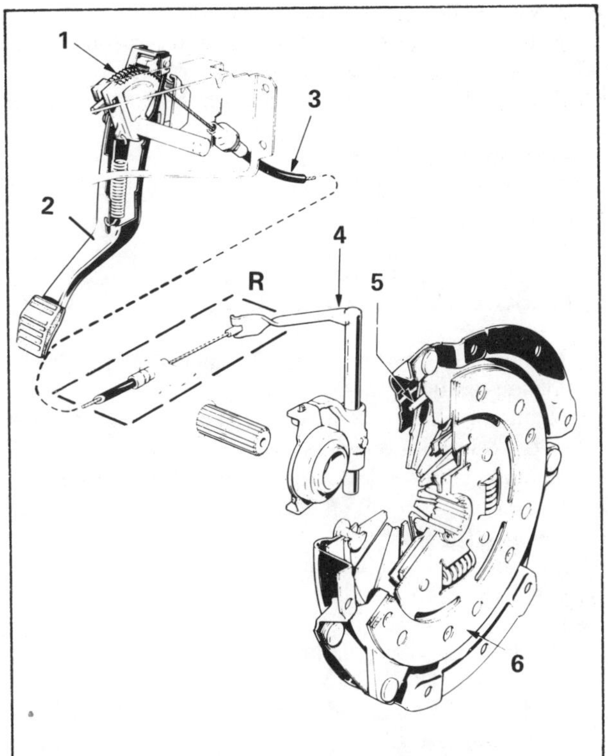

1. *Automatic adjuster*
2. *Clutch pedal*
3. *Clutch cable*
4. *Release bearing and shaft*
5. *Clutch pressure plate*
6. *Clutch driven plate*

Fig. H:1 The clutch and pedal operating mechanism

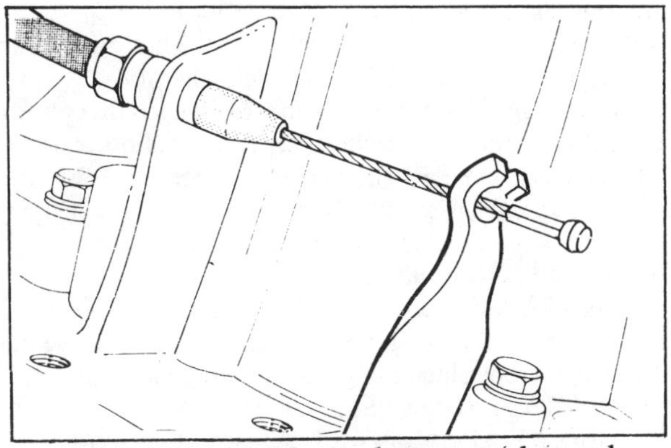

Fig. H:2 Clutch cable attachment - gearbox end

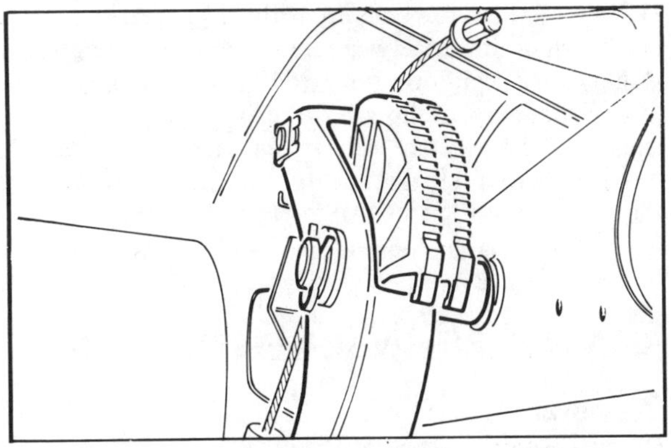

Fig. H:3 Clutch cable attachment - pedal end

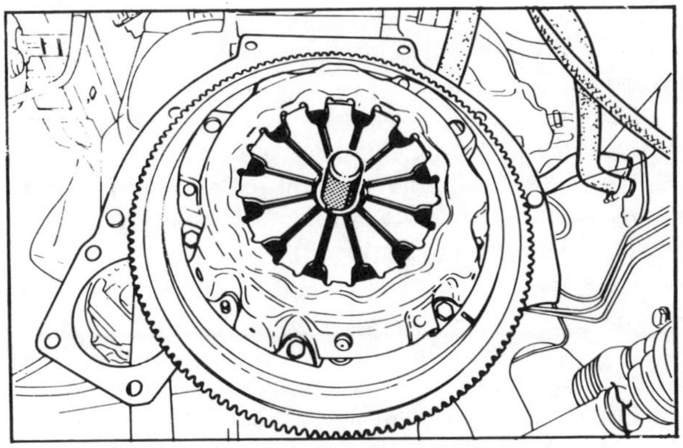

Fig. H:4 Clutch plate showing aligning mandrel

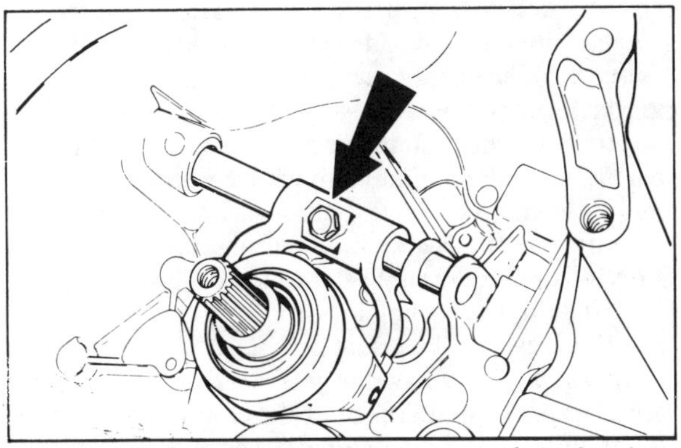

Fig. H:5 Undo bolt arrowed to remove release bearing

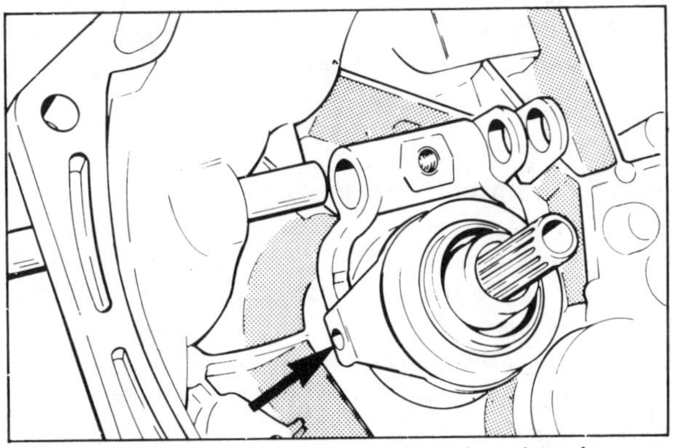

Fig. H:6 Release bearing bracket locating pin

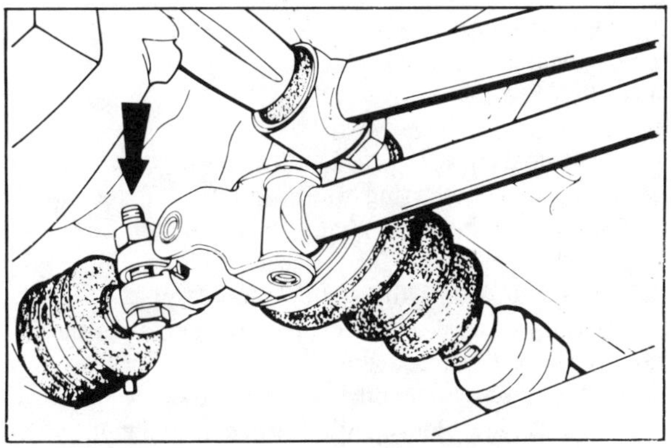

Fig. H:7 Gearshift rod clamp bolt arrowed

4. Smear the selector shaft and the inside of the seal with gearbox oil and fit the seal into the housing by tapping it into position with a small hammer.
5. Refit the rubber gaiter and reconnect the gearshift rod. Adjust the linkage play after reconnection as detailed in the CVH ENGINE chapter, under 'Removal & Installation'. Check the gearbox oil level and top-up if necessary.

Driveshaft Oil Seals

1. Disconnect one or both driveshafts from the gearbox (depending on which oil seal is to be replaced), as detailed in the FRONT SUSPENSION chapter.
2. Using a blunt screwdriver, carefully lever the seal from the gearbox housing to remove it.
3. Clean the seal housing to remove all dirt and oil.
4. Smear the oil seal lip with clean gearbox oil and then install the seal with the open side facing inwards using the special Ford tool 16-018 (Fig. H:14).
5. Fit a new circlip to the driveshaft inner CV joint splines and smear the oil seal contact area on the shaft with oil before refitting the driveshaft(s) to the gearbox as detailed in the FRONT SUSPENSION chapter.
6. Check the gearbox oil level and top-up as necessary.

GEARBOX REMOVAL & INSTALLATION . [6]

Removal

1. Raise and support the front of the car, drain the gearbox oil and disconnect both driveshafts from the gearbox as detailed in the CVH ENGINE chapter under 'Engine Removal & Installation' operations 8 to 18. Support the engine.
2. Referring to Fig. H:15, undo the four bolts securing the front gearbox mounting to the gearbox, and the single bolt securing the mounting to the bearer.
3. Unscrew the single bolt securing the gearbox rear mounting to the bearer and then unscrew the three bolts securing the mounting bearer to the car body (Fig. H:16).
4. Undo the two bolts securing the lower part of the gearbox flange to the engine.
5. Carefully withdraw the gearbox away from the engine until the input shaft is clear of the clutch.
6. Lower the gearbox and remove it from under the car.

NOTE: It may be necessary to raise or lower the engine to give extra clearance when removing the gearbox on 5 - speed models.

Installation

1. The installation procedure is basically a reversal of removal. Before starting the installation, make sure that the splines of the input shaft and the clutch release bearing guide sleeve are lightly smeared with graphite grease.
2. Make sure that the engine adaptor plate (Fig. H:17) is correctly located on the engine block guide bushes.
3. Raise the gearbox and insert the input shaft splines into the clutch, then push the gearbox into position so that the engine block guide bushes are aligned with the gearbox clutch housing flange.
4. Refit and tighten the two lower bolts to the correct torque. Raise the engine to its original position.
5. Attach the mountings to the gearbox and the car body.
6. Reconnect the remainder of the components as detailed in the CVH ENGINE chapter under 'Engine Installation'.

GEARBOX OVERHAUL . . . . [7]

Remove the gearbox from the engine as detailed in the previous section.

1. Drain any oil remaining in the gearbox through one of the drivejoint openings.
2. Remove clutch release shaft and thrust bearing with fork by undoing the bolt (Fig. H:5).
3. Remove the selector shaft locking mechanism cap nut together with interlock pin and spring (Fig. H:19).
4. Undo the 8 bolts and detach the gearbox housing end cover.
5. Remove the snap rings from the main and input shaft bearings (Fig. H:18).
6. Undo the 14 bolts and separate small housing section from the main casing by carefully tapping housing section loose with a plastic headed mallet before lifting the housing off.
7. Lift out the magnetic disc (Fig. H:20).
8. Withdraw the selector fork shaft, then remove the selector forks, shift locking plate and spring (Fig. H:21).
9. Lift the mainshaft and input shaft with reverse gear simultaneously out of the gearbox casing.
10. Now lift the differential assembly out of the gearbox casing.
11. Remove the mainshaft roller bearings by breaking the plastic bearing cage (A, Fig. H:22). Remove the bearing rollers and cage. Withdraw the oil slinger with pliers and remove retainers. Extract bearing race using special tools 16-021 and 15-053 (B, Fig. H:22).

Note that a new oil slinger should be fitted on reassembly.

12. Drift out the input shaft oil seal rearwards through the gearbox housing using a screwdriver.
13. Lever out the CV joint output shaft oil seals.
14. Remove the differential gear bearing cup using a drift.
15. Using a suitable drift, drive the bearing cup and diaphragm springs out of the small housing section.

NOTE: As the bearing race is a sliding fit, it is secured by deforming the edge of the bearing seat, (this serves solely as an aid to assembly).

Gearbox Mainshaft Assembly (Fig. H:18)

16. Release the mainshaft bearing circlip and insert the outer snap ring in the peripheral groove in the outer bearing race. Remove the bearing with a universal two-jaw puller. Remove the snap ring again.

NOTE: After removal, the bearing must be discarded and replaced with a new one.

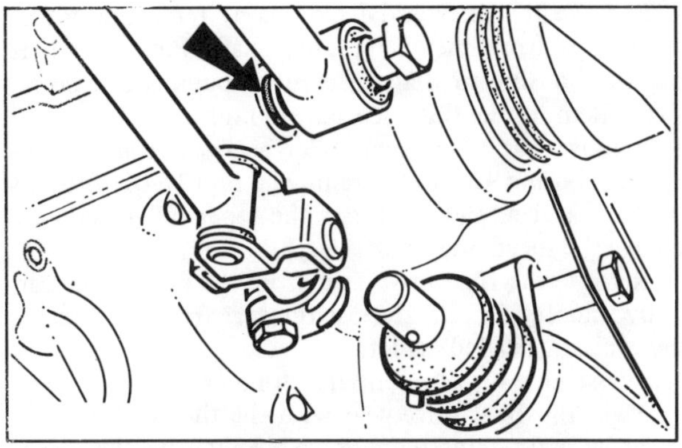

Fig. H:8 Details of stabiliser rod to gearbox

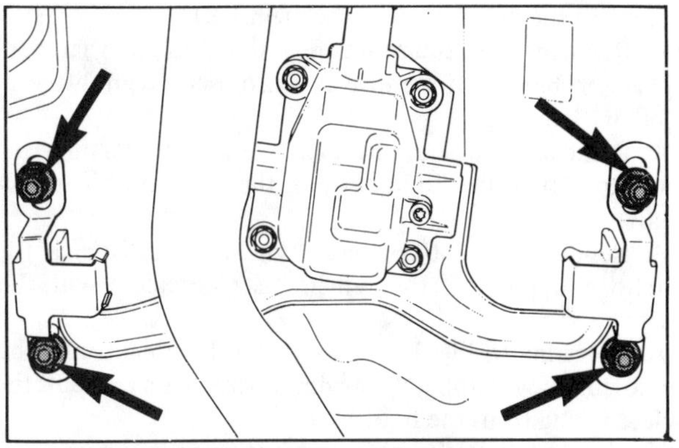

Fig. H:9 Gearshift housing bracket bolts

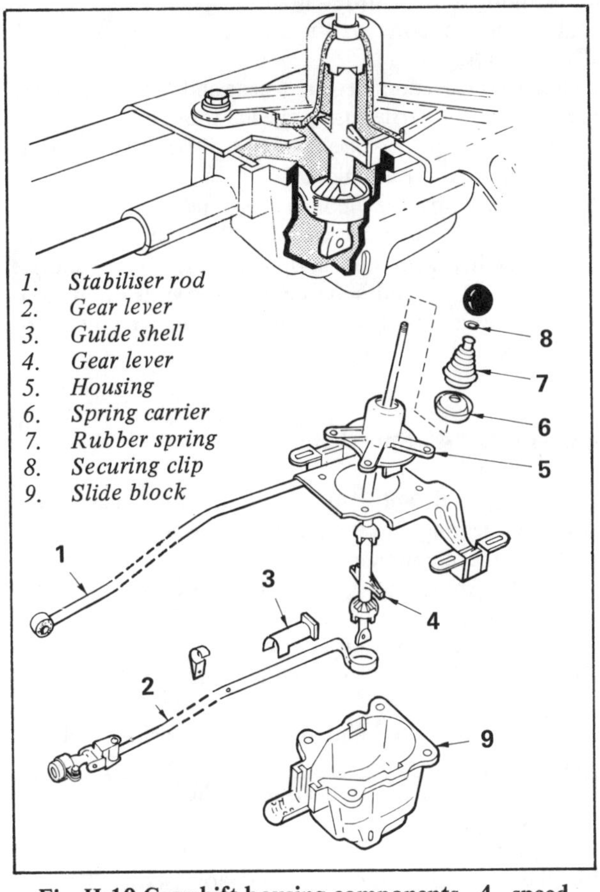

Fig. H:10 Gearshift housing components - 4 - speed

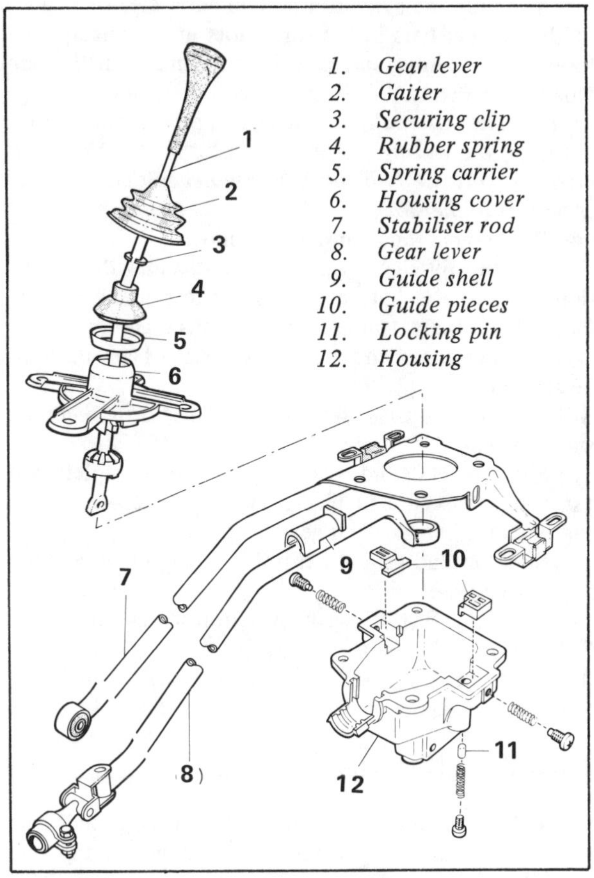

Fig. H:11 Gearshift housing components - 5 - speed

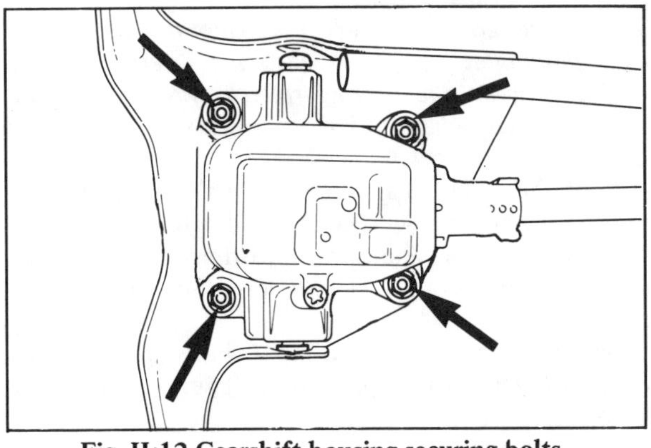

Fig. H:12 Gearshift housing securing bolts

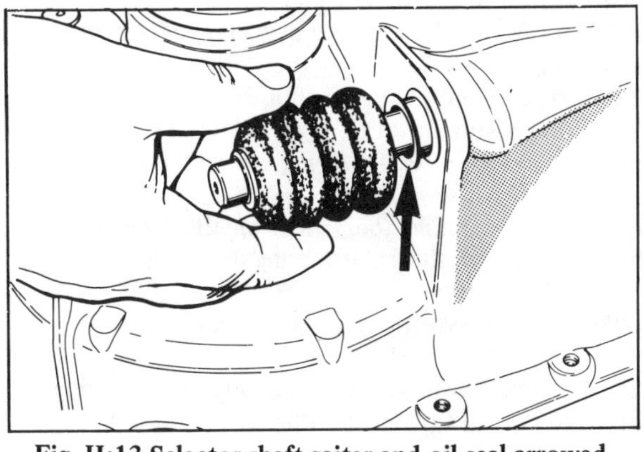

Fig. H:13 Selector shaft gaiter and oil seal arrowed

17. Slide the 4th gear from the mainshaft.
18. Remove the circlip securing the 3rd/4th gear synchroniser hub then remove synchroniser assembly complete with 3rd gear.
19. Remove 2nd gear, then remove the supporting ring and thrust half rings before sliding the gear from the mainshaft.
20. Remove the 1st/2nd gear synchroniser. Detach the circlip and pull off the synchroniser complete with 1st gear.
21. Referring to Fig. H:18, dismantle the synchroniser by removing the retaining rings with blocker bars and pull the selector ring from the hub.
22. Check all parts for wear and replace as necessary.
23. Reassemble synchroniser as follows:

Assemble the synchroniser hub with selector ring and fit the blocker bars in the three slots in the selector ring. Insert the retaining springs offset, starting with the same blocker bar Fig. H:23. Fit the synchroniser rings.

NOTE: The internally splined face of the selector ring has slots, the width of 3 splines milled at 3 points (120^0 intervals). Fit the hub and ring together so that the blocker bars lie in the slots.

24. Reassemble the mainshaft as follows:

Thoroughly clean and lightly lubricate with transmission oil all parts before assembly, the running and side faces of the gears, splined shaft sections together with the synchroniser cones and mainshaft bearing faces must also be greased.

NOTE: Do not overstretch circlips when fitting them with snap ring pliers.

a) Slide on 1st/2nd gear synchroniser complete with 1st gear and secure in place. Ensuring selector groove in the synchroniser faces 1st gear.
b) Slide on 2nd gear, fit thrust half ring and secure with supporting ring.
c) Slide the 3rd/4th gear synchroniser complete with 3rd gear onto the mainshaft.
d) Slide on 4th gear, press on mainshaft ball bearing (with groove facing outwards) using suitable tool.

Input Shaft (Fig. H:18)

25. Release input shaft large ball bearing and insert outer snap ring in peripheral groove in bearing outer race. Withdraw bearing using a universal two-jaw puller.
26. Remove input shaft small ball bearing with the aid of a press or universal two-jaw puller.

NOTE: Discard the ball bearings and replace them with new ones.

27. Reassemble the input shaft as follows:
a) Using a suitable tool, press on input shaft small ball bearing (with lettering facing upwards) as far as the stop.
b) Using a suitable tool, press on the input shaft large ball bearing with the groove facing outwards.

Differential Assembly (Fig. H:18)

28. Remove both axle drive pinions by swivelling them both laterally out of the housing.
29. Remove the differential pinions by removing one of two circlips from the differential shaft. Press the shaft from the housing so that differential pinions can be removed. Remove the thrust cage from housing.
30. Remove both the taper roller bearings from the differential housing. Clamp the removing tool 15-026A by its handle in a vice, and withdraw the taper roller bearings. Remove the speedometer drive gear.
31. Remove the final drive gear from the differential carrier by undoing 6 bolts. Release the gear from the housing using a plastic headed mallet.
32. Reassemble the differential as follows:
a) Align the final drive gear with the threaded holes in the carrier and tighten them down evenly using the original bolts. The deeply chamfered edge of the inner ring of the gear must face the differential.

Tighten the bolts to the correct torque see TECHNICAL DATA.

b) Refit the speedometer drive gear and press both taper bearings onto the carrier using special tool 15-025A.
c) Insert the thrust cage into the differential housing. Refit differential drive pinions.

NOTE: Do not damage the thrust cage when installing.

d) Refit the axle drive pinions. With the differential pinions parallel with one another, bring the two drive pinions into mesh and simultaneously swivel the pinions inwards. Prevent the drive pinions from moving by inserting a plastic or wood plug.

Reassembly - 4 - Speed Gearbox

1. Using special tool 16-020, drift the outer cup of the taper roller bearing into the large housing.
2. Insert the diaphragm springs into the small housing. The diaphragm springs should be positioned with the internal diameters next to one another.
3. Insert the bearing cup and secure it with a light blow from a punch. This prevents the cup falling out when the housing is reassembled.
4. Press in the new input shaft oil seal using a suitable tool or pipe.
5. Install the mainshaft bearing in the housing, with a new oil slinger.
6. Press in the roller bearing using special tool 16-020 with the rear of the cage facing downwards. Using a punch, stake the mainshaft bearing with light blows on the edge of the bearing seat.

NOTE: Take great care when staking the bearing to avoid aluminium splinters becoming detached and jamming between the bearing rollers. Also, do not use excessive blows when staking as this will distort the bearing cup.

7. Refit the differential assembly in the large housing. Lubricate the bearings with gearbox oil before fitting.
8. Slide the reverse idler gear onto the shaft, and at the same time guiding the selector lever into the selector groove.
9. Before the mainshaft and input shafts are fitted, lift the selector lever with idler gear and support it on the selector shaft detent.

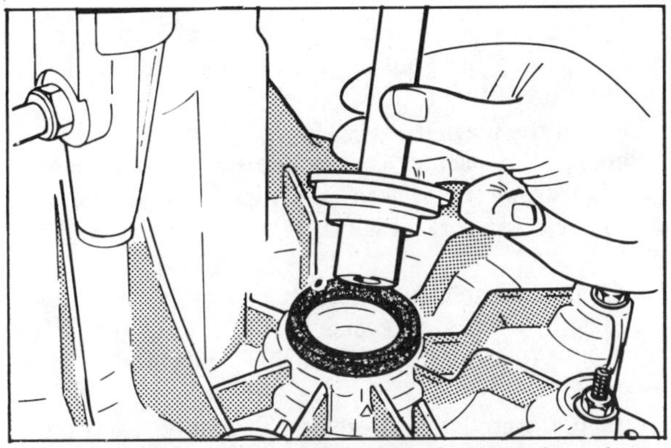

Fig. H:14 Fitting driveshaft oil seal with tool

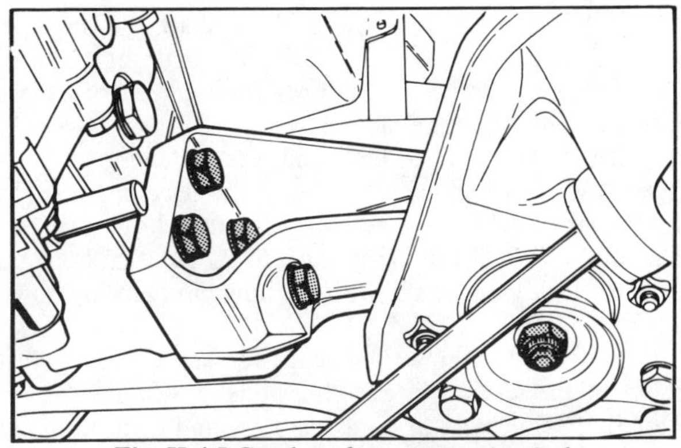

Fig. H:15 Gearbox front mounting bolts

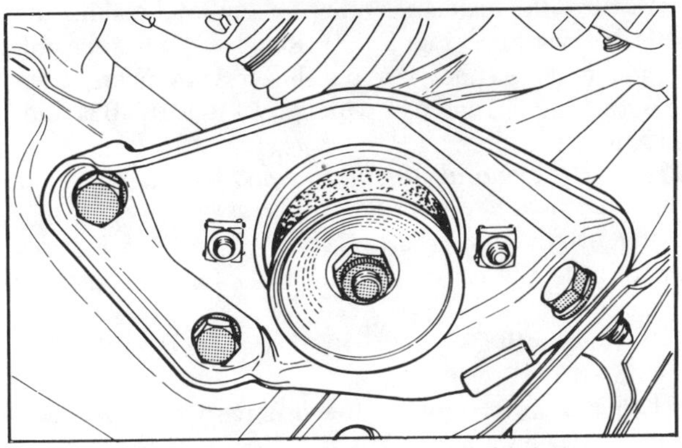

Fig. H:16 Gearbox rear mounting bolts

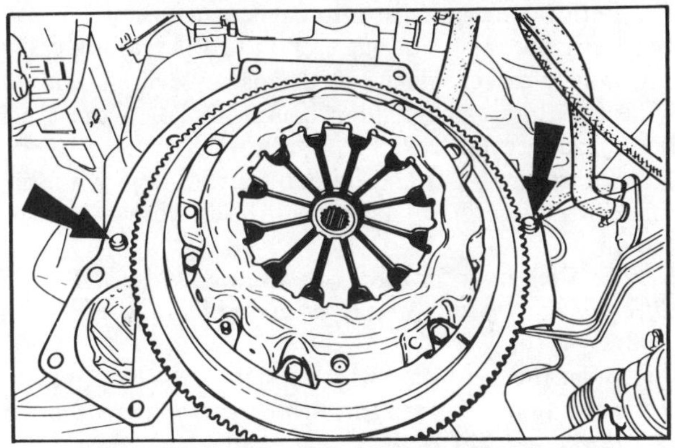

Fig. H:17 Engine guide block bushes arrowed

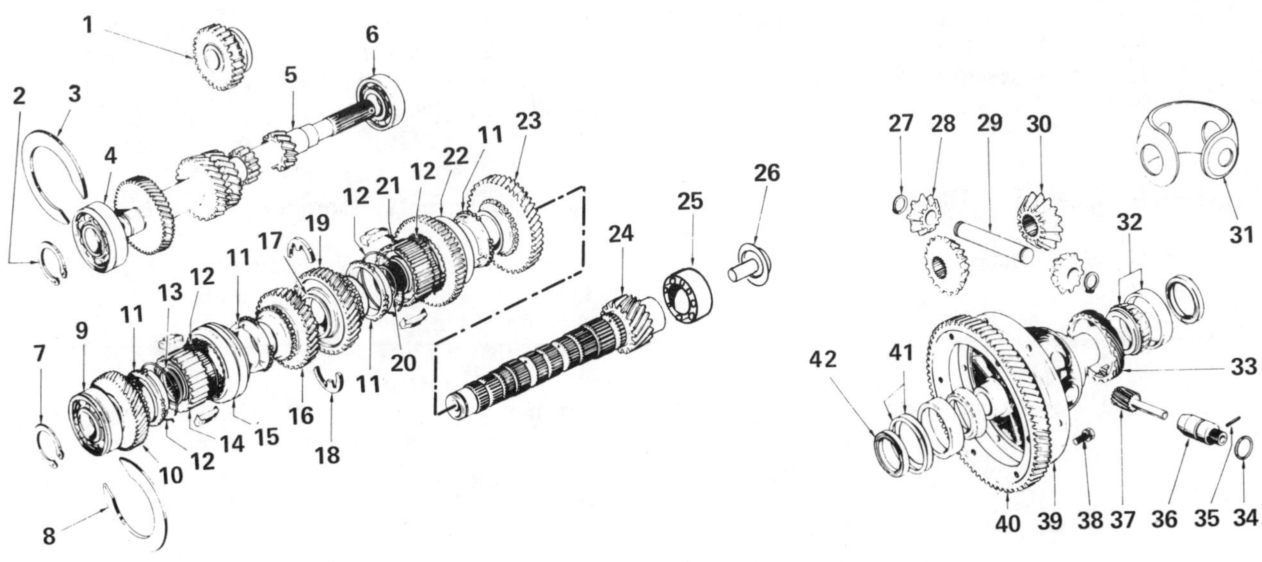

1. *Reverse idler gear*
2. *Circlip*
3. *Snap ring*
4. *Large ball bearing*
5. *Input shaft and gear cluster*
6. *Small ball bearing*
7. *Circlip*
8. *Snap ring*
9. *Ball bearing*
10. *4th gear*
11. *Synchroniser ring*
12. *Synchroniser spring*
13. *Circlip*
14. *3rd/4th gear synchroniser*
15. *3rd/4th gear selector ring*
16. *3rd gear*
17. *Retaining ring*
18. *Half thrust rings (2)*
19. *2nd gear*
20. *Circlip*
21. *1st/2nd gear synchroniser hub*
22. *1st/2nd gear selector ring with reverse gear*
23. *1st gear*
24. *Mainshaft*
25. *Ball bearing*
26. *Oil slinger*
27. *Circlip*
28. *Differential pinion*
29. *Differential shaft*
30. *Axle shaft pinion*
31. *Thrust cage*
32. *Taper roller bearing*
33. *Speedometer drive worm*
34. *O-ring*
35. *Retaining pin*
36. *Speedometer drive pinion bearing*
37. *Speedometer drive pinion*
38. *Flange bolts*
39. *Differential casing*
40. *Drive gear*
41. *Diaphragm spring*
42. *Oil seal*

Fig. H:18 Exploded view of 4 - speed gearbox internal component arrangement

10. Install the mainshaft and input shaft by the same method as used on removal. Bring the mainshaft and input shaft gears into mesh and carefully guide both shafts into the housing at the same time.
11. Install the selector lever and reverse idler gear. Then insert the shift locking plate and fit the selector forks (1st /2nd gear has the short actuating lever, 3rd/4th gear fork has long lever). Refit the spring into the housing bore and selector fork guide shaft, with the long pin pointing downwards.
12. Select 4th gear at this stage for subsequent adjustment of the selector lever This is done by turning the selector shaft clockwise as far as the stop, and then pushing it inwards as far as it will go.
13. Refit the magnet disc into the housing recess.
14. Refit the small housing section to the gearbox casing with a new gasket. Make sure the magnetic disc is correctly seated. Insert and tighten housing bolts to the specified torque see TECHNICAL DATA.
15. Secure the input and output shaft bearings with snap rings, lifting the shafts slightly with a screwdriver to insert the ring.
NOTE: Snap rings for these two bearings are available in 3 different thicknesses. Select a snap ring that just fits in the groove.
16. Align the snap rings with the step in the cover gasket. Refit the housing cover and tighten the bolts to the correct torque see TECHNICAL DATA.
17. Refit the selector shaft locking interlock pin, spring and cap nut. Coat the cap nut with sealant.
18. Press new driveshaft oil seals into the gearbox housings using special tool 16 - 018.
19. Refit the clutch release shaft, lever and thrust bearing in the reverse order of removal.

Overhaul 5 - speed (Fig. H:27)

1. Drain any oil remaining in the gearbox through one of the shaft openings.
2. Remove clutch release lever and release bearing with fork by undoing the bolt (Fig. H:5).
3. Remove both cap nuts on the selector shaft interlock mechanism and remove the interlock pin and springs (Fig. H:19).
4. Undo the screw and detach the selector pin from the selector shaft.
5. Remove the snap ring and detach 5th gear complete with synchroniser clutch and selector fork from the mainshaft (Fig. H:27).
6. Remove snap ring and detach the 5th gear driving gear from the input shaft with a puller.
NOTE: This snap ring must be discarded after removal and a new one used on reassembly.
7. Undo the 9 bolts and detach the 5th gear casing from the main gearbox casing.
8. Remove the snap rings from the main and input shaft bearings.
9. Detach the gearbox mounting from the gearbox casing by undoing the three nuts.
10. Undo the 14 bolts and detach the small housing half, carefully tapping the housing section loose with a plastic hammer before lifting it off.
11. Lift out the magnetic disc (Fig. H:20).
12. Remove the selector shaft guide sleeve, then remove the circlips from the selector shaft guide sleeve and 1st/2nd gear selector fork (Fig. H:24). Withdraw the guide sleeve.
13. Lift the complete mainshaft assembly and input shaft, selector forks and reverse gear the out of the housing together. Remove the selector shaft and the shift locking plate.
14. Lift the complete differential assembly out of the gearbox large housing section.
15. Remove the mainshaft roller bearing by breaking the plastic bearing cage (Fig. H:22). Remove the bearing rollers and cage. Extract the oil slinger with pliers, then knock out the bearing cup with special tool 16 - 021 and, slide hammer 15-053.
NOTE: After removal, the bearing and oil slinger should be discarded and new ones fitted on reassembly.
16. Remove the input shaft oil seal by drawing it inwards from outside the casing using a drift. Take care not to damage the seal housing.
17. Remove both CV joint shaft oil seals by levering them out with a screwdriver.
18. Using a drift, remove the differential bearing cup from the large housing section.
19. Drive the bearing cup and spring washers out of the small housing section using a suitable drift.
NOTE: Because the bearing is a sliding fit in its housing, the bearing cup is secured by deforming the edge of the bearing seat with a blow from a punch (this serves solely as an aid to assembly).

Gearbox Mainshaft Assembly (Fig. H:27)

Overhaul of the 5 - speed gearbox mainshaft assembly is the same as that on 4 - speed gearboxes detailed previously, operations 18 - 24.

Note that the blocker bars of the 5th gear synchroniser are held by a retaining plate (Fig. H:25). The retaining spring must press against the blocker bars and be located by the retaining plate and synchroniser hub.

Input Shaft (Fig. H:27)

20. Refit the snap ring in the peripheral groove in the outer bearing ring the (larger ball bearing), and use it to detach the bearing with a universal two leg puller.
21. Remove the smaller ball bearing from the shaft using a universal two leg puller.
NOTE: Discard the bearings after removal and fit a new ones on reassembly.
22. Press the new smaller ball bearing onto the shaft, with the lettering facing upwards, as far as the stop.
23. Press on the new larger ball bearing to the shaft, with the groove facing outwards and secure it with the snap ring.

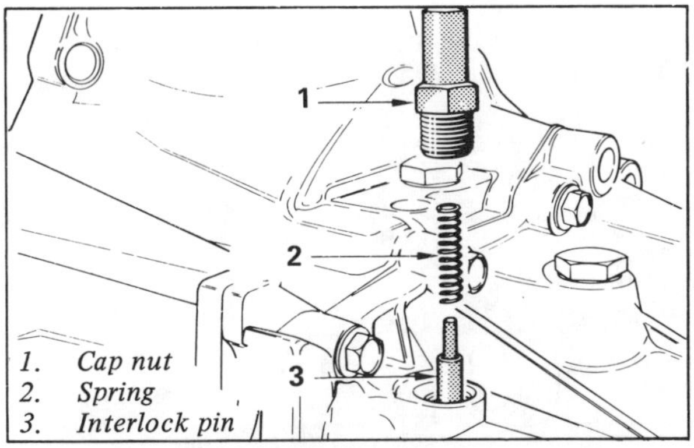

Fig. H:19 Selector interlock cap components

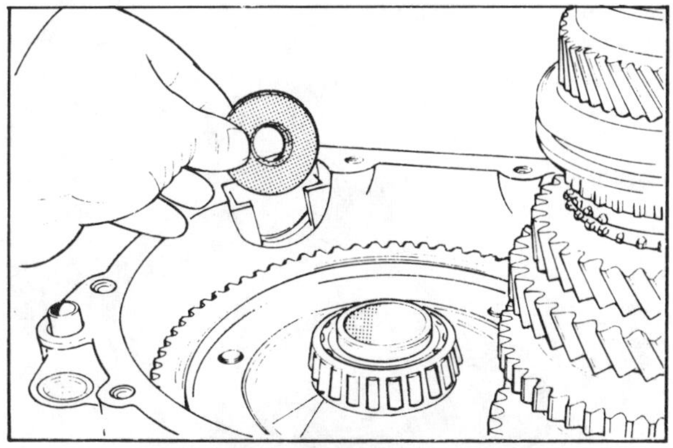
Fig. H:20 Position of magnetic disc

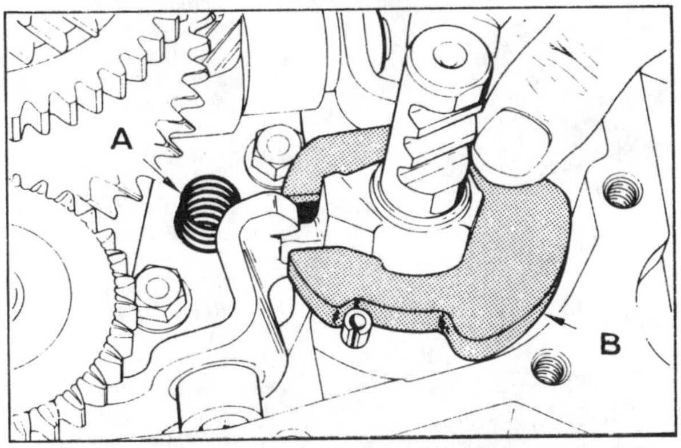

Fig. H:21 Guide shaft spring (A), locking plate (B)

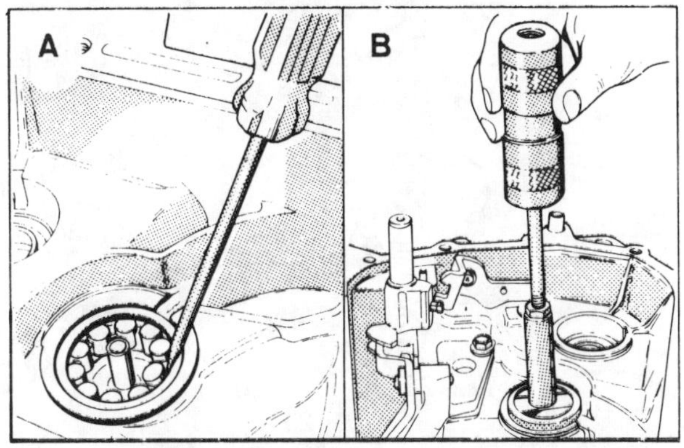

Fig. H:22 Method of removing mainshaft bearing

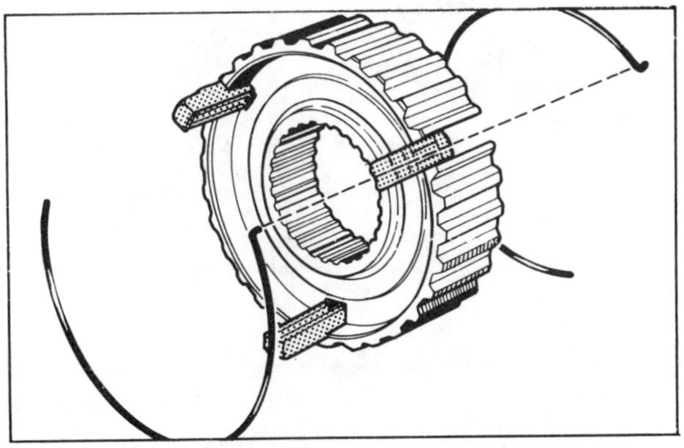
Fig. H:23 Correct position of synchroniser springs

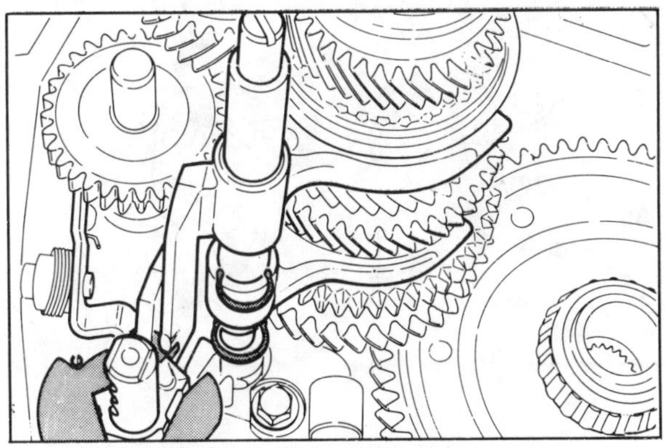
Fig. H:24 Selector shaft guide sleeve - 5 - speed

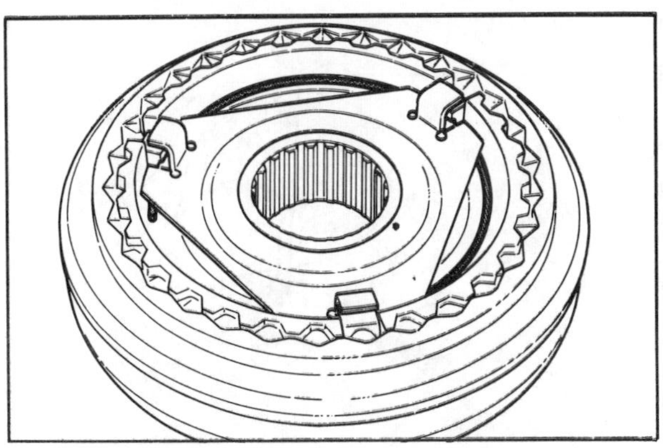
Fig. H:25 Details of the 5th gear synchroniser

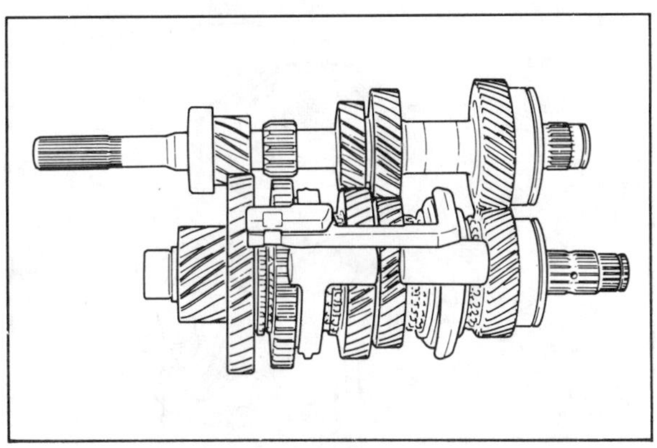
Fig. H:26 Gear and selectors ready for refitting

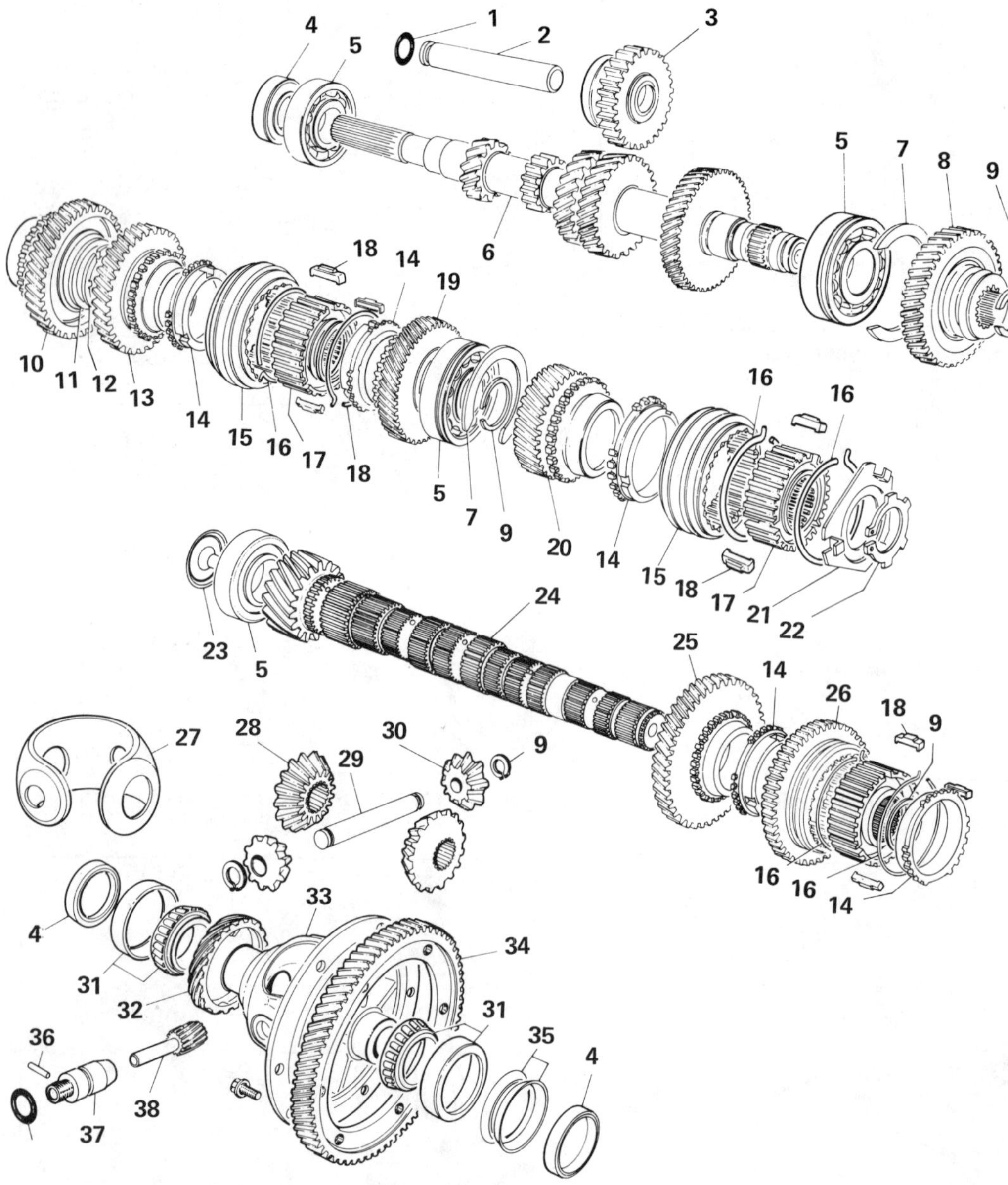

1. O-ring
2. Shaft - reverse idler gear
3. Reverse idler gear
4. Radial oil seal
5. Bearing
6. Input shaft
7. Snap ring
8. 5th gear
9. Circlip
10. 2nd gear (driven)
11. Supporting ring
12. Thrust half washers
13. 3rd gear (driven)
14. Synchroniser ring
15. Selector ring
16. Retaining spring
17. Synchroniser hub
18. Blocker bar
19. 4th gear (driven)
20. 5th gear (driven)
21. Retaining plate
22. Circlip
23. Oil slinger
24. Mainshaft
25. 1st gear (driven)
26. Selector ring with reverse gear
27. Thrust cage
28. Axle shaft pinion
29. Differential shaft
30. Differential pinion
31. Taper roller bearing
32. Speedometer drive worm
33. Differential housing
34. Final drive gear
35. Spring washers (2)
36. Locking pin
37. Speedometer drive pinion bearing
38. Speedometer drive pinion

Fig. H:27 Exploded view of 5 - speed gearbox internal component arrangement

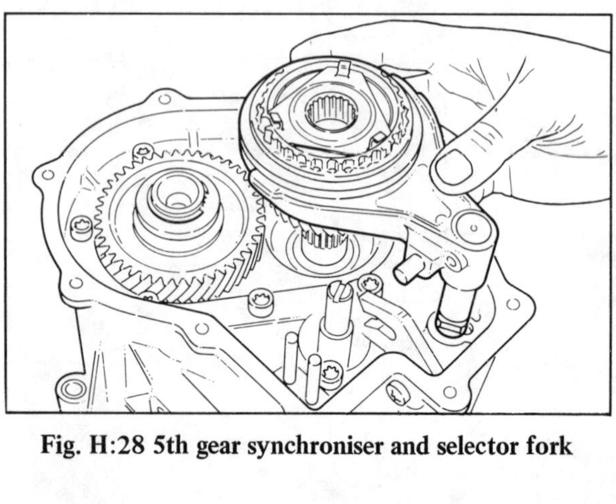

Fig. H:28 5th gear synchroniser and selector fork

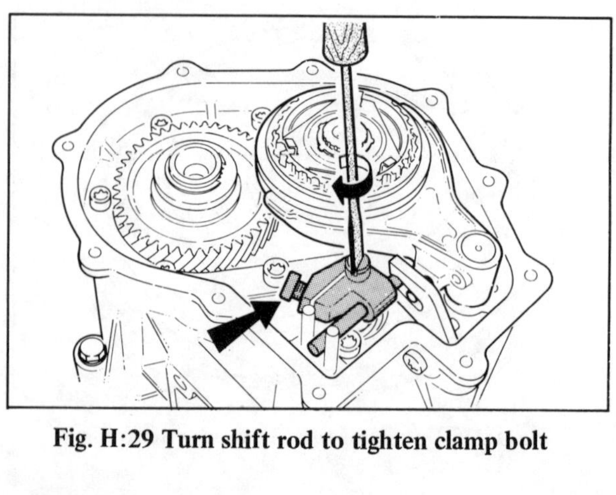

Fig. H:29 Turn shift rod to tighten clamp bolt

Differential Assembly (Fig. H:27)

The differential assembly overhaul procedure on 5 - speed gearboxes is the same as that detailed previously for 4 - speed gearboxes, operations 28 - 32.

Reassembly - 5 - Speed Gearbox

1. Using special tool 16-020, drift outer cup of the taper roller bearing into large housing.
2. Insert the diaphragm spring washers into small housing section with their internal diameters next to one another.
3. Insert the bearing cup and secure it in position with a light blow from a punch. This prevents the cup falling out when the housing is reassembled.
4. Press in the new input shaft oil seal using a suitable tool or pipe.
5. Fit a new oil slinger, then install the mainshaft roller bearing in the housing.
6. Press in the roller bearing using special tool 16-020 with the rear of the cage facing downwards. Using a punch, stake the mainshaft bearing with light blows on the edge of the bearing seat.
NOTE: Take great care when staking the bearing to avoid aluminium splinters becoming detached and jamming between the bearing rollers. Also, do not use excessive blows when staking as this will distort the bearing cup.
7. Refit the differential assembly in the large housing. Lubricate the bearings with gearbox oil before fitting.
8. Slide the reverse idler gear onto the shaft, and at the same time, guiding the selector lever into the selector groove.
9. Install the selector shaft and refit the shift locking plate to guide shaft.
10. Insert the mainshaft, input shaft and selector forks in the housing together as shown in Fig. H:26, bringing the gears into mesh and guiding the selector forks past the shift locking plate. Note that the shift locking plate must be turned clockwise to bear on the dowel.
11. Insert selector shaft guide sleeve and secure 1st/2nd gear selector fork on the guide sleeve using circlips, Fig. H:77.
12. Refit the magnetic disc into the housing recess.
13. Refit the small housing section to the gearbox casing using a new gasket. Make sure that the magnetic disc is correctly seated. Refit and tighten the bolts to the specified torque. See TECHNICAL DATA.
14. Refit the gearbox mounting bracket and tighten the nuts to the correct torque.
15. Secure the mainshaft and input shaft ball bearings with snap rings, lifting the shafts slightly with a screwdriver to insert the ring.
NOTE: Snap rings for both bearings are available in 3 different thicknesses. Select the snap ring that just fits the groove.
16. Refit the 5th gear housing and turn the snap rings to align the ends with stepped parts in the cover gasket. Refit the housing and tighten the bolts to the specified torque see TECHNICAL DATA.
17. Coat the splines of the 5th gear and shaft with grease, then refit 5th gear on the input shaft. To do this, heat the gear to about 80°C and drive it onto the input shaft.
NOTE: The marks on the web of the 5th gear and the input shaft centering bore must be the same colour.
18. Fit the circlip with special tool 16 - 031.
NOTE: A new circlip MUST be used. The circlip must not be fitted with pliers as these will overstretch it and make it unusable.
19. Fit the 5th gear complete with synchroniser and selector fork (Fig. H:28).
20. Fit the cap nut with interlock pin and spring (5th gear selector fork shaft) and tighten to the specified torque.
NOTE: Coat thread of selector shaft locking assembly with sealer prior to installation.
21. Insert the 5th gear selector pin and adjust as follows:
a) Engage 5th gear with selector shaft by turning selector shaft clockwise as far as the stop from the neutral position and then pulling it right out.
b) Slide selector ring (with selector fork) onto the 5th gear.
c) Turn shift rod clockwise as far as the stop with a screwdriver and hold in this position (Fig. H:29).
d) Tighten selector pin clamping bolt to the specified torque in this position.
NOTE: Use "Loctite" when inserting the clamp bolt.
22. Refit the housing cover, using a new gasket, and tighten bolts to the specified torque see TECHNICAL DATA.
23. Refit the interlock pin 1st - 4th and reverse gear, with spring and cap nut and secure to the specified torque (Fig. H:19).
NOTE: Coat cap nut thread with sealing compound (see TECHNICAL DATA).
24. Check the operation of the selector mechanism by going through all the gears with the selector shaft.
25. Using special tool 16 - 018 fit the new driveshaft C. V. joint oil seals (Fig. H:14).
26. Refit the clutch release shaft and lever with release bearing Fig. H:5.

CLUTCH & GEARBOX
Trouble Shooter

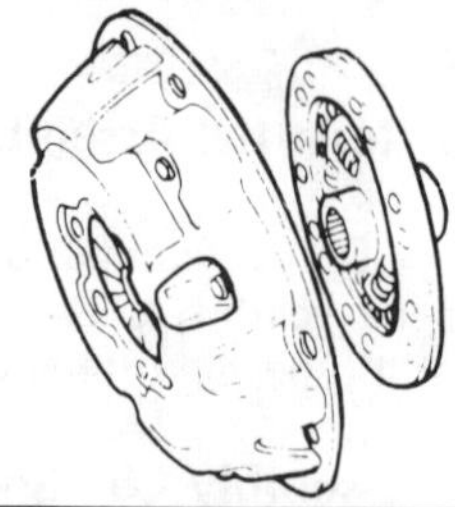

FAULT	CAUSE	CURE
CLUTCH		
Clutch slips	1. Driven plate lining worn 2. Driven plate lining contaminated with oil 3. Clutch incorrectly adjusted	1. Replace clutch assembly 2. Check cause and replace driven plate or clutch assy. 3. Adjust clutch (if adjustable)
Clutch drags - i.e. gear fails to engage smoothly	1. Faulty clutch operation on hydraulic type system 2. Clutch incorrectly adjusted 3. Driven plate sticking on hub shaft splines 4. Broken clutch pressure plate fingers 5. Broken release bearing or release arm	1. Check/overhaul/replace clutch hydraulic components 2. Adjust clutch (if adjustable) 3. Remove clutch, clean and lubricate splines with graphite 4. Check and replace pressure plate assembly 5. Check and replace components
Clutch grabs	1. Oil on driven plate lining 2. Driven plate cushion springs broken 3. Warped, or damaged pressure plate 4. Weak or broken engine mountings	1. Replace driven plate 2. Replace driven plate 3. Check pressure plate and replace if necessary 4. Check/replace mounting rubber
Scream from clutch pedal fully depressed	1. Worn release bearing or broken pressure plate/driven plate springs	1. Remove gearbox and check/replace spigot or release bearing
Constant noise/chatter from clutch - pedal released, disappearing when pedal depressed	1. Worn spigot bearing in flywheel or worn release bearing	1. Remove gearbox and check/replace clutch components
GEARBOX		
Gearchange difficult when cold	1. Incorrect grade of oil in gearbox 2. Clutch driven plate not clearing 3. Normal condition on some models. 4. Worn or incorrect adjustment on gearchange linkage	1. Check/replace gearbox oil with known correct grade 2. See Clutch drag 3. Check possible causes 1 and 2 4. Check for play in linkage joints. Replace and/or adjust.
Gearchange difficult when hot - gears grate	1. Worn synchromesh unit 2. Incorrect clutch adjustment	1. Overhaul or replace gearbox. 2. Check clutch adjustment
Constant moan or grating noise from gearbox	1. Insufficient oil in gearbox 2. Worn layshaft bearings 3. Worn front and rear mainshaft bearings	1. Check/top up gearbox oil 2. Overhaul or replace gearbox 3. Overhaul or replace gearbox
Oil leakage from bottom of clutch housing	1. Front bearing oil seal worn	1. Remove gearbox, clutch release mechanism and fit new oil seal
Oil leakage from rear of gearbox	1. Output shaft oil seal worn	1. Remove propshaft and replace output shaft oil seal

Steering

INTRODUCTION .... [1]

The steering system fitted to all Escort models is of the rack and pinion type. A collapsible tube in the steering column provides some protection in the event of an accident.

Removal and replacement of the steering system components is straightforward, although you will need a ball joint taper-breaking tool to separate the track rod end joint from the suspension steering arm. This tool can either be bought quite cheaply, or hired locally - see Yellow Pages under 'Tools & Equipment Hire'.

Both the steering column and the steering rack can be overhauled: However, several special tools are necessary to adjust the steering rack assembly. These are described with their Ford part numbers in the Steering Rack 'Overhaul' section.

After replacing or adjusting any part of the steering system, it is important that the front wheel alignment is checked afterwards by a Ford dealer or local Tyre Centre using proper measuring equipment.

However, after any repairs or adjustments, the wheel alignment can be roughly set as detailed in the following section prior to driving the car to a specialist.

When dealing with any steering or suspension components it is most important that any bolts or nuts are tightened to the specified torque figures. Any fastening devices such as tab washers or split pins should be replaced with new parts if disturbed.

FRONT WHEEL ALIGNMENT .... [2]

Toe Setting

The toe setting of the front wheels, affects the stability and controllability of the car, also the wear characteristics of the front tyres.

In practice, the toe setting is the amount by which the wheels point inwards or outwards at the front in relation to each other (Fig. I:1). This is measured from the wheel rims hub height and may be given either as a total dimension for both wheels or as an angular measurement for each wheel in degrees. When the wheels point inwards they are said to have toe-in, and toe-out when they point outwards.

Ideally, the front wheel toe setting should be checked after any repair or overhaul operation in which is steering or suspension component has been removed, or its location altered. It should also be checked if the front wheels have been subjected to heavy impact, such as hitting the kerb - even at low speeds (when parking for example).

A reasonably accurate indication of whether or not the front wheels are correctly aligned can be gained by examining the wear pattern of the tyre treads. The presence of 'fins' or 'feathers' on the tread surface are an indication of incorrect alignment. This condition takes the form of a sharp 'fin' on the edge of each pattern rib, and the position of this indicates the direction of misalignment. Fins on the outboard edges indicate excessive toe-in, whereas fins on the inboard edges of the pattern ribs indicate excessive toe-out.

Preliminary Conditions

Before attempting to check or adjust the front wheel toe setting, the following items should first be carried out:

1. Check that there is no excessive wear or looseness in any of the steering or front suspension ball joints, bushes or mountings, or in the front wheel bearings.
2. Check that all tyres are inflated to their correct pressures.
3. Ensure that the car is in its normal unladen condition (i.e. spare wheel in place, luggage compartment empty, etc).
4. Position the car on level, horizontal ground. Bounce the car a few times at each corner to settle the suspension. Once this has been done the ride height must not be disturbed either by jacking up the car or by someone sitting in it.
5. With the front wheels in the straight-ahead position, roll the car forwards a few feet and stop it without using the brakes. It must not be moved backwards after this.

Checking & Adjusting

As stated previously, the front wheel toe setting can be checked and roughly set using a mechanical tracking bar. This should normally be treated only as a temporary measure, but if great care is taken a fairly accurate setting can be obtained.

For those unacquainted with this piece of equipment, a tracking bar is basically a telescopic rod which fits between the front wheel rims with a pair of stands to set it at the correct height. The length of the rod is adjustable and can be locked in position by a wing nut. It should be possible to obtain one from a tool hire shop.

1. Position the tracking bar in front of the front wheel axis and adjust its length so that the bar ends are at the wheel rims at hub height. Note the reading on the bar scale.
2. Similarly, take a second reading of the distance between the wheel rims, this time behind the wheel axis.
3. The difference by which the first reading exceeds the second reading will give the existing toe setting figure. The toe setting is given in the TECHNICAL DATA.
4. If necessary, to minimise the possibility of inaccuracy due to wheel rim run-out, the car can be moved forwards sufficiently to rotate the front wheels through 180^0 and the measuring procedure repeated. An average of the two readings can then be taken.
5. If adjustment is necessary, the track rods on each side should be adjusted qually.
6. Slacken the clips on the outer end of the bellows. To adjust the toe setting, slacken the locknut adjacent to the track rod end. Use one spanner to hold the body of the ball joint and another spanner to loosen the nut. Do not use only one spanner as this will tend to force the ball pin from its socket. Rotate the track rod, using a pair of pipe grips or a spanner on the rod until the correct setting is obtained (Fig. I:3).
7. When adjustment is completed, tighten the locknut at the track rod end. Both track rods should be approximately the same length after adjustment. Tighten the clips on the steering rack bellows.

TRACK ROD END JOINTS . . . . [3]

Replacement

Wear in the track rod end balljoint cannot be removed by adjustment. Renewal of the complete track rod end is necessary when worn.

1. Remove the trim from the relevant front wheel and slacken the wheel nuts. Raise and support the front of the car and make sure the handbrake is fully applied.
2. Using two spanners, one supporting the track rod joint sleeve, and the other on the locknut, slacken the locknut about half a turn.
3. Remove the split pin and undo the nut (Fig. I:3) securing the joint pin to the steering arm.
4. Using a ball joint taper-breaking tool (Fig. I:4) separate the track rod end joint from the steering arm. A suitable tool can either be purchased quite cheaply, or hired for the occasion - see Yellow Pages under 'Hire Services - Tools and equipment.
5. Unscrew the joint from the track rod, noting the exact number of turns required to remove it. This is essential to ensure correct positioning of the new joint. Do not disturb the position of the locknut.
6. Apply a smear of grease to the track rod threads, then screw the new joint onto the rod, the same number of turns needed to remove the old one.
7. Connect the joint stud to the steering arm and tighten the stiffnut to the correct torque, then continue to tighten until the split pin hole aligns with a nut hole.
8. Tighten the joint locknut. Refit the road wheel and wheel trim, then lower the car to the ground. Check the front wheel alignment.

STEERING RACK BELLOWS . . . . [4]

Replacement

If the rubber bellows is damaged or split is should be replaced as soon as possible afterwards. As a temporary measure, a plastic bag can be placed over the bellows to prevent dirt and water from getting inside and damaging the steering rack unit (Fig. I:5).

1. Remove the track rod end joint as described in the previous section. Count the number of turns required to remove the joint from the tract rod. Unscrew the locknut from the rod.
2. Unscrew the clip securing the outer end of the bellows to the track rod, then repeat for the inner end clip.
3. Pull the bellows over the track rod and remove it. If the bellows has been damaged and allowed water to enter, expel the lubricant. If the pinion end bellows has been removed, turn the steering wheel from side to side to allow the fluid to escape. If the inner end bellows has been removed, it will be necessary to disconnect the pinion end bellows. Place a tray beneath the bellows to collect the expelled grease.
4. Clean the track rod to remove any dirt and check that there is no dirt around the ball joint at the inner end of the track rod. If there is, it should be carefully wiped clean.
5. Smear the inside surface of the new bellows with clean lubricant, see TECHNICAL DATA for specification. Slide the bellows into position over the track rod.
6. Refit the screw clip to the inner side bellows inner end only. Refit the track rod end joint as described previously.

Twisted wire clips should not be reused, fit worm drive clips instead.

7. Refill the rack with lubricant through the pinion end bellows. See the TECHNICAL DATA for fluid type and quantity. Do not overfill. Replace the bellows clips. Slowly turn the steering wheel back and forth to distributor the lubricant.
8. Check the front wheel alignment as described at the start of the chapter.

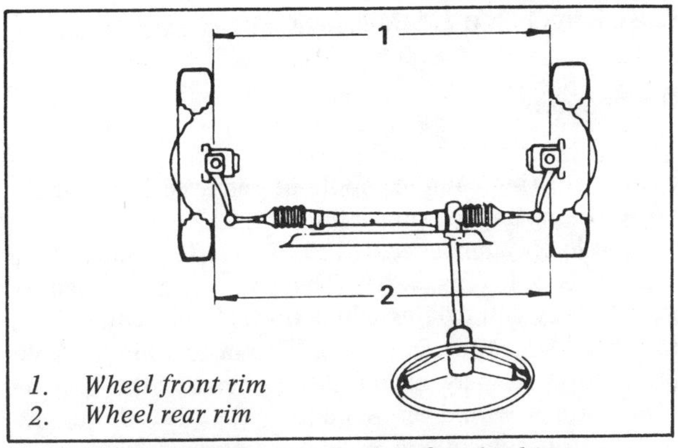

Fig. I:1 Measurement points for checking front wheel alignment

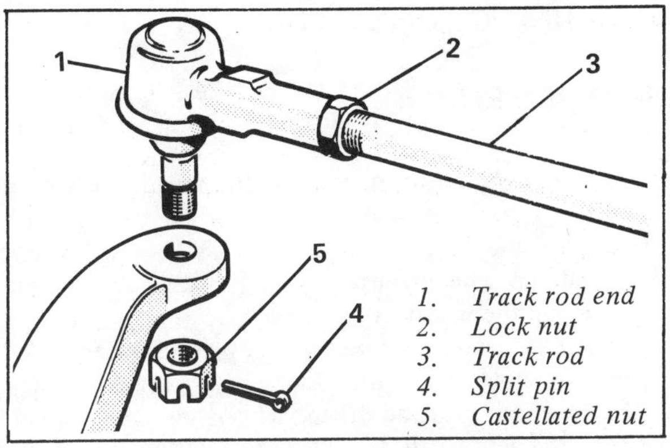

Fig. I:2 Track rod assembly

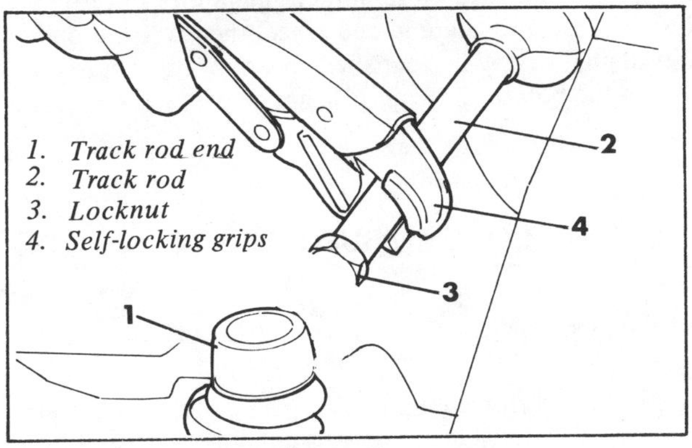

Fig. I:3 Track rod adjustment

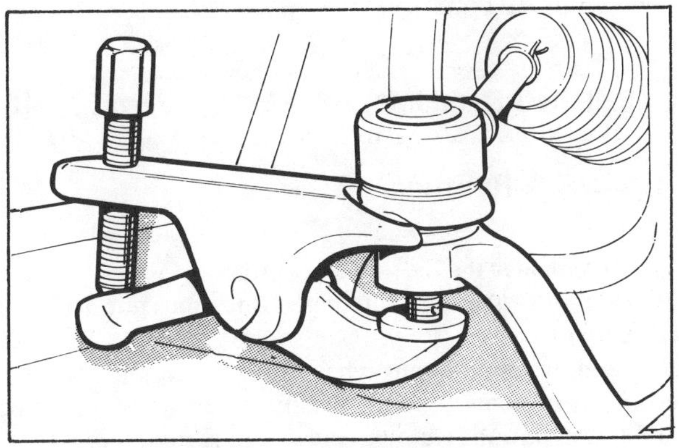

Fig. I:4 Using a ball joint taper breaking tool

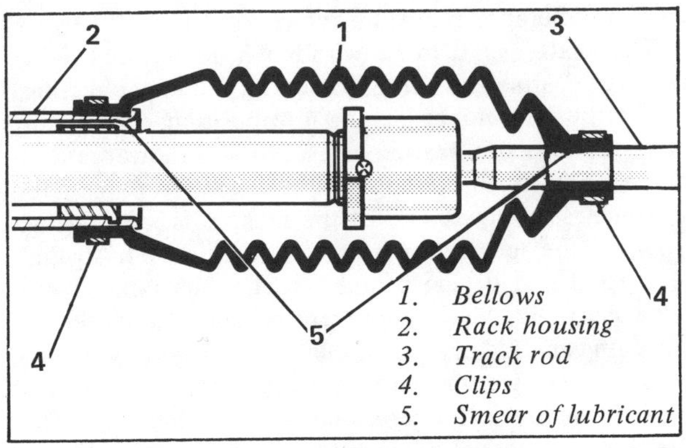

Fig. I:5 Section through rack bellows

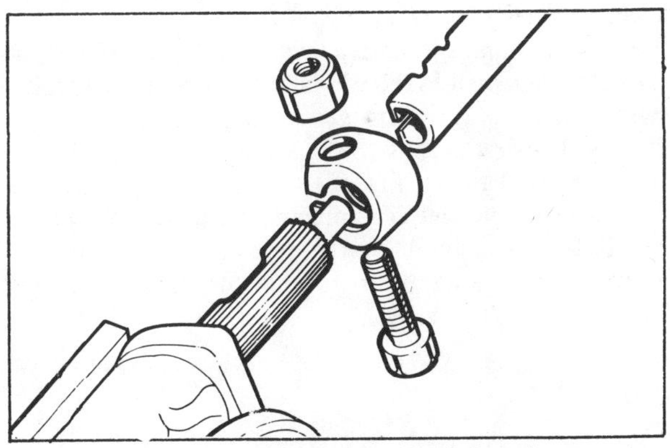

Fig. I:6 Steering shaft to rack pinion clamp

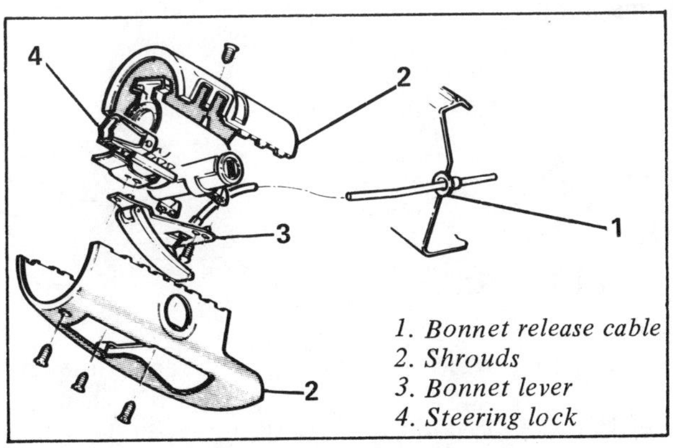

Fig. I:7 Steering column shroud

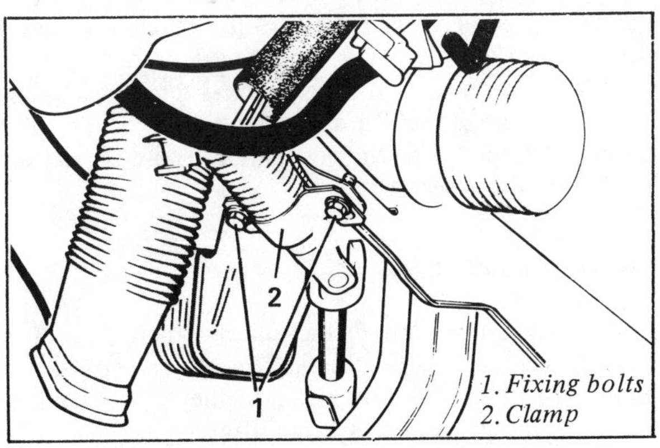

Fig. I:8 Steering column lower mounting

STEERING WHEEL . . . . [5]

Removal & Refitting

Before attempting to remove the steering wheel, insert the ignition key and turn the switch to position I. Disconnect the battery earth strap. With a screwdriver lever out the wheel centre insert. Remove the retaining nut and pull the wheel from the column.

Refit the steering wheel in the reverse order. Before fitting the nut, check that the wheel is in the straight ahead position with the driving wheels pointing forward. Tighten the nut to the specified torque, see the TECHNICAL DATA.

STEERING COLUMN ASSEMBLY . . . . [6]

Removal & Refitting

1. Disconnect the battery earth strap.
2. Remove steering coupling clamp and nut and bolt (Fig. I:6).
3. Remove the steering wheel from the column as detailed in the previous section. Remove steering column upper and lower shroud retaining screws and shrouds (Fig. I:7).
4. Remove lower dash insulation panel, where fitted. Remove bonnet release lever mounting screw, and detach lever.
5. Remove multi-function switches from column, see the GENERAL ELECTRICS chapter. Disconnect ignition switch multi-plug. Remove steering column retaining clamps. Upper clamp is stud and nut fixing, lower clamp is nut and bolt fixing (Fig.I:8).
6. Remove the steering column and shaft from the car.
7. Refit the column in the reverse order of removal. Tighten the fixing bolts to the specified torque, see TECHNICAL DATA.

Adjustment

If free play can be felt in the upper bearing, adjust as follows. Turn the steering wheel until the position of maximum free play can be felt. Holding the wheel in this position, slacken the lower mounting bolts and pull the column upwards until the free play is removed. Tighten the bolts to the specified torque and recheck for free play. If play is still felt, find the position of maximum play and repeat the adjustment.

Steering Column Joints

If play can be felt in the universal couplings in the steering shaft assembly, the shaft must be replaced complete (Fig. I:9). No kits are available to overhaul the joints, which are sealed. Do not attempt to separate the joints or repair.

STEERING COLUMN LOCK . . . . [7]

Replacement

The steering column should be removed from the car before attempting to remove the lock.

The lock clamp is secured to the steering column by a single shear-bolt (Fig. I:10). This should be removed by dot punching and drilling a hole through the centre. Then use a left handed thread screw extractor to undo the bolt. Alternatively it may be possible to remove the bolt by tapping round with a centre punch to unscrew it. Remove the lock from the column.

Position the new steering lock around the tube (Fig. I:11). Fit a new shear-head bolt. Tighten until finger tight. Refit the steering column and wheel then test the operation of the lock. When satisfied that is working correctly, tighten the bolt until the head snaps off.

STEERING COLUMN BUSHES . . . . [8]

Replacement

If the steering column bushes are worn or if rattling is heard coming from the steering column, replace or refit the bearings as described below.

Remove the steering column as described above. Remove the tolerance ring from the top of the shaft (Fig. I: 9). The shaft can then be withdrawn downwards with the lower bearing and spring. Lever out the upper bearing.

Fit the new upper bearing into position, taking care not to damage the bearing surface. At the lower end of the column, fit the spring so that the larger diameter end touches the lower end of the collapsible section (Fig. I:10). Fit the new lower bearing onto the shaft so that the tapered end of the bearing touches the tapered end of the shaft. Fit the shaft into the column. Refit the tolerance ring and indicator cam ring.

Refit the steering wheel and tighten the bolt to ensure that the lower bearing is pulled into the column. The slots should engage in the locating pegs in the tube (Fig. I: 13). Refit the column assembly to the car and check the column free play.

STEERING RACK . . . . [9]

Removal & Refitting

1. Slacken the wheel nuts and raise the front of the car as described in the ROUTINE MAINTENANCE chapter, under Jacking Procedures. Remove the front wheels.
2. Inside the engine bay undo the nut and remove the

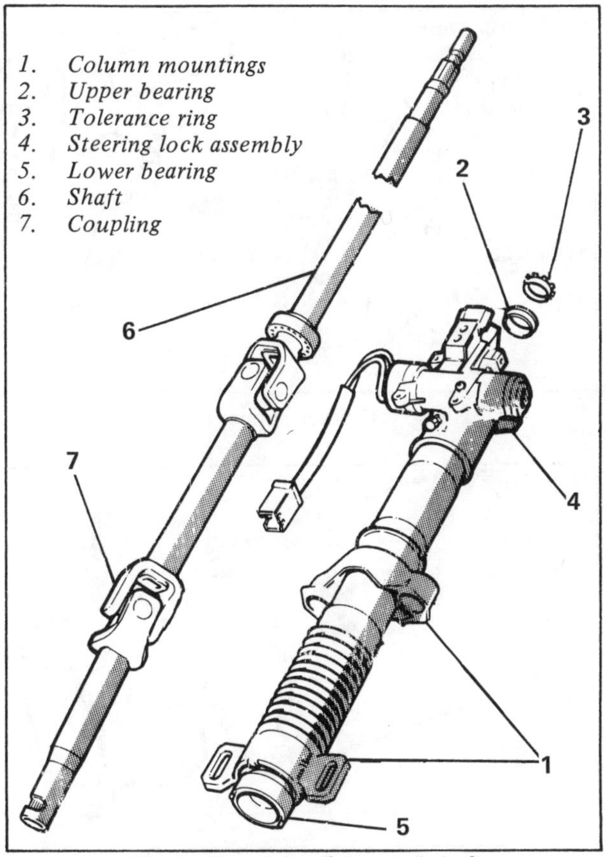

Fig. I:9 Steering column and shaft

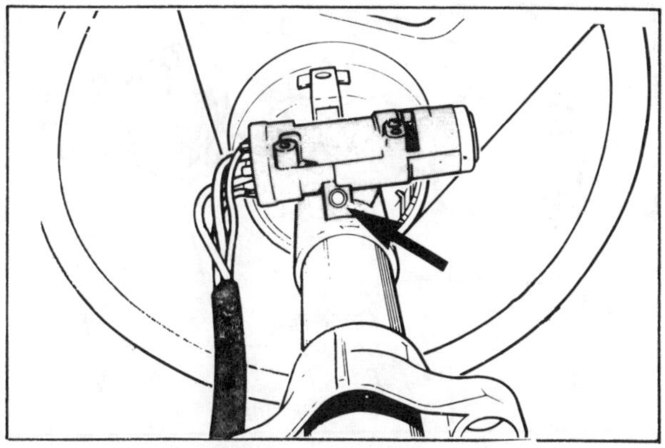

Fig. I.10 Column lock shear bolt

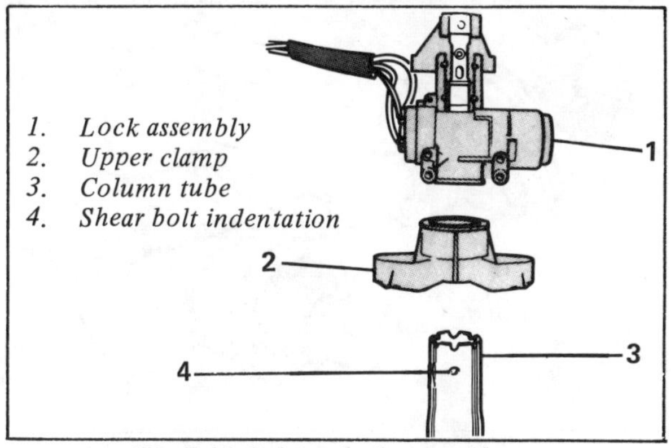

Fig. I:11 Column lock assembly

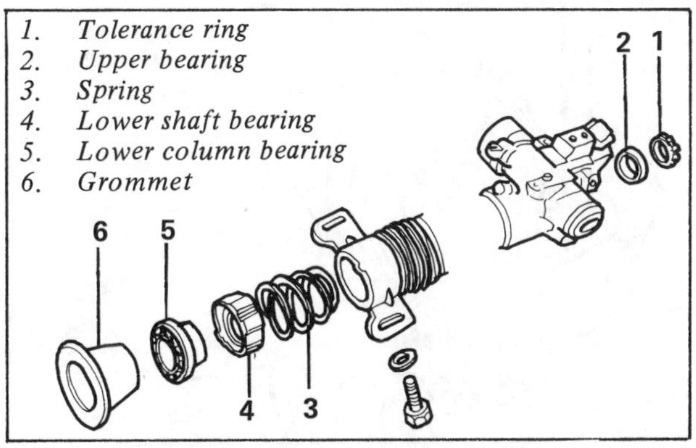

Fig. I:12 Column bearing assemblies

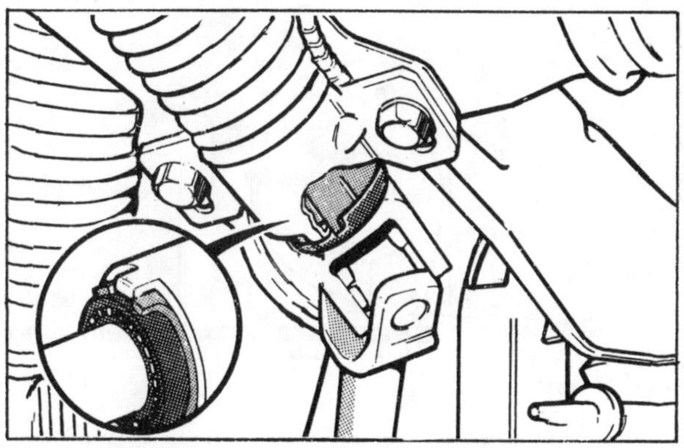

Fig. I:13 Lower column bearing location

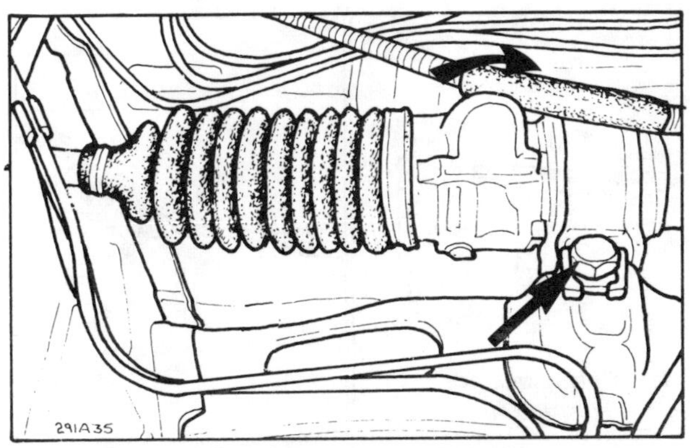

Fig. I:14 Steering rack mounting clamps

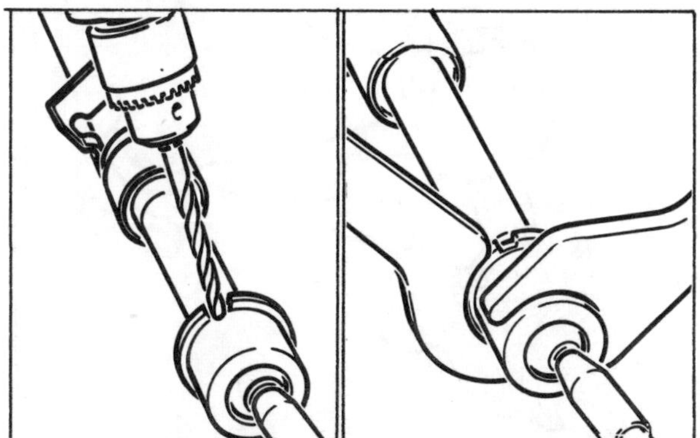

Fig. I:15 Remove track rod inner ball joints

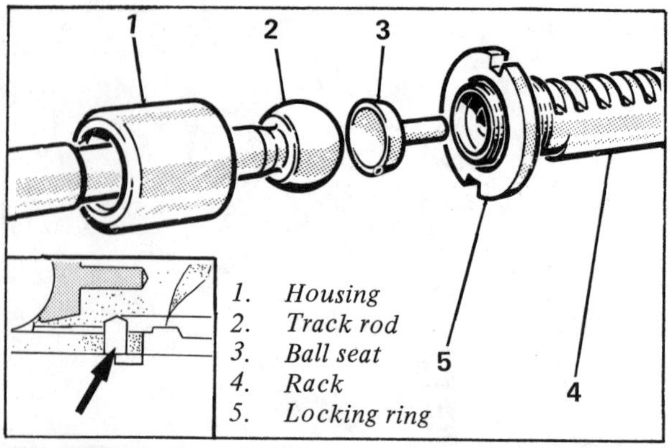

Fig. I:16 Track rod ball joint assembly, showing locking pin hole

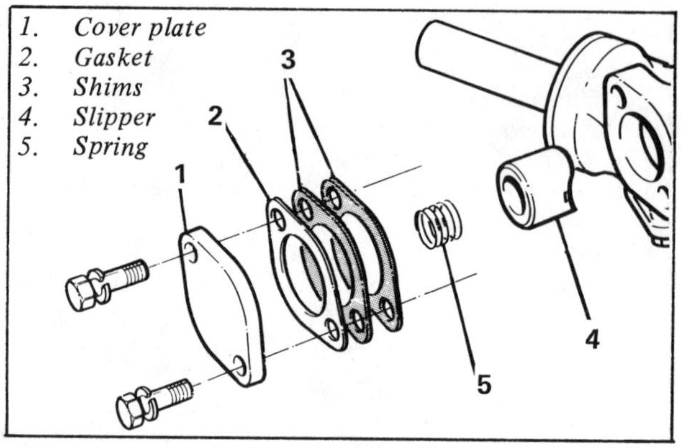

Fig. I:17 Rack slipper assembly

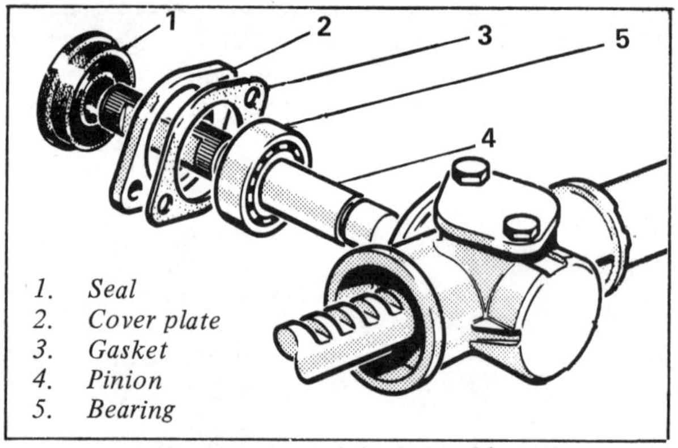

Fig. I:18 Gear pinion assembly

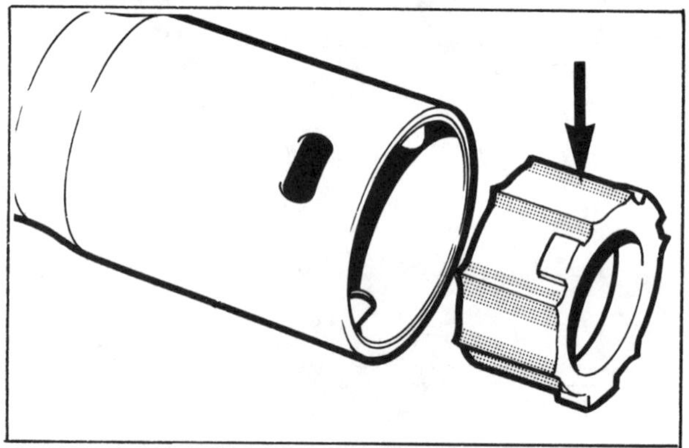
Fig. I:19 Rack support bush

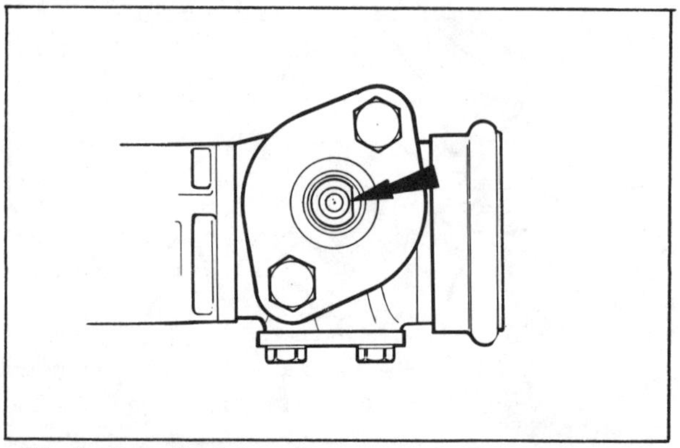
Fig. I:20 Rack pinion alignment, wheels straight ahead

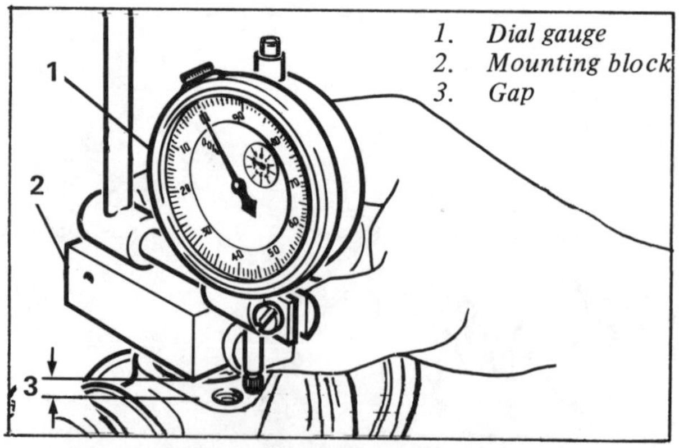

Fig. I:21 Rack slipper clearance measurement

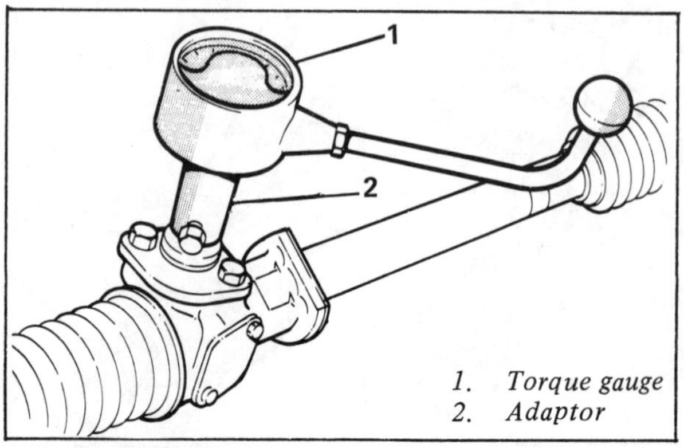

Fig. I:22 Pinion turning torque measurement

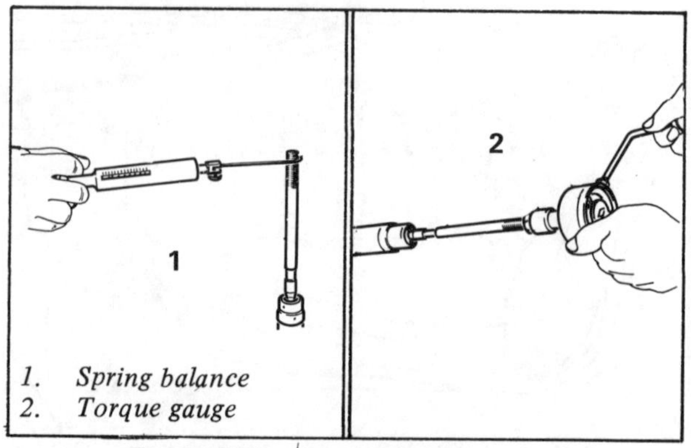

Fig. I:23 Track rod articulation check

pinch bolt holding the coupling clamp to the rack pinion shaft (Fig. I:8).

3. Under each wheel arch, remove the split pin and undo the castellated nut securing each track rod end joint to the swivel hub (Fig. I:2). Use a ball joint taper breaking tool to separate each joint from the hubs (Fig. I:4).

4. Under the car undo the lock washers and bolts securing the rack clamps to the engine bay bulkhead (Fig. I: 14). Remove the clamps and rubber pads.

5. Withdraw the rack unit from the car through the driver's side.

Refit the steering rack in the reverse order to removal. Before refitting the bolts, ensure that the rack and wheels are in the straight ahead position. Refit the rack clamp bolts with new lock washers (Inset, Fig. I:14). Tighten the bolts to the specified torque, see the TECHNICAL DATA. Fit new split pins to the track rod end ball joint nuts. Check the front wheel alignment as described at the start of the chapter.

Rack Overhaul

It is possible to obtain all component parts of the rack from a Ford dealer if overhaul is necessary. Kits are available to overhaul the track rods and steering gear bearing. Before overhauling the assembly, however, check the cost of the necessary components against the cost of a factory overhauled exchange unit.

1. Remove both bellows and drain fluid lubricant from gear. Mount steering gear in a vice with protected jaws.

2. Using 5/32 in (4,9 mm) diameter drill, centre punch and drill out pins securing track rod ball joint housings (Fig. I:15). Do not drill deeper than 0.4 in (9,5 mm) maximum. Using suitable spanners, separate ball joint housings from locking rings (Fig. I:16). Remove housings, track rods and ball seats.

3. Remove rack slipper cover plate, gasket shim pack spring and slipper (Fig. I:17). Remove the pinion bearing cover plate, gasket and pinion seal (Fig. I:18). Withdraw pinion and bearing assembly. Withdraw rack from housing.

4. Clean and inspect all parts for wear or damage. Includes checking condition of rack support bush which is replaceable and pinion lower bearing. The pinion bearing is only serviced complete with pinion.

5. If necessary, fit new rack support bush (Fig. I:19) and new lower pinion bearing. Position rack in housing. Assemble pinion and bearing assembly to housing. This must be carried out with rack positioned centrally in housing. After assembly, the flat on pinion must be at 90° to centre line of rack and facing outside of vehicle (Fig. I: 20).

6. Fit pinion bearing gasket, cover plate and seal. Apply sealer to bolt threads. Ensure cover plate is fitted centrally in relation to pinion. Tighten bolts to the specified torque, see TECHNICAL DATA.

7. Assemble rack slipper, spring shims, gasket and cover plate selecting shim pack to give correct clearance. Adjust shim pack and assemble as follows.

a) Fit dial gauge to mounting block. Position on flat surface and with stylus of dial gauge on surface, set dial to zero. Hold mounting block on slipper with stylus of dial gauge on main body of pinion housing (Fig. I:21). Record maximum dial gauge deflection from zero as rack is traversed from lock to lock.

b) Assemble shim pack. Including gasket, thickness must be 0.0006 to 0.006 in (0,015 to 0,15 mm) greater than dimension recorded. Pack must be set accurately, since incorrect thickness will result in heavy, stiff steering or 'knocking' from steering.

8. Fit shim pack gasket and cover plate to housing. Measure steering pinion turning torque. Use torque gauge 15-041 and adaptor, tool no. 13-008 (Fig. I:22). Rotate gauge at approximately 4 rev/min. The correct torque is given in the TECHNICAL DATA. Incorrect torque indicates incorrect shim pack (check as above) or faulty components within steering gear e.g. tight bearing, lack of lubricant, etc.

9. Lubricate and assemble the track rod assemblies in the reverse order of removal. Check track rod articulation pull. With wire hook fitted to a spring balance and attached to track rod 0.25 in (6,0 mm) from end, measure effort to move track rod when in line with rack (Fig. I:23).

Adjust ball housing to give pull of 5.1 lb (2,3 kg). Lock housing with locknut using spanner and recheck.

Alternatively, lock two nuts together on end of track rod. Using torque gauge 15-041 and socket turn the track rod (Fig. I:23). Tighten ball housing until turning torque is as specified in TECHNICAL DATA. Lock housing locknut and re-check.

10. Centre punch and drill holes on housing to locknut interface (Fig. I:16). Holes to be 5/32 in (4,0 mm) diameter, 0.4 in (9,5 mm) deep and positioned on break line between housing and locknut. New holes must be drilled even if existing hole aligns. Fit new retaining pins. Hold in position by peening over surrounding metal using pin punch.

11. Fit one rack bellows and secure with clamps, see Steering Rack Bellows.

12. Check pinion turning torque, as described earlier.

13. Remove rack assembly from vice and refit to car.

STEERING

Trouble Shooter

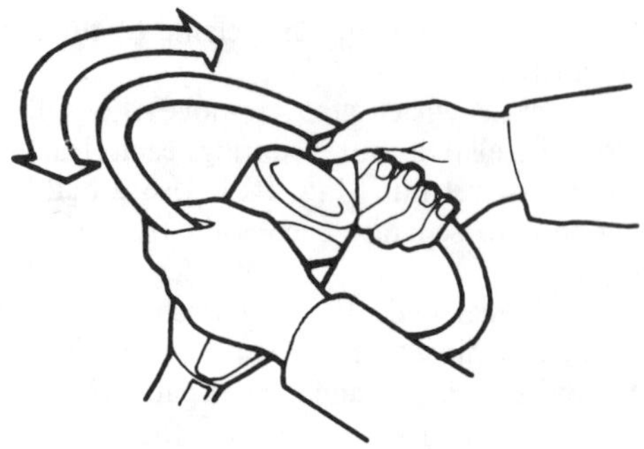

FAULT	CAUSE	CURE
Steering feels stiff	1. Low tyre pressures 2. Incorrect wheel alignment 3. Stiff track rod ends 4. Steering rack needs adjustment	1. Correct tyre pressures. 2. Correct wheel alignment. 3. Check and replace if necessary. 4. Adjust if necessary.
Steering wheel shake	1. Wheels and tyres need balancing 2. Tyre pressures incorrect 3. Incorrect wheel alignment 4. Wheel hub nut loose 5. Wheel bearings damaged 6. Front suspension distorted 7. Steering rack needs adjustment 8. Shock absorbers faulty	1. Balance as necessary. 2. Correct. 3. Correct alignment. 4. Adjust wheel bearings. 5. Replace wheel bearings. 6. Check, repair or replace. 7. Adjust as necessary. 8. Check and rectify.
Steering pulls to one side	1. Uneven tyre pressure 2. Wheel alignment incorrect 3. Wheel bearings worn or damaged 4. Brakes improperly adjusted 5. Shock absorbers faulty 6. Suspension distorted 7. Steering rack worn	1. Correct. 2. Correct. 3. Replace and adjust. 4. Adjust brakes. 5. Check and rectify. 6. Check and rectify. 7. Adjust or replace.
Wheel tramp	1. Over-inflated tyres 2. Unbalanced tyre and wheel 3. Defective shock absorber 4. Defective tyre	1. Correct pressure. 2. Check and balance if necessary. 3. Check and rectify. 4. Repair or replace.
Abnormal tyre wear	1. Incorrect tyre pressure 2. Incorrect wheel alignment 3. Excessive wheel bearing play 4. Improper driving	1. Check pressures. 2. Check wheel alignment. 3. Adjust wheel bearings. 4. Avoid sharp turning at high speeds, rapid starting and braking, etc.
Tyre noises	1. Improper tyre inflation 2. Incorrect wheel alignment	1. Correct tyre pressures. 2. Correct wheel alignment.

Front Suspension

INTRODUCTION .... [1]

The front suspension on all Escort models is of the MacPherson strut type with a bolt-on swivel hub carrier at the lower end. The swivel hub carrier contains two sets of non-adjustable taper roller bearings supporting the driveshaft.

The swivel hub carrier is connected by a ball joint to the lower arm which controls lateral movement of the front wheel. On 1.1 litre models, the lower arm is of the pressed steel type, all other models have a solid forged type of lower arm.

Fore and aft movement of the strut and lower arm is controlled by a tie bar on 1.1 litre models, and on 1.3 and 1.6 models by a stabiliser bar, connecting the outer end of the lower arms to mounting brackets under the front of the car.

The driveshafts are of unequal length with sliding inner CV joints and fixed length outer CV joints. The longer (tubular) driveshaft has a greater diameter.

The front suspension components are quite easy to remove to replace any parts or rubber bushes. Note that on solid forged type lower arms, the swivel hub ball joint is an integral part of the arm and cannot be replaced separately. If the suspension strut is worn or leaking, it can be removed by unbolting it from the swivel hub carrier and at the top mounting, without disturbing the remainder of the suspension.

In addition to a metric spanner and socket set, a number of special tools will be needed when working on the front suspension, as follows: Coil spring clamps and a 6 mm Allen key for removing the coil spring from the strut assembly, a universal two legged puller and a torque wrench with 200 lb ft capacity for removing the driveshaft or swivel hub assembly.

The lower arm ball joint is held by a Torx type clinch bolt which requires a special Torx drive adaptor to hold the bolt stationary while the nut is slackened and tightened.

For removing and fitting hub bearings and oil seals, a selection of sockets and pieces of tube should be sufficient.

Whenever any work has been carried out on the front suspension, the front wheel alignment (toe-setting) should be checked afterwards by a local Ford dealer or tyre specialist, and adjusted if necessary to avoid possible excessive tyre wear.

SWIVEL HUB CARRIER .... [2]

Removal

1. Remove the centre cover from the relevant wheel, then unlock the hub nut staking using a thin punch.
2. With an assistant applying the brakes, slacken the hub nut using a torque wrench and socket.
3. Slacken the wheel bolts before raising and supporting the car on the relevant side as detailed in the ROUTINE MAINTENANCE chapter. Remove the road wheel.
4. Remove the brake caliper mounting bolts (Fig. J:2) and detach the caliper from the brake disc. Tie the caliper to one side, avoiding placing any strain on the brake flexible hose.
5. Unscrew the hub nut, then remove the brake disc from the hub flange after undoing the small disc locating screw.
6. Using a two-legged puller, remove the hub from the carrier and driveshaft (Fig. J:3).
7. Disconnect the lower arm from the hub carrier by removing the clinch bolt to free the ball joint. Push the lower arm down to separate the joint pin from the hub carrier.
8. Remove the bolt securing the lower part of the suspension strut to the hub carrier. Lever the suspension strut housing slot apart to detach the carrier from the strut (Fig. J:11).
9. Undo the nut and separate the steering track rod from the suspension steering arm using a ball joint taper breaking tool.
10. Support the driveshaft under the outer CV joint and carefully withdraw the hub carrier from the driveshaft and suspension.

Installation

Install the hub carrier in the reverse order of removal, noting the following points:

a) Thoroughly clean the outer driveshaft and lightly grease the driveshaft splines.
b) Fit a new hub nut and tighten it to the correct torque

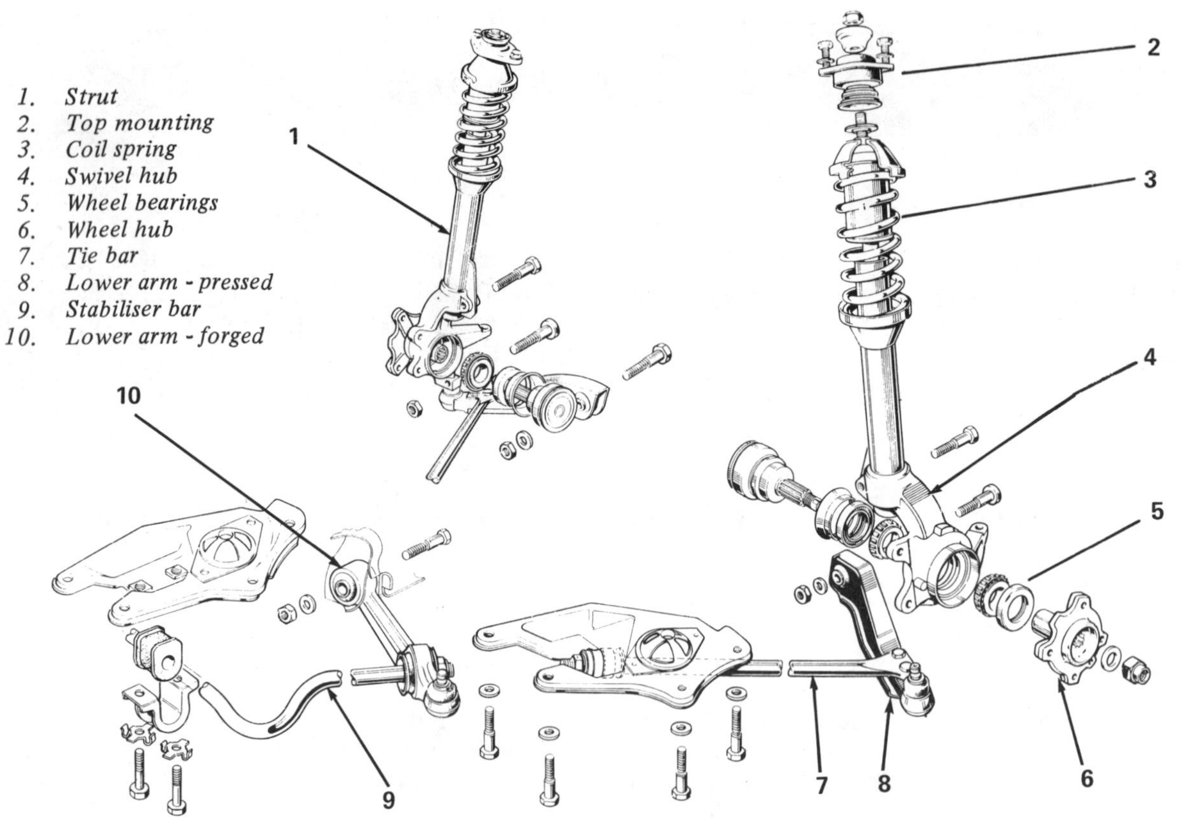

Fig. J:1 Front suspension assembly

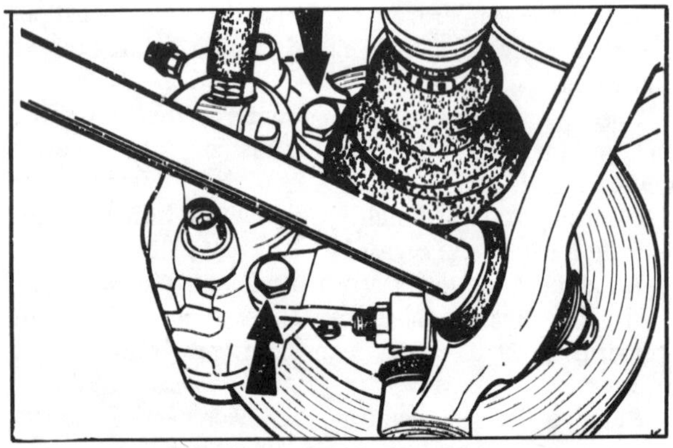

Fig. J:2 Brake caliper mounting

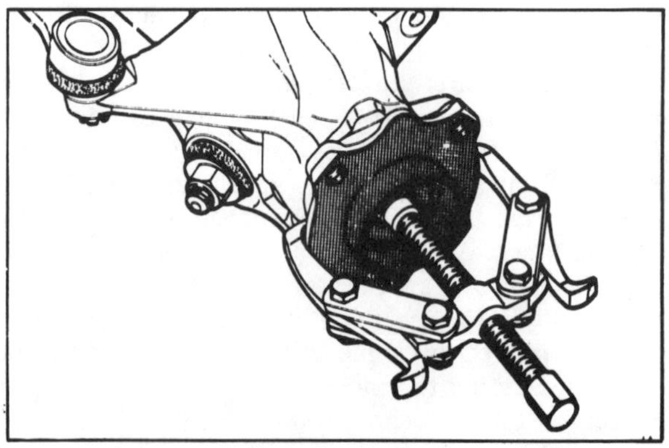

Fig. J:3 Wheel hub removal using puller

Fig. J:4 Track rod end ball joint separation

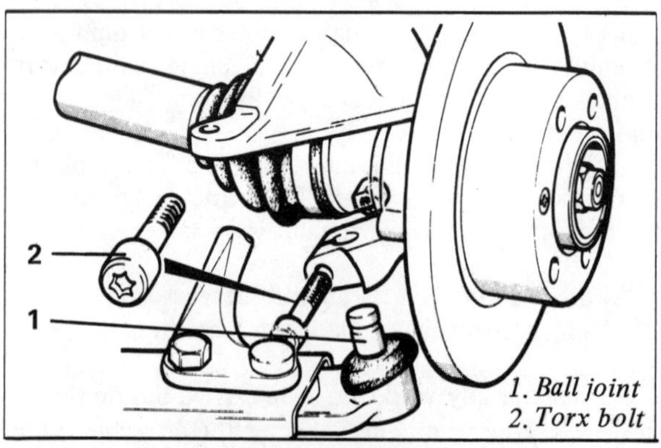

Fig. J:5 Lower arm ball joint pinch bolt

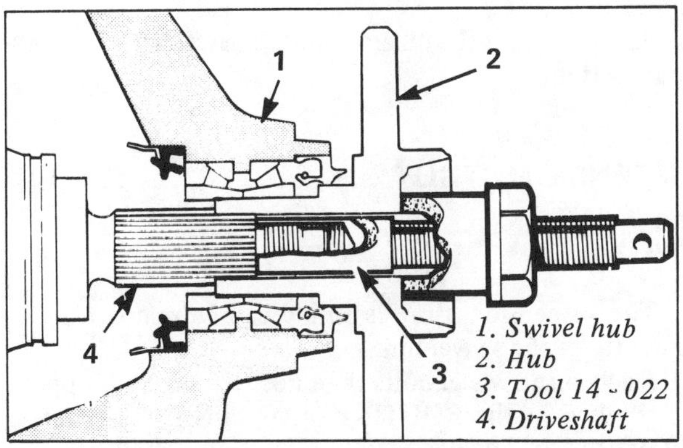

Fig. J:6 Refitting hub using tool

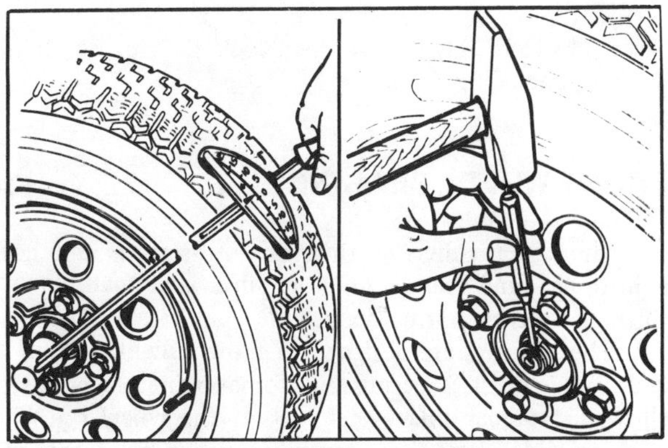

Fig. J:7 Fitting new hub nut, then staking

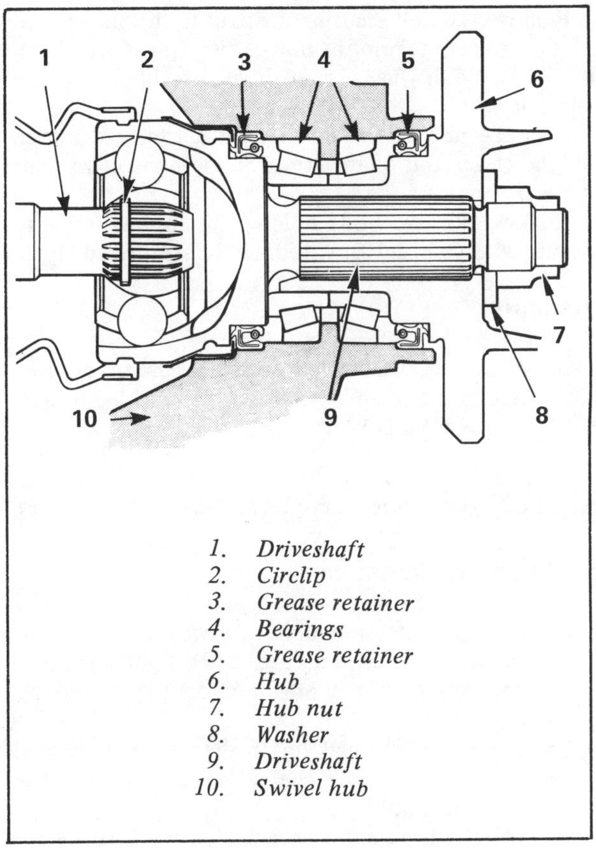

Fig. J:8 Section through front wheel hub

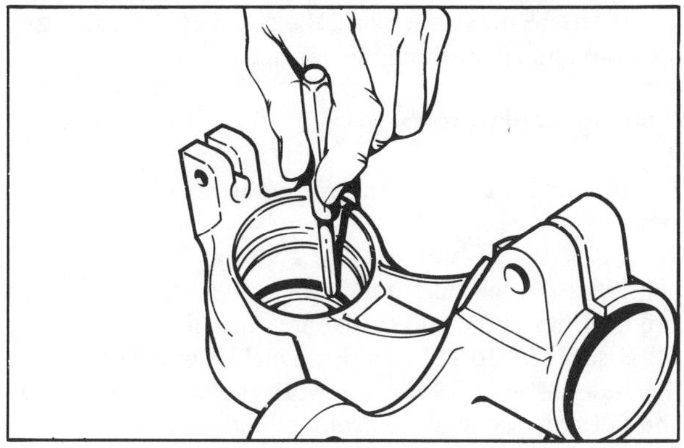

Fig. J:9 Drift out the bearing caps

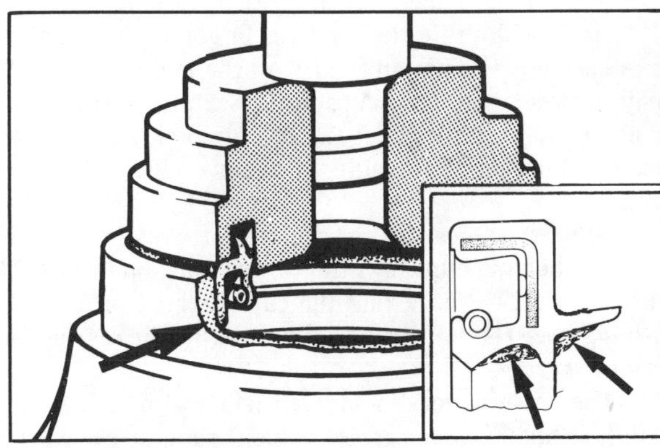

Fig. J:10 Grease retainer fitting, showing grease cavity

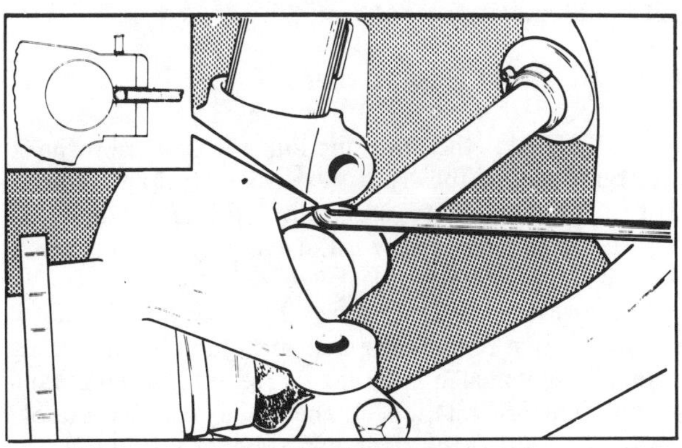

Fig. J:11 Swivel hub clamp spreading

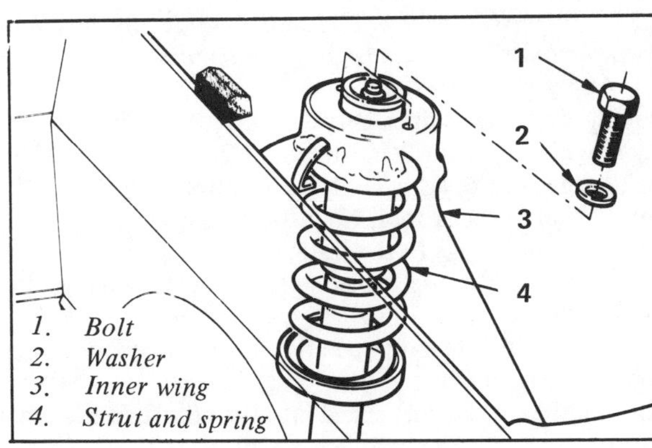

Fig. J:12 Strut top mounting

- see TECHNICAL DATA, then lock the nut by staking with a punch (Fig. J:7).

FRONT WHEEL BEARINGS .... [3]

The taper roller bearings are self-setting and non-adjustable, and should not require any maintenance throughout their normal service life.

The bearings are designed to allow a slight amount of end-float, and it is permissible for even new bearings to allow just perceptible wheel rock at the wheel rim. The only possible check for bearing condition is to spin the wheel and feel for any roughness or hear any sound indicating bearing wear.

If the bearings are worn, the complete bearing assembly (cup and cone) should be replaced.

Bearing Replacement

1. Remove the swivel hub carrier as detailed in the previous section.
2. Fit a pair of 'soft' jaws to a vice and mount the suspension assembly horizontally in the vice so that the seals and bearings in the hub carrier are accessible.
3. Use pliers to pull the dust shield from the groove in the hub carrier. Using a screwdriver, lever out the inner and outer grease retainers from the hub carrier.
4. Lift out the inner and outer bearings. Place them in removal order for examination.
5. Using a flat ended punch, drift out the bearing cups (Fig. J:9). Make sure the punch is in good condition and take care not to raise any burrs on the cup seats as these may prevent the new cups from seating properly and hence will adversely affect the bearing preload. Clean the bearing housing thoroughly to remove all traces of old grease.
6. Support the carrier on a block of wood, then drive the new bearing cups into the carrier using one of the old cups as a drift. Check that the cup is seated properly by turning over the hub carrier and checking the seating face from the rear.
7. The new bearings should be packed with high melting point lithium based grease which should be worked thoroughly between the rollers. Insert one bearing in one side of the spindle carrier.
NOTE: Do NOT pack the chamber inside the hub between the inner and outer bearings with grease.
8. Apply grease to the sealing lips of the grease retainer, making sure the cavities between the sealing lips are also filled with grease. When fitting the new grease retainer make sure the open side is facing inwards towards the hub. A Ford special tool, No. 14-024 is available (Fig. J: 10) to avoid damage to the sealing lip.
Alternatively, use a piece of tubing that fits around the seal lip.
9. Repeat the previous operation for the remaining bearing and grease retainer.
10. Gently tap the dust shield into the groove in the inner face of the hub carrier with a block of wood.
11. Refit the swivel hub assembly as described in the previous section.

SUSPENSION STRUT .... [4]

Removal

The suspension strut assembly can be removed independantly of the swivel hub.

1. Slacken the wheel bolts then raise the car and support as described in the ROUTINE MAINTENANCE chapter. Remove the road wheel.
2. Place a jack or suitable block beneath the swivel hub and lower arm to support it.
3. Remove the bolt securing the strut to the hub carrier. To detach the strut from the hub carrier, spread the clamp (Fig. J:11) by driving a screwdriver or chisel into the clamp slot.
4. Push the hub carrier down until it can be detached from the lower end of the strut. Support the drive shaft and lower arm.
5. Remove the two bolts holding the top mounting to the inner wing (Fig. J:12). Withdraw the strut assembly.

Installation

Install the suspension strut in the reverse order of removal. Tighten the retaining nuts to the specified torque, see the TECHNICAL DATA.

COIL SPRING & UPPER MOUNTING...... [5]

Coil Spring Replacement

Mount the suspension unit in a vice and fit spring compressors to the coil spring (Fig. J:13). Tighten gradually and evenly to remove any pressure from the top mounting assembly.

Use a 6 mm hexagonal key in the socket at the end of the piston rod to hold the whilst undoing the top mounting retaining nut.

Remove the top mounting assembly followed by the coil spring (Fig. J:14).

Assemble the spring and top mounting to the new strut in the reverse order of removal.

Shock Absorber Replacement

As detailed in the Introduction , the shock absorber is an integral part of the front suspension strut and replacement of the complete unit minus the spring and top mounting will be necessary if the damping action is defective or if there are signs of leakage.

Remove the strut assembly from the car as detailed previously, then remove the coil spring and top mounting assembly as detailed in the following section. When obtaining replacement units, check the colour code on the old component. Note that strut units should be replaced as

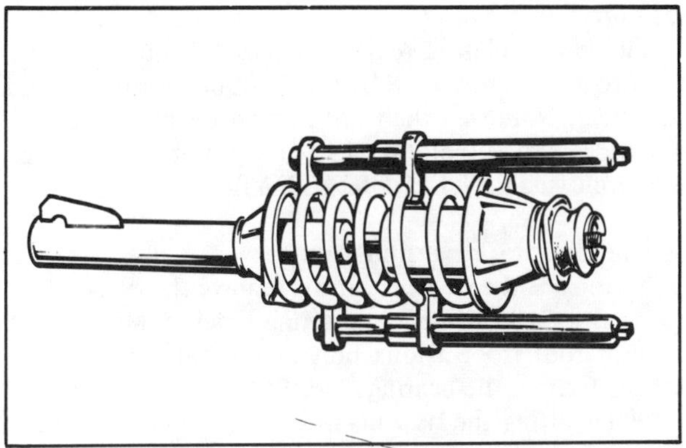

Fig. J:13 Compress strut coil spring

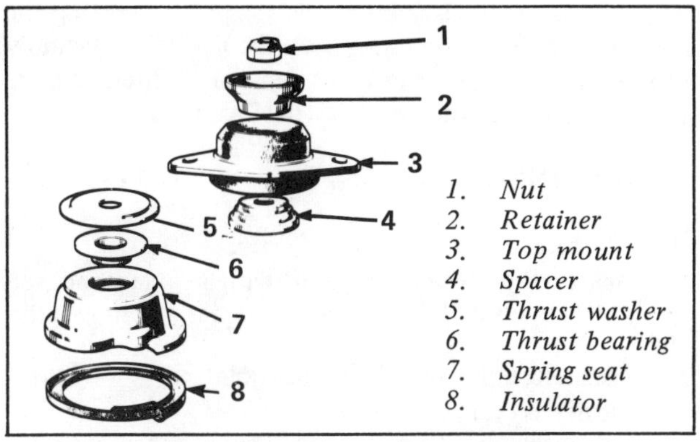

Fig. J:14 Top mounting assembly

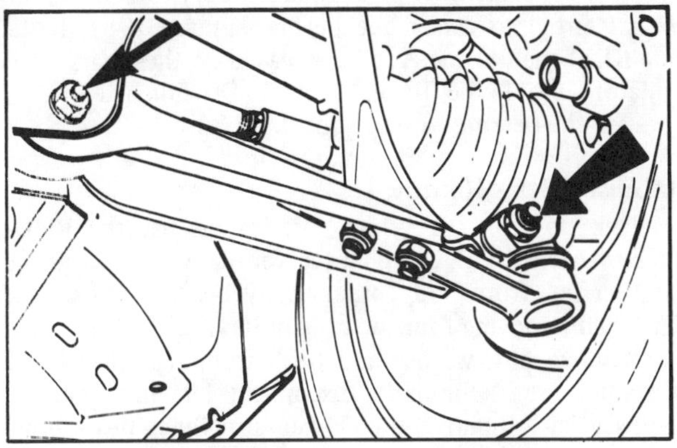

Fig. J:15 Lower arm with tie bar

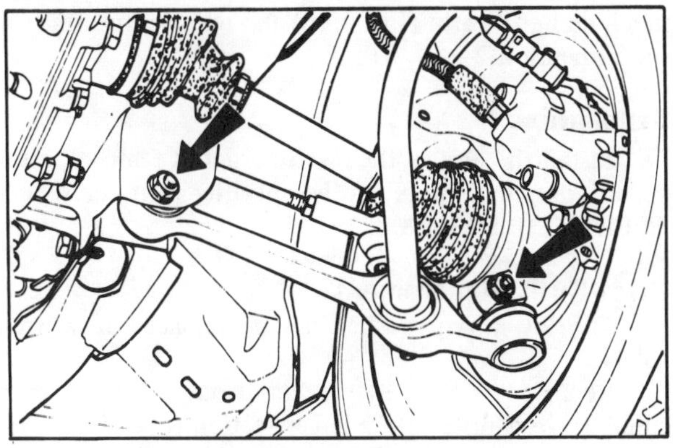

Fig. J:16 Tie bar to lower arm fitting

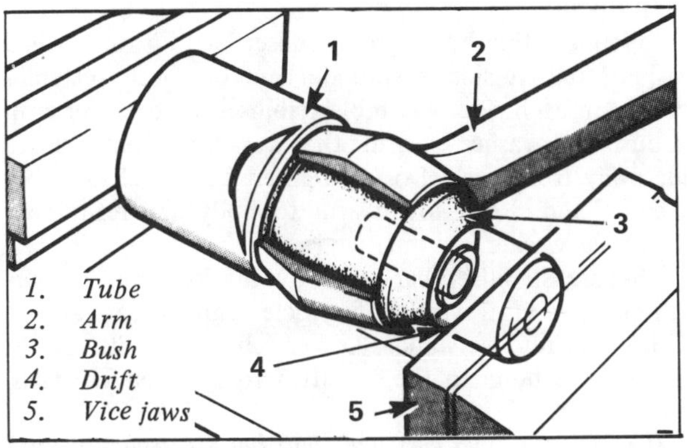

Fig. J: 17 Removing lower arm bush

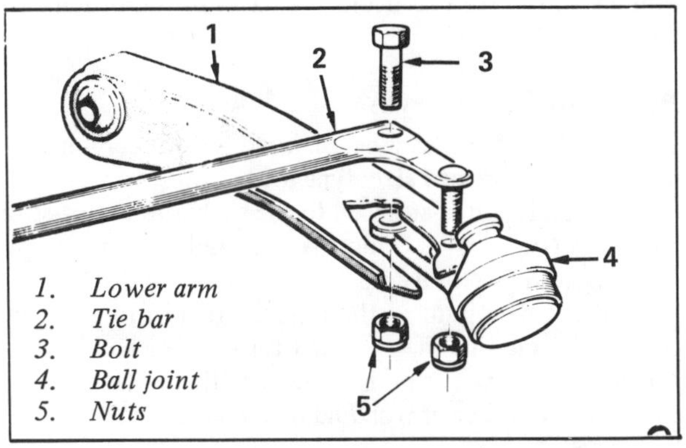

Fig. J:18 Lower arm assembly tie bar only

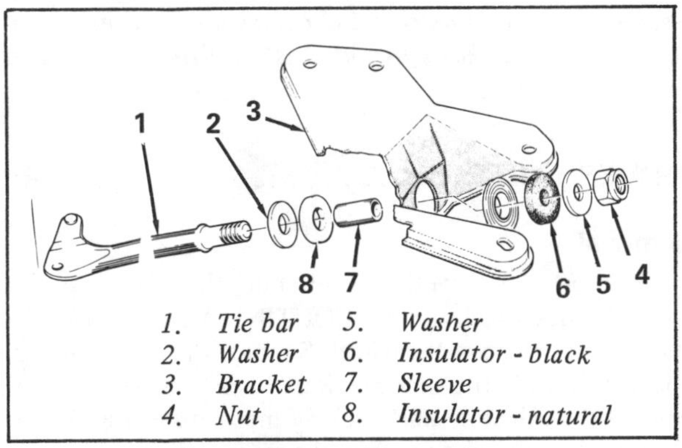

Fig. J:19 Tie bar front mounting

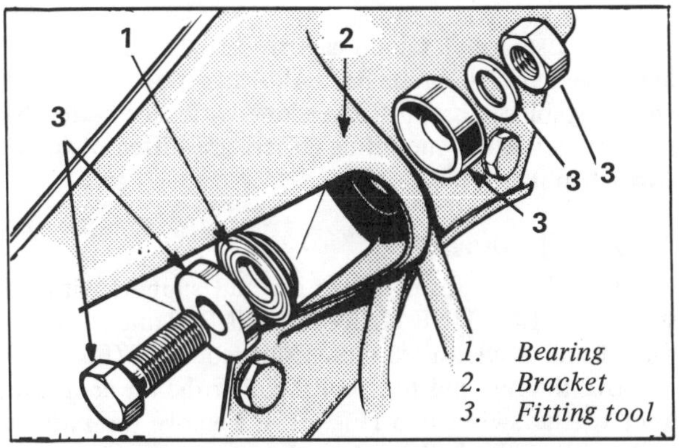

Fig. J:20 Tie bar bearing installation

pairs to maintain suspension balance. After refitting the coil spring and top mounting, refit the new strut assembly to the car in the reverse order of removal. Tighten all bolts to the specified torque, see TECHNICAL DATA.

LOWER ARM . . . . [6]

Removal

1. Jack up the wheel and support with an axle stand, see the ROUTINE MAINTENANCE chapter. Remove the wheel.
2. Disconnect the swivel hub ball joint, see the 'Swivel Hub Carrier' section (Fig. J.5).
3. If a tie bar is fitted, remove the two retaining nuts and bolt (Fig. J:15). If a stabiliser bar is fitted, remove the retaining nut and washer. Undo the inner pivot bolt and nut holding the arm (Fig. J:16) and remove the arm from the suspension.

Installation

Install the arm in the reverse order of removal. Tighten the retaining nuts to the specified torque, see the TECHNICAL DATA.

Pivot Bush Replacement

The lower arm bushes can be replaced if worn as follows:

Remove the bush by pressing it out using a vice and suitable sized drift into a larger section tube (Fig. J:17). Clean the bush eye and lubricate with rubber grease or vaseline. Fit the new bush by pressing it in, in one continuous movement to avoid permanently deforming the bush.

Ball Joint Replacement

The ball joint can be replaced separately only on models with a pressed steel type lower arm.

The ball joint is attached to the tie bar retaining studs and bolts (Fig. J:18). Undo the nuts and detach the joint retaining arm to remove it.

Fit the new joint in the reverse order of removal and tighten the nuts to the specified torque. The ball joint on forged type lower arms is integral with the arm, and if worn, the complete arm should be replaced.

TIE BAR/STABILISER BAR . . . . [7]

The Escort 1.1 saloon cars are fitted with tie bars to hold the swivel hubs in line. All other models are fitted with a stabiliser bar, XR3 versions with an uprated bar. Note that the bar mounting panels are different for the two assemblies.

Tie Bar - Removal

Jack up the front of the car and support with axle stands, see ROUTINE MAINTENANCE. Remove the nut, washer and insulator from the front mounting (Fig. J:19). Remove the nut and torx bolt holding the lower arm ball joint to the swivel hub (Fig. J:5). Remove the nuts and bolt holding the ball joint to the lower arm (Fig. J:18). The tie bar can now be removed.

Installation

Fit new insulators to the tie bar and refit in the reverse order to removal. Refit the Torx bolt with the head to the rear. Note that the front insulator is black, the rear insulator is natural colour. Tighten the bolts to the specified torque, see the TECHNICAL DATA.

Tie Bar Bearing - Replacement

To replace the tie bar bearing, remove the tie bar. The bearing is a pushfit in the mounting bracket. Make up an extractor tool from a long bolt and suitable tubing and washers. Extract the bearing (Fig. J:20).

When fitting the new bearing use no lubricants. Push into position using a tool similar to the removal tool. Refit the tie bar.

Note that the bearing supplied in service is different to that fitted to cars made before January 1981, build code BJ. If the bearing is to be replaced on these cars, new insulators should be fitted as well. Do not attempt to lubricate the bearing to stop squeaking.

Stabiliser Bar - Removal

Jack up front of vehicle, fit stands. See the ROUTINE MAINTENANCE chapter for details. Disconnect stabiliser bar from body brackets by bending back lock plates and removing four securing bolts (Fig. J:21).

Remove nut, washer and bush from right hand end of stabiliser bar. Similarly disconnect left hand end of stabiliser bar. Remove one lower arm inner pivot bolt. Remove mounting bushes from stabiliser bar.

Installation

Fit mounting bushes to stabiliser bar. Fit one dished washer (concave side to shoulder) and one bush to stabiliser bar on each side. Assemble stabiliser bar to lower arm, fit bush and washer ensuring that washer is fitted concave side away from bush. Do not tighten nuts until assembly is completed. Refit lower arm to body and refit inner pivot bolt, washer and nut.

Secure stabiliser bar to body brackets but do not tighten bolts until weight of vehicle is on wheels. Jack up vehicle and remove stands. Lower vehicle onto its wheels. Tighten the bolts to the specified torque, see TECHNICAL DATA.

NOTE: Early model cars had right hand threads on both ends of the stabiliser bar. Later models have a left hand thread on the right hand side. These can be identified by a green or blue paint mark, with a black insert in the retaining nut. The left hand thread must be fitted to the right hand side of the car.

DRIVESHAFTS . . . . [8]

Removal

1. Slacken the driveshaft hub nut, then jack up the wheel as described in the ROUTINE MAINTENANCE chapter, and remove the wheel. Support the body with an axle stand. Drain the transmission oil. Place a suitable tray beneath the selector shaft locking mechanism (Fig. J:23). Unscrew the cap nut and allow the oil to drain.
2. Free the lower arm ball joint by removing the clamp

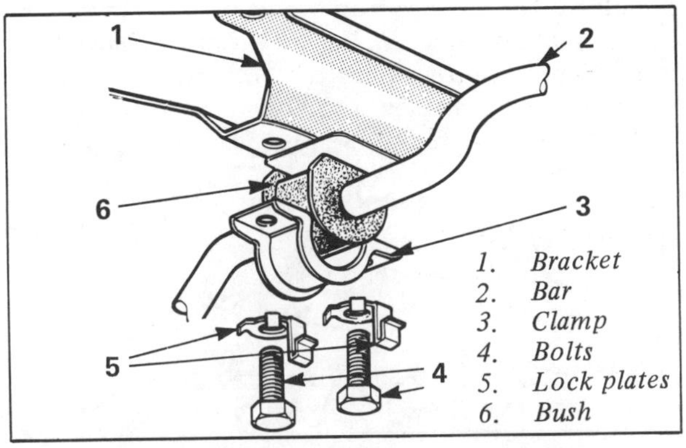

Fig. J:21 Stabiliser bar body mounting

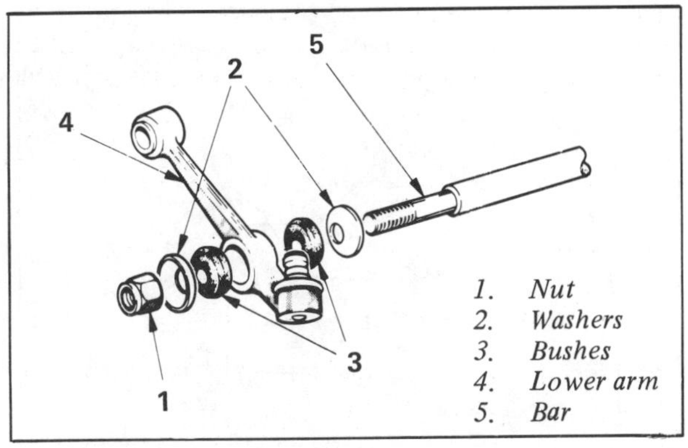

Fig. J:22 Stabiliser bar lower arm mounting

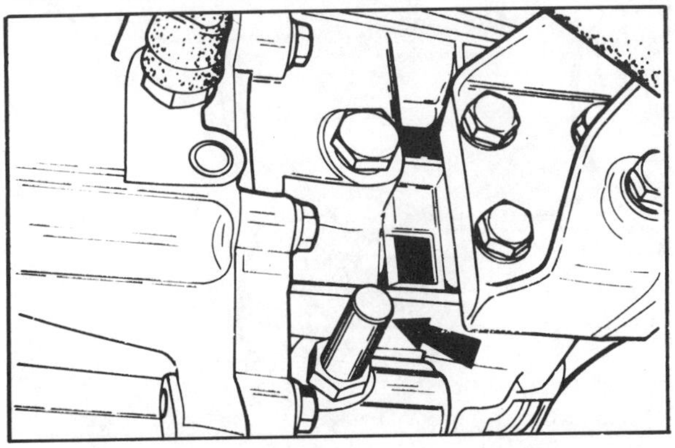

Fig. J:23 Drain gearbox by unscrewing selector lock

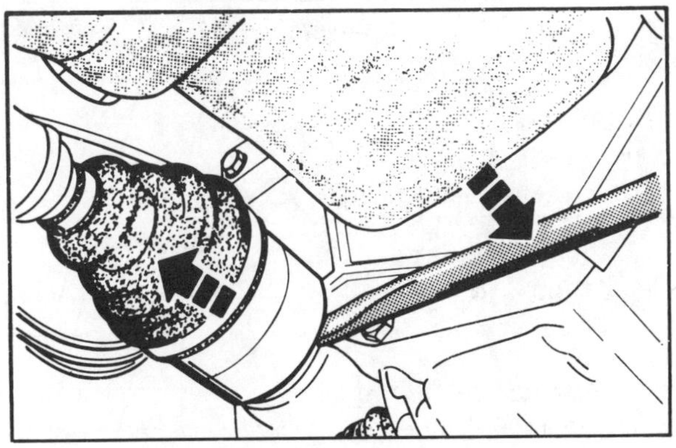

Fig. J:24 Lever drive shaft from differential

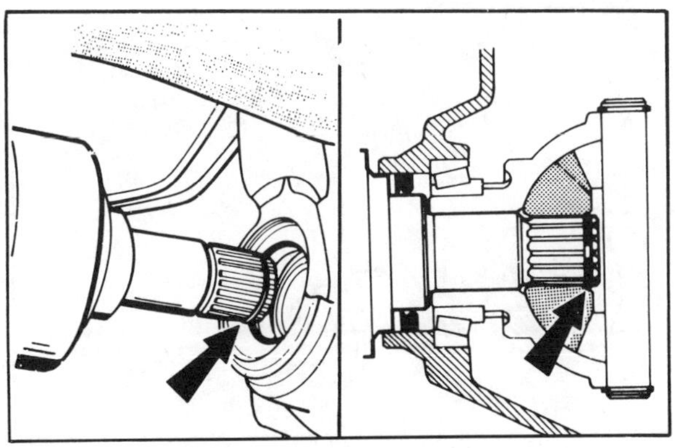

Fig. J:25 Driveshaft inner circlip

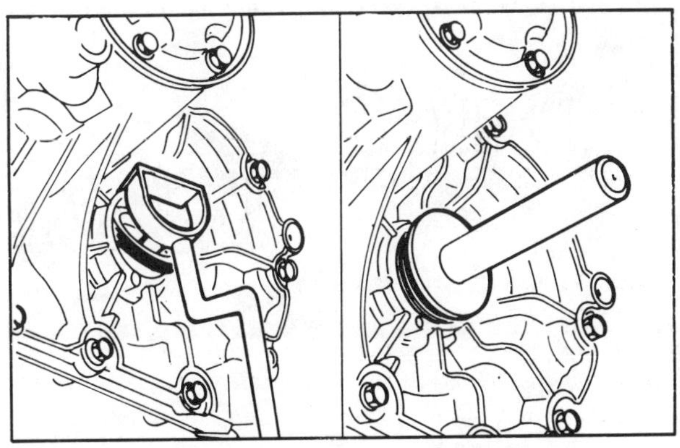

Fig. J:26 Driveshaft inner oil seal replacement

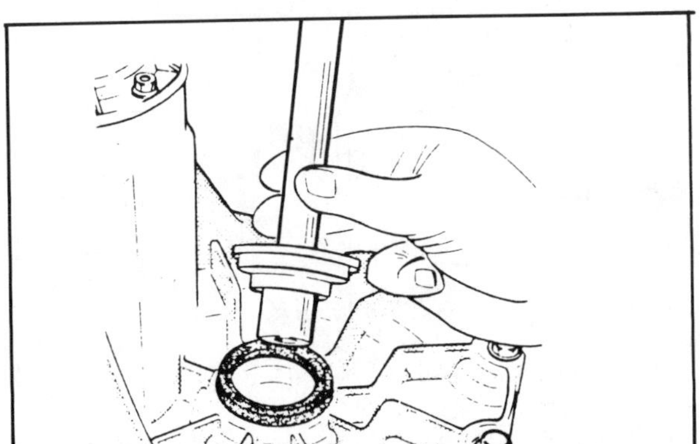

Fig. J:27 Fit shaft inner oil seal with tool 21 - 051

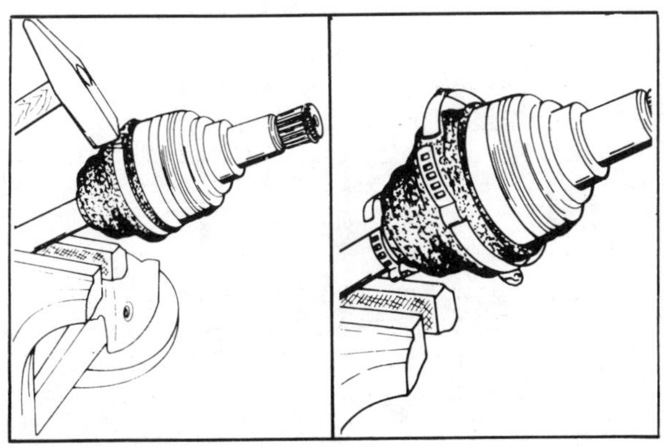

Fig. J:28 Release bellows by tapping clamp

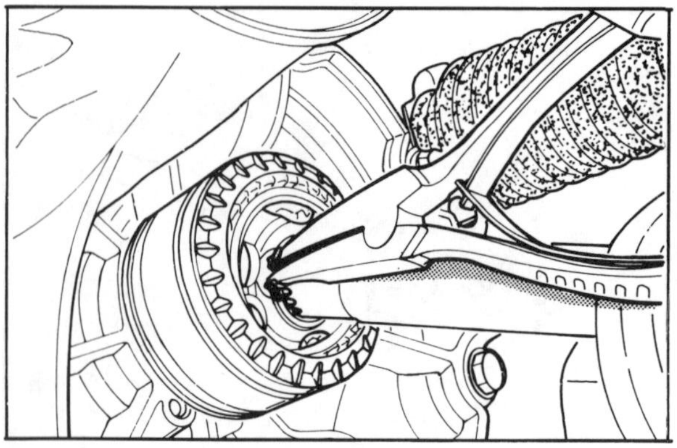

Fig. J:29 Open inner driveshaft circlip

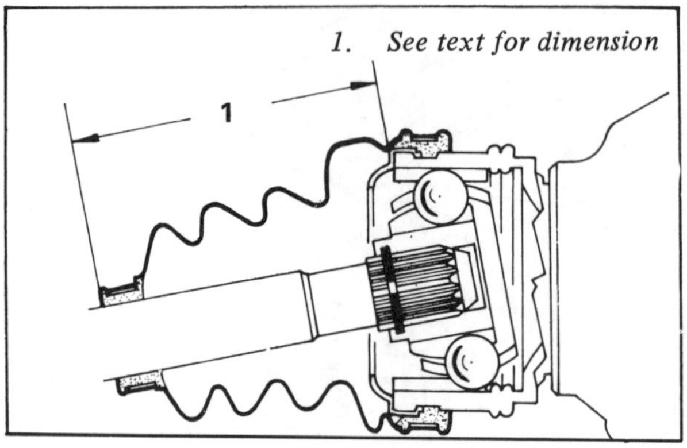

Fig. J:30 Bellows refitting dimension

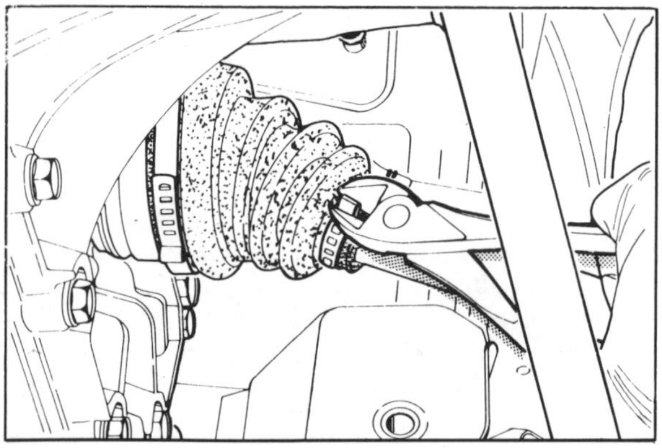

Fig. J:31 Crimp bellows clamp

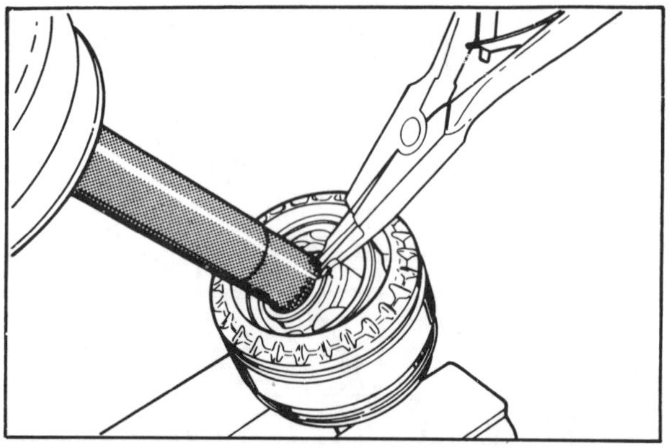

Fig. J:32 Outer joint circlip removal

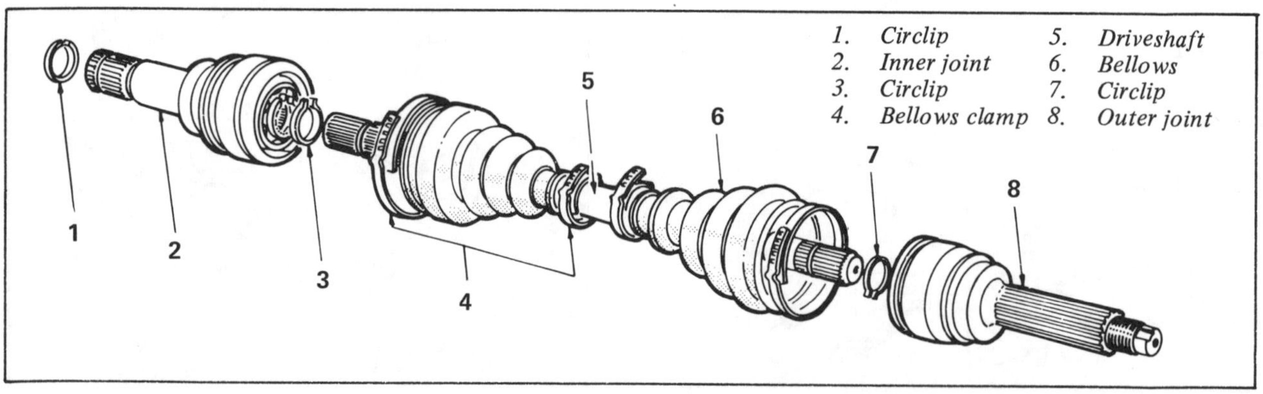

Fig. J:33 Drive shaft bellows assembly

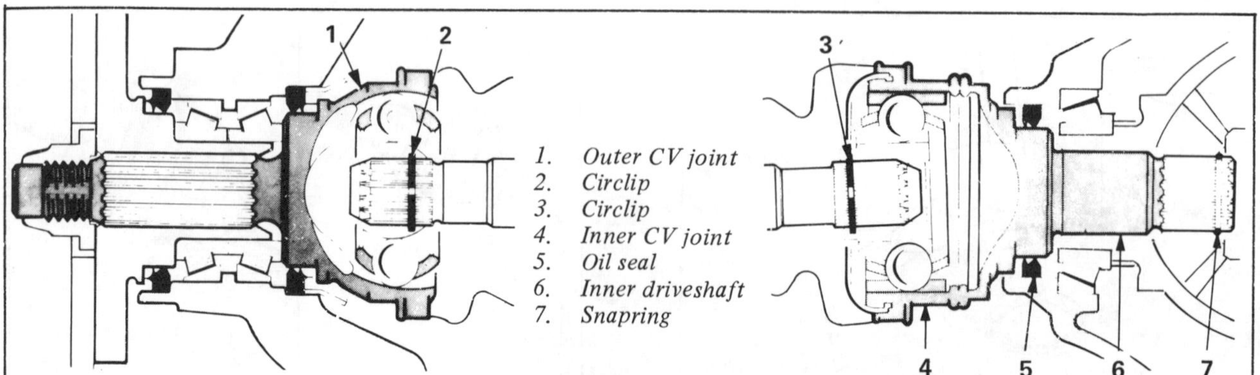

Fig. J:34 Section through driveshaft joints

bolt (Fig. J:5). If a stabiliser bar is fitted, detach the lower arm from the body mounting. Temporarily refit the road wheel.

3. Insert a lever between the differential housing and the drive joint. Pull the road wheel outward. When all the slack is taken up, hammer the end of the lever inwards (Fig. J:24). Support from below using suitable boxes. Remove the wheel and hub nut with washer.

NOTE: The driveshafts must not be allowed to hang free or the final drive oil seals will be damaged.

4. Detach the brake caliper from the hub carrier. Remove the brake disc. See the BRAKES chapter for details. Remove the hub from the assembly, see the 'Swivel Hub' section. The driveshaft assembly can now be withdrawn from the swivel carrier (Fig. J:25). Remove the circlip from the inner shaft splines.

5. If both driveshafts are to be removed together, the differential pinions should be immobilised. Place an old driveshaft or suitable plug into the final drive to prevent the pinions from turning.

Installation

1. Smear the shaft splines with suitable grease, see the TECHNICAL DATA for details of specification. Push the hub onto the drive shaft, pulling the shaft in as far as possible. A special tool is available (Fig. J:6).

2. Fit the washer and a new hub nut. Refit the brake disc and the caliper to the swivel. Fit a new circlip to the drive shaft inner spline. Push the shaft into the differential, until the circlip can be heard to engage.

3. Refit the rest of the assembly in the reverse order of removal. All bolts must be tightened to their specified torque, see the TECHNICAL DATA. Refill the transmission with the specified lubricant. See the ROUTINE MAINTENANCE chapter for refilling details.

4. After lowering the car to the ground, tighten the nut to the specified tightening torque. See the 'Swivel Hub' section (Fig. J:7).

Final Drive Oil Seals

The oil seals can be replaced after detaching the inner end of the shaft. It is not necessary to detach the shaft from the hub assembly. See the 'Driveshaft Removal & Installation' section for details of the removal and refitting procedure.

With the shaft supported using suitable boxes or stands, lever the old oil seal from the transmission casing. Although a special tool is made for this job (Fig. J:26) a screwdriver can be used. Take care not to scratch the seal housing bore.

The new oil seal should be fitted using Ford special tool 21-051. The seal should be driven into the recess a pre-set depth, (Fig. J:27). Alternatively, measure the oil seal's depth in the recess before removal. Drive the new oil seal into the measured depth using a suitable sized drift. Refit the driveshaft inner end as described previously.

Driveshaft Bellows - Replacement

If the joint bellows are damaged, the lubricant will be allowed to escape and dirt and water will be able to enter, ruining the joint within. The bellows should be examined whenever working beneath the car. Should there be any signs of the bellows cracking or the smallest cut or split, the rubber should be renewed immediately. As a temporary measure, a plastic bag can be tied around the bellows to keep dirt and water out.

The inner bellows can be removed without disconnecting the shaft from the swivel hub. Replacement of the outer bellows requires removal of the hub (Fig. J:33).

To replace the inner bellows on either side, jack up the appropriate wheel and remove, see the ROUTINE MAINTENANCE chapter. Detach the lower ball joint from the swivel hub (Fig. J:5). Do not turn the torx head bolt, but turn the nut.

Detach the bellows from the inner joint. The bellows is held in place with a clamp. Tap the clamp on the locking latch to separate (Fig. J:28). Slide the bellows along the shaft and release the joint circlip (Fig. J:29). Pull the shaft from the joint and release the bellows.

Repack the joint with grease as necessary. Refit the shaft using a new bellows and circlip. The smaller end of the bellows should be 5 in (127 mm) from the joint (Fig. J:30), on 1.1 litre cars, 5.19 in (132 mm) on 1.3 and 1.6 cars.

Tighten the bellows clamp by hand, fit the hook into the nearest hole. Crimp the clamp using a pair of pliers (Fig. J:31). Reassemble the drive shaft assembly in the reverse order. Tighten the lower ball joint nut to its specified torque, see the TECHNICAL DATA.

To replace the outer bellows, slacken the hub nut and disconnect the lower arm ball joint from the hub carrier. Separate the drive shaft from the hub, see 'Swivel Hub Carrier' section previously. Separate the drive shaft from the joint as described for the inner joint. Remove the bellows and refit a new component as described previously. Reassemble the joint in the reverse order of removal.

Driveshaft Joints - Replacement

1. The drive shaft joints can be replaced after removing the shafts. Release the joint bellows retaining clamps and slide the bellows along the shaft. Wipe all surplus grease from the joint, prise open the retaining circlip (Figs. J:29 & 32) and withdraw the driveshaft from the joint.

2. Thoroughly clean the joint, removing all traces of old grease and dirt. Inspect the condition of the circlips and replace them if necessary. Repack each joint with the specified grease, see TECHNICAL DATA.

3. Slide the bellows along the shaft and refit the joints to the shaft, pushing the shaft into each joint until the circlip is felt to engage (Fig. J:33).

4. Refit the bellows to the joints. See the 'Driveshaft Bellows' section for details of refitting procedure (Fig. J: 34).

Rear Suspension

INTRODUCTION....................... [1]

The rear suspension fitted to Escort Saloon and Estate models is fully independant with coil springs transverse arms and strut type shock absorbers. Tie bar link rods provide fore and aft location for the transverse arms (Fig. K:1). Gas filled Bilstein shock absorber units are fitted to the XR3 models.

Rear wheel alignment is set by spacers on the end of each tie bar. All suspension pivot bushes are of the rubber voided type, and are designed to flex. These bushes can be checked for wear by levering the bush to one side, then examining for signs of cracking in the supporting webs. Only the bushes on the tie bar can be replaced, those on the lower arms cannot, and a replacement lower arm must be fitted if the bushes are worn.

Van versions are fitted with a transverse beam type axle and single leaf type springs with double-acting shock absorbers (Fig. K:2). Stub axles are welded to each end of the axle beam with taper roller bearings supporting the rear hubs. Note that when replacing bearings or shock absorbers, although the components may appear similar, they are different to those used on the car and estate models.

Whenever replacing shock absorbers, springs or any other components on either type of suspension, it is most important to state the exact model to obtain correct replacement parts - see TECHNICAL DATA section.

Apart from a comprehensive Metric spanner and socket set, the only tools necessary to overhaul or replace the rear suspension components are suitable sized drifts and sockets for removing and refitting the bearings, grease retainers and replaceable bushes. A heavy duty vice will also be necessary.

REAR WHEEL BEARINGS .... [2]

Bearing Adjustment

The rear wheel bearings on saloon and estate cars are integral with the brake drum, and will require adjustment every time the drum is removed. The bearings on van versions are mounted in a separate hub and will not normally need to be adjusted except after a high mileage, or if the hub has been removed for bearing replacement.

The rear wheel bearings on all models should be adjusted as follows:

1. Slacken the wheel bolts, then raise and support the rear of the car as detailed in the ROUTINE MAINTENANCE chapter.
2. Remove the road wheel and lever off the grease cap using a screwdriver.
3. Using pliers, withdraw the split-pin and detach the nut retainer (Fig. K:3), to expose the bearing retaining nut.
4. Tighten the nut to a torque of 15 - 18 lb ft (2,0 - 2,5 kg m) whilst, at the same time, turning the brake drum anti-clockwise.
5. Now slacken the nut off by 180°, then retighten the nut finger-tight. This will ensure that the bearings are seated correctly.
6. Check that the brake drum revolves freely. If not, slacken the nut and repeat the adjustment.
7. When the adjustment is correct, refit the nut retainer and secure it with a new split pin. Make sure that the grease cap is clean and that the small breather hole is clear before refitting it. Do not fill the cap with grease.
8. Refit the road wheel and lower the car to the ground.

Bearing Replacement - Saloon & Estate Models

If, after adjustment, the bearings are rough or noisy as the wheel is revolved, or if the bearing cone rollers or cups are pitted, then the complete bearing assembly should be replaced together with a new grease seal.

Jack the rear of the car and support it on stands. Remove the rear wheel, grease cup, split pin and nut retainer. Remove the adjusting locknut, washer and outer bearing. Making sure the handbrake is fully released, pull the drum

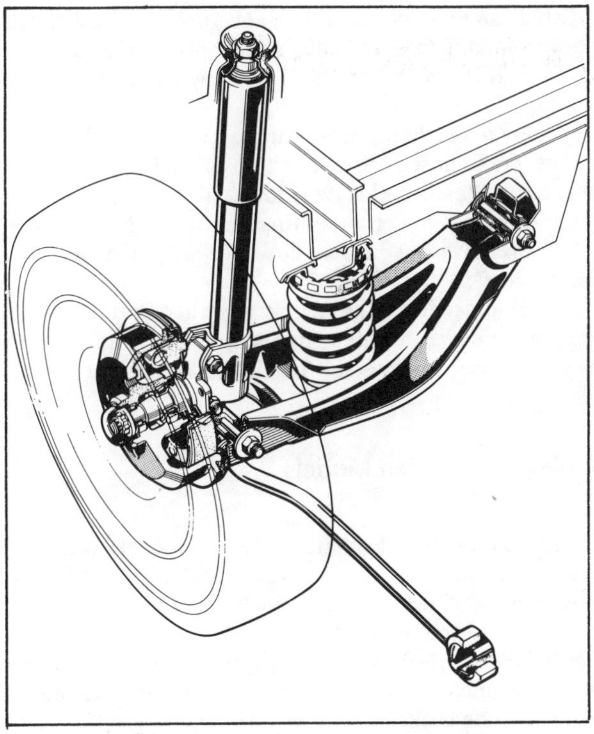

Fig. K:1 Saloon and estate rear suspension, right hand side shown

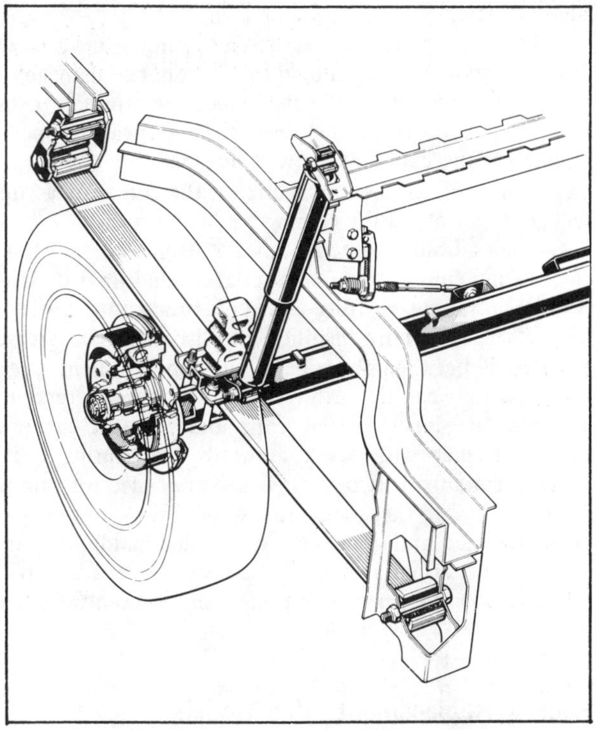

Fig. K:2 Van rear suspension, right hand side shown

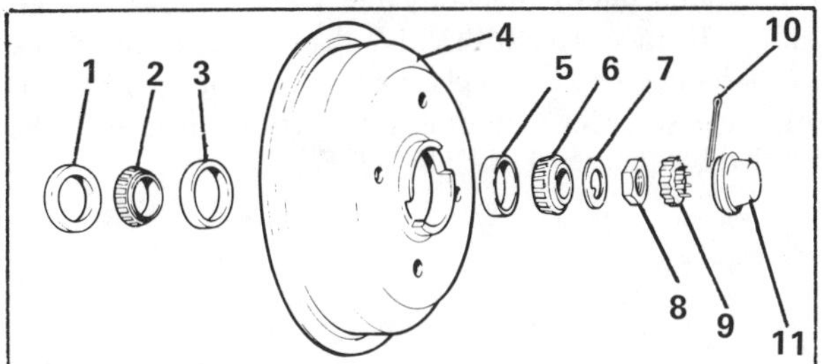

1. *Grease retainer*
2. *Bearing cone*
3. *Inner cup*
4. *Drum and hub*
5. *Outer cup*
6. *Bearing cone*
7. *Tab washer*
8. *Locknut*
9. *Nut retainer*
10. *Split pin*
11. *Grease cup*

Fig. K:3 Saloon and estate bearing assembly

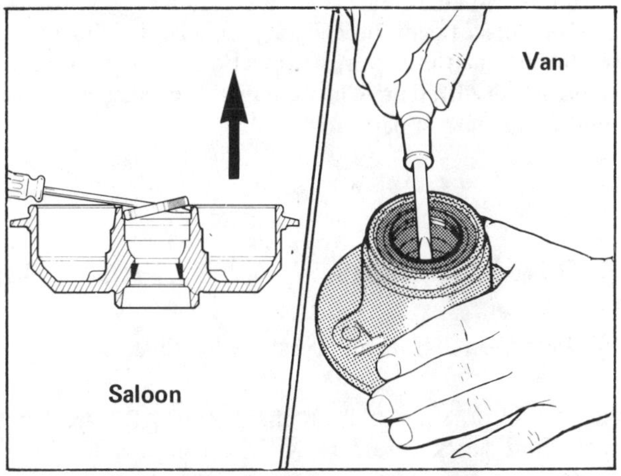

Fig. K:4 Grease retainer removal

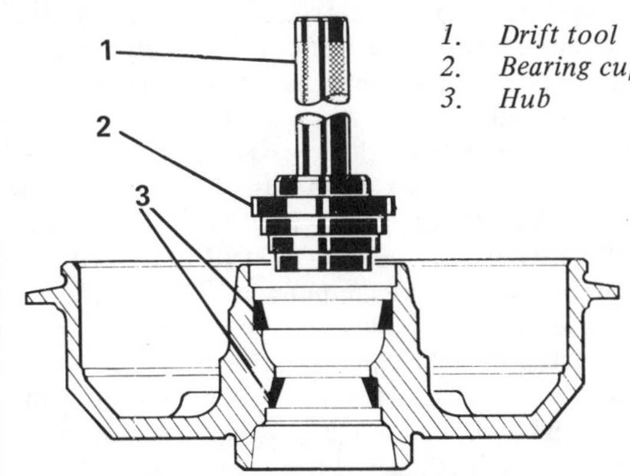

1. *Drift tool*
2. *Bearing cups*
3. *Hub*

Fig. K:5 Bearing fitting

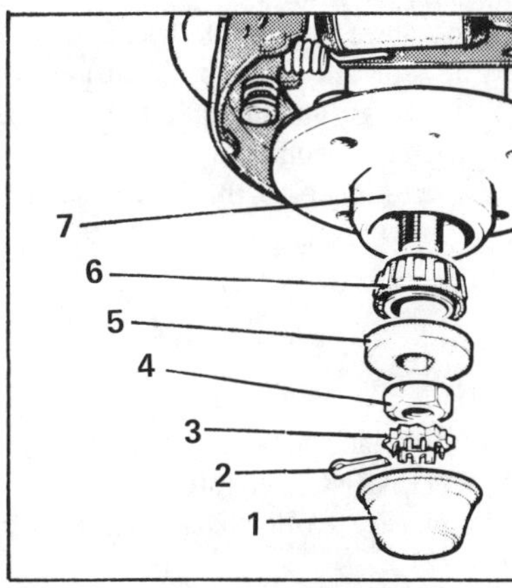

1. *Grease cap*
2. *Split pin*
3. *Retainer*
4. *Nut*
5. *Washer*
6. *Bearing*
7. *Hub*

Fig. K:6 Van wheel bearing assembly

and hub assembly from the stub axle.

Lever the grease retainer from the hub using a screwdriver (Fig. K:4). Care should be taken not to damage the seal bore. Lift out the bearing cone. Use a punch to tap out the bearing cups, tapping at diametrically opposed points on the cup. Make sure it does not tilt in the hub. Take care not to raise any burrs on the cup seats as this will adversely affect the bearing pre-load.

Clean all old grease and any dirt from the hub. Refit the bearing cups with an appropriate round drift or large socket or tubing, drifting each one firmly into position (Fig. K:5). Pack both bearings with lithium based grease, working it between the rollers. Also, apply a light smear of grease between the sealing lips of a new grease retainer. Position the inner bearing cone and then fit the new grease retainer with its lip towards the bearing, using either a suitable drift or a large socket or piece of tubing. Refit the hub to the spindle followed by the outer bearing cone, washer and locknut. Apply the handbrake and adjust the bearings as described above. Replace the roadwheel, lower the car to the ground and tighten the wheel bolts.

Bearing Replacement - Van Models

The wheel bearings on van models are mounted in a separate hub with detachable brake drum (Fig. K:6). The bearing replacement procedure is similar to the saloon models detailed in the previous section, however the components are different. When obtaining replacement parts, ensure that correct parts are fitted.

SHOCK ABSORBERS. . . . . [3]

Replacement - Saloon & Estate Models

All versions apart from the Escort XR3 are fitted with Ford shock absorbers. XR3 versions are fitted with Bilstein gas filled shock absorbers. The removal and replacement procedures are similar for all Escort saloons and estates.

When replacing shock absorbers due to wear or leakage, both units should be replaced to maintain suspension balance. Care must be taken to ensure that the replacement units are of the correct specification level. This is because of the many changes made to the Escort suspension specifications.

Replacement

Slacken the wheel bolts, jack up the car as described in the ROUTINE MAINTENANCE chapter and remove the wheel. If available place a trolley jack beneath the lower suspension arm.

Inside the luggage compartment, raise the parcel tray (if fitted) and remove the shock absorber cap. Slacken the top nut, holding the threaded rod in place using an Allen key (Fig. K:7). Withdraw the insulating rubber and metal cup.

From beneath the wheel arch, slacken the clip holding the brake hose to the shock absorber body then free the hose. Remove the two nuts and bolts holding the shock absorber lower mounting to the stub axle (Fig. K:11). Remove the shock absorber with the cup and bump rubber.

Refit or replace the shock absorber in the reverse order of removal. Tighten the securing nuts and bolts to the specified torque, see the TECHNICAL DATA.

Replacement - Van Models

The shock absorbers fitted to van models with leaf spring rear suspension are different in construction to saloon models. Replace the shock absorbers as follows.

Jack up the rear of the van as described in the ROUTINE MAINTENANCE chapter, ensuring that the van is unladen. Place suitable supports or axle stands under the axle beam.

Remove the nut and bolt holding the lower mounting (Fig. K:8). Remove the two bolts holding the shock absorber bracket to the body bracket. Withdraw the assembly. Remove the bolt holding the bracket to the shock absorber and fit to the replacement unit. Fit the new unit in the reverse order of removal, tightening the nuts to the specified torque, see TECHNICAL DATA.

TIE BAR. . . . . [4]

Removal & Installation - Saloon & Estate Only

Each rear stub axle is retained fore-and-aft by a tie bar (Fig. K:9). Raise the appropriate wheel, see ROUTINE MAINTENANCE. Remove the nut and bolt holding the front end of the bar to the underbody. Remove the nut from the rear end of the tie bar (Fig. K:10). When removing the washers and bushes carefully note the order of components, as the washers determine rear wheel alignment.

Refit the tie bar in the reverse order of removal, ensuring that the washers on the stub axle mounting are replaced in the correct order. Tighten the nuts to the specified tightening torque, see TECHNICAL DATA.

If the front end bush requires replacement, a vice will be necessary to extract and fit a new bush. Use a socket or piece of tubing smaller than the internal diameter of the rod to push the rubber into a socket or tube larger than the bush. Use vaseline or a suitable rubber lubricant to aid fitting the new bush, which can be pushed straight in.

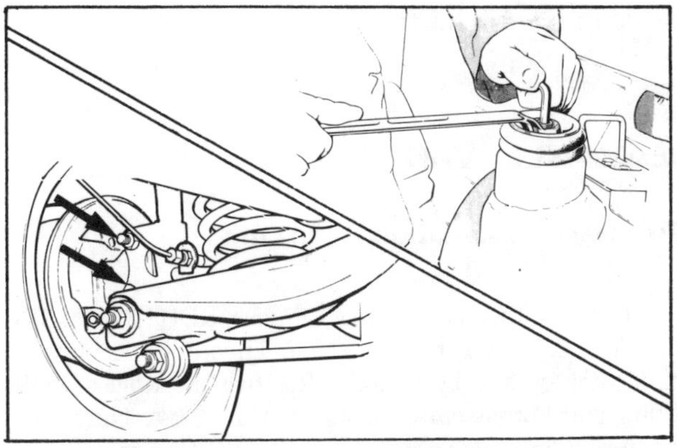

Fig. K:7 Saloon and estate shock absorber mountings

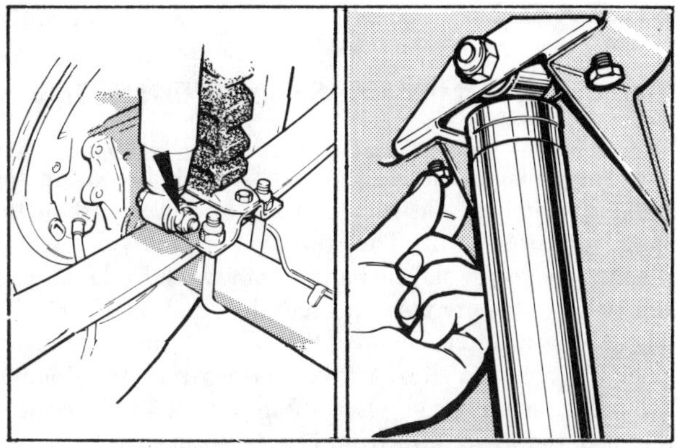

Fig. K:8 Van shock absorber mountings

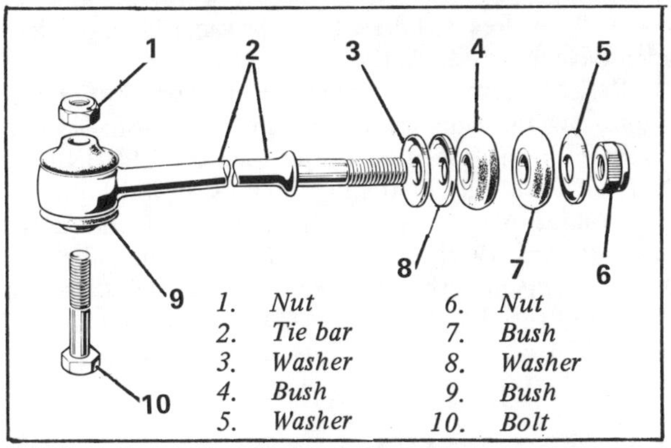

Fig. K:9 Tie bar assembly

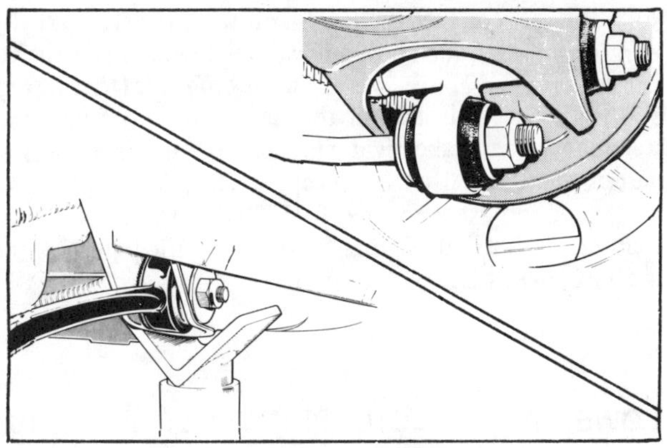

Fig. K:10 Tie bar mountings

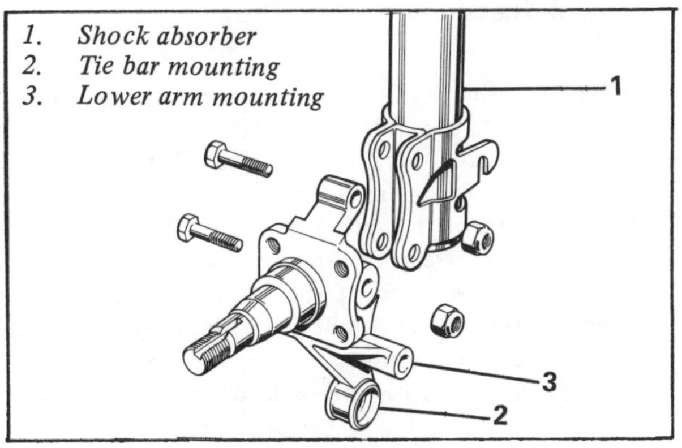

Fig. K:11 Rear stub axle

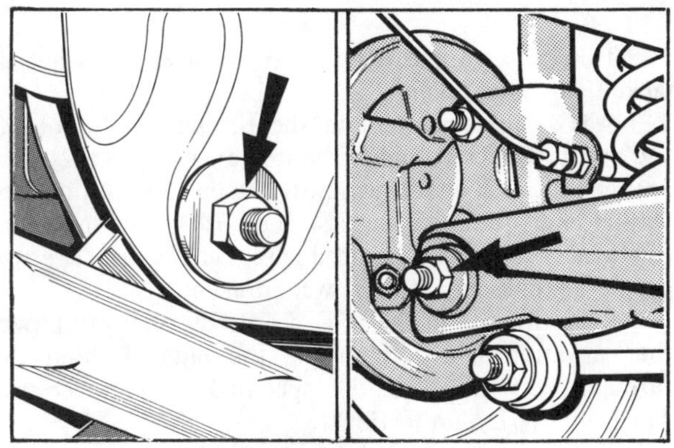

Fig. K:12 Lower arm mountings

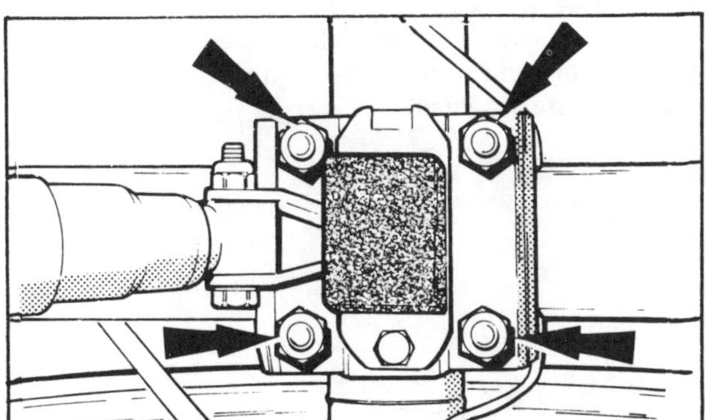

Fig. K:13 Beam axle U bolt nuts

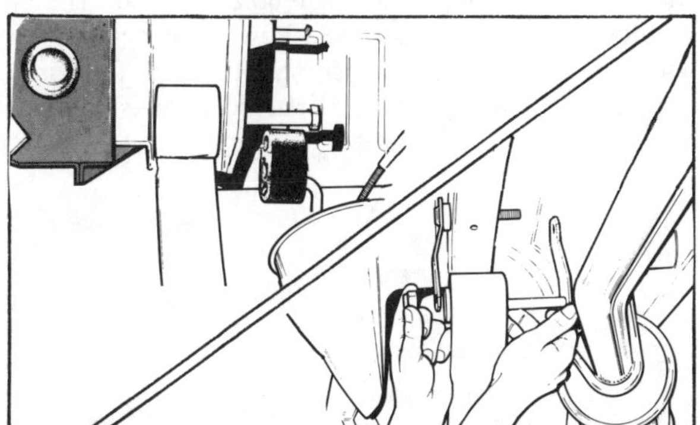

Fig. K:14 Van leafspring front and rear mountings

STUB AXLE . . . . [5]

Removal & Installation - Saloon & Estate Only

The stub axle holds the rear wheel hub bearing and brake assemblies, and is bolted to the lower part of the shock absorber strut. The stub axle assembly is located by the lower arm and tie bar. To remove the axle, jack up the wheel and remove it, see the ROUTINE MAINTENANCE chapter. Remove the brake drum and bearing assembly, see the 'Rear Wheel Bearings' section. Remove the brake back plate as described in the BRAKES chapter. Support the lower arm from beneath using a trolley jack or blocks.

Remove the bolts holding the shock absorber to the stub axle. Disconnect the lower arm from the stub axle by removing the through bolt and washers. Disconnect the tie bar as described above. Withdraw the stub axle (Fig. K:11).

Refit the stub axle in the reverse order of removal. Reconnect the brake rigid pipe and bleed the system as described in the BRAKES chapter. Reconnect the brake cable and adjust, see the ROUTINE MAINTENANCE chapter. Tighten all fixings to their specified torque, see the TECHNICAL DATA.

LOWER ARM & COIL SPRING . . . . [6]

Removal & Installation - Saloon & Estate Only

The lower arm can be removed after the wheel is jacked up and removed, see the ROUTINE MAINTENANCE chapter. Remove the through bolts holding the arm to the body and to the stub axle (Fig. K:12). The arm can then be withdrawn. Note the position of the large washer and the rubber bushes. The bush on the inner end of the arm cannot be replaced in service if worn. The arm should be replaced with a new component.

Refit the arm in the reverse order of removal. Lower the car after lightly tightening the nuts. Tighten the through bolt nuts to the specified torque, see the TECHNICAL DATA section.

The coil spring can be removed after removing the shock absorber upper mounting, see the 'Shock Absorber' section, and the outer through bolt removed. The arm should be supported by a jack. Lower the jack and arm. Withdraw the spring and upper mounting pad. Refit in the reverse order. Note that the plastic sleeve on XR3 models should be fitted at the top.

BEAM AXLE - VAN . . . . [7]

Removal & Installation - Van Only

Jack up the rear of the van, see ROUTINE MAINTENANCE. Support the vehicle underbody using axle stands-not on the beam axle. Remove the wheels, brake drums and hub assemblies. Remove the brake back plate, see the BRAKES chapter.

Remove the nuts holding the U bolts (Fig. K:13) while supporting the axle from below using stands or blocks. When free withdraw from the van, sliding the light laden valve linkage from the spacer tube.

Refit the axle assembly in the reverse order of removal, greasing the light laden valve link rod bush. Tighten the nuts to the specified torque, see the TECHNICAL DATA. Before fully tightening the nuts, lower the van to the ground. After refitting the brake drum bleed the brakes system, see the BRAKES chapter. Adjust the light laden valve. Adjust the rear wheel bearings as described at the start of the chapter.

LEAF SPRINGS - VAN. . . . . [8]

Removal & Installation - Van Only

Raise the rear of the car and support the axle tube with axle stands, see the ROUTINE MAINTENANCE chapter. Remove the four nuts holding the U bolt through the axle retaining plate (Fig. K:13).

The rear end of the spring is held to the underbody by a shackle plate assembly. Remove the nuts at the end (Fig. K:14). The rear end of the spring should then be free. Remove the nut and bolt from the front end. The spring can then be withdrawn.

The spring bushes can be replaced using a vice and tubing of similar diameter to the bush metal sleeve. Push the bush from the spring eye into a larger tube diameter. Clean the inner eye and regrease. Push the new bush into position using the vice.

Note that two similar shackle plate assemblies may be fitted. If replacement parts are required, ensure that the correct type are obtained.

Brakes

INTRODUCTION . . . . [1]

All Escort models have a dual-circuit braking system providing two separate hydraulic circuits for each pair of diagonally opposed wheels, thus providing optimum emergency braking in the event of one circuit failing.

The front brake calipers are of single piston, sliding caliper design, with self-adjusting, drum type rear brakes. Certain Escort variants are fitted with a direct-acting brake servo.

A brake pressure control valve is also fitted in the hydraulic circuit on saloon and estate car variants to minimise the tendency for the rear brakes to lock prematurely under conditions of heavy braking. On Escort van versions, the 'light laden' valve is actuated by a linkage attached to the rear axle and controls brake pressure to the rear brakes by reacting to changes in the suspension height resulting from varying load conditions.

On certain Escort variants, brake pad wear sensors are fitted to the front brake pads. When the pad lining material has worn to a predetermined thickness, the sensor contacts the brake disc and operates a warning light on the instrument panel.

The handbrake is cable operated on the rear brakes, and can be adjusted to compensate for wear and cable stretch.

Apart from the brake pressure control valve mentioned previously, the remainder of the Escort braking system components can be overhauled or replaced by the experienced DIY mechanic - see 'Hydraulic System Overhaul' section. Although it is possible to overhaul the brake servo unit, special tools are essential for the job which the DIY mechanic is unlikely to obtain. In this case, we recommend that a replacement unit is fitted.

Health Warning

Brake shoes and pads contain asbestos, and after use leave asbestos dust around the drum and caliper. When working or cleaning components covered with this dust, allow no dust to become air-borne, or it may be inhaled. Wash any asbestos dust from the skin.

FRONT BRAKE PAD REPLACEMENT . . . . . [2]

The brake pads should be renewed if they are worn below the minimum safe thickness of 0.060 in (1,5 mm), or if they are the cause of braking problems.

Whenever one or more brake pads require replacement, BOTH pads at BOTH front brakes should be replaced to maintain even braking balance.

1. Slacken the wheel bolts, jack up the front of the car, fit axle stands and remove the front wheels.
2. Remove the retaining clip from the caliper, see Fig. L:1.
3. Slacken the caliper piston housing bolts until they are free from the anchor brackets using a 0.27 in (7 mm) allen key (Fig. L:2).
4. Remove the piston housing (Fig. L:3) and suspend it from the body with a piece of wire. Make sure the piston assembly is securely suspended.
5. Withdraw the inner brake pad from the piston housing and the outer pad from the anchor bracket.
6. Brush any dirt or dust from the caliper, taking care not to inhale any since it is asbestos based and is harmful. Ideally use a vacuum cleaner to remove the dust, or use a brush or rag with care.
7. Check that the replacement pads are of the correct type and that the pads, retaining clips and disc are free from grease, oil or dirt.
8. Before fitting the new pads, push the caliper piston back into its bore to provide adequate clearance. This should be done by carefully applying pressure to the face of the piston. It will be necessary to either open the caliper bleed valve as the piston is being retracted, or to suck out some of the brake fluid from the master cylinder reservoir during this operation since the level will rise as the piston is pushed back into its bore. Take care not to remove too much otherwise air will enter the system and

the brakes will have to be bled as described later in this chapter.

9. Fit the new inner brake pad into the piston housing, locating the spring on the back of the pad before pushing it home. If fitted, reconnect up the pad wear sensor wire at the multi-plug and check the circuit. Fit the outer pad to the anchor bracket.

10. Replace the piston housing and secure with the mounting bolts. Tighten the bolts to the correct torque - see TECHNICAL DATA.

11. Refit the retaining clip (Fig. L:1).

12. Operate the brake pedal several times to bring the brake pads into their correct operating position.

13. Repeat the operation for the other front brake.

14. Refit the road wheels and lower the car to the ground. Fully tighten the road wheel bolts and check the fluid level in the master cylinder reservoir. Top-up as necessary.

REAR BRAKE SHOE REPLACEMENT..... [3]

The brake shoes in the rear brakes should be renewed if the linings are worn to or approaching the minimum wear thickness of 0.06 in (1,5 mm), or if they are the cause of braking problems - see ROUTINE MAINTENANCE chapter.

Brake shoes should be renewed in axle sets, i.e. BOTH sides of BOTH rear brakes to maintain braking balance.

NOTE: The lining thickness of a new trailing shoe is considerably less than that on the leading shoe to compensate for the trailing shoe's lower wear rate. Only the leading shoe lining should be used to assess the wear rate and lining thickness.

Removal

1. Slacken the rear wheel bolts, jack up the rear of the car, fit axle stands and remove both rear wheels.

2. Fully release the handbrake.

3. On saloon/estate cars, remove the hub nut dust cap, split pin, nut retainer, adjusting nut, washer and outer bearing from the spindle.

4. Slide the brake drum/hub assembly from the spindle. On van models, undo the countersunk screw and pull off the brake drum.

5. Use a pair of pliers to depress the shoe hold down spring washer on the leading shoe. Turn the spring washer through 90° so that the slots in the pin hole coincide with the lugs on the retaining pin. Remove the washer and spring and then remove the pin from the back plate. Take care not to lose the spring.

7. Twist the leading shoe outwards and upwards away from the back plate until it is possible to detach the shoe and remove the springs (Fig. L:4).

8. Remove the trailing shoe hold down spring in the same manner as the leading shoe spring assembly.

9. Move the trailing shoe assembly upwards and away from the back plate (Fig. L:5). Disconnect the handbrake cable from trailing shoe lever.

10. Remove the spacer strut from the trailing shoe by twisting it and removing the spring (Fig. L:6).

11. Thoroughly clean the brake backplate and stub axle.

NOTE: Harder lining material brake shoes can be fitted, identified by code number M-M 180 HH together with revised strut and pull off spring fitting as shown in Fig. L: 7.

Installation

1. Apply a smear of high melting point grease to the shoe raised platforms on the back plate.

2. Fit the handbrake lever return spring to the trailing shoe. Then hook the spacer strut to the spring and lever it into position on the shoe (Fig. L:6).

3. Locate trailing brake shoe against the wheel cylinder and the backplate lower abutment position making sure that the lower end of the handbrake lever is correctly located on the face of the plastic plunger and not trapped behind it.

4. Refit shoe hold down spring to trailing shoe.

5. Referring to Fig. L:8, refit the larger of the two springs, between leading and trailing shoes at the back plate abutment position.

6. Refit the spring between the leading shoe and the strut. With shoe web at right angles to the backplate, enter lower end of the shoe behind abutment retainer plate and fit the aperture at the upper end of the shoe, over the quadrant lever. Press the shoe onto the backplate.

7. Refit the shoe hold down spring to the leading shoe.

8. On saloon and estate cars, refit the hub/drum assembly to the spindle, replacing the outer bearing, retaining washer and adjusting nut. Tighten the latter to the correct torque - see TECHNICAL DATA, whilst rotating the drum. Then back off the nut 90°. Fit the nut retainer, a new split pin and the dust cap.

On van models, refit the brake drum in the reverse order of removal.

9. Depress the foot brake pedal several times to adjust the brakes. This is important.

10. Repeat the procedure for the other brake assembly.

BRAKE HYDRAULIC CIRCUIT .... [4]

Overhaul of the components of the hydraulic system should always be carried out under conditions of scrupulous cleanliness. Clean all dirt and grease from the exterior of components before removal or dismantling.

After dismantling, wash all parts in commercial alcohol, methylated spirit or clean brake fluid. Do NOT use mineral based oils such as petrol, paraffin or carbon tetrachloride. Blow out all internal passages with compressed air.

Inspect pistons and cylinder bores of the master cylinder, looking for scores, ridges or corrosion pits. The unit MUST be discarded if any of these conditions are present. The calipers and wheel cylinders should also be inspected for leaks and for signs of wear after dismantling. If any signs of wear are present, the unit must be replaced.

It is essential that only new seals are used when reassembling. These are normally available in the form of a repair kit containing all the necessary parts required for

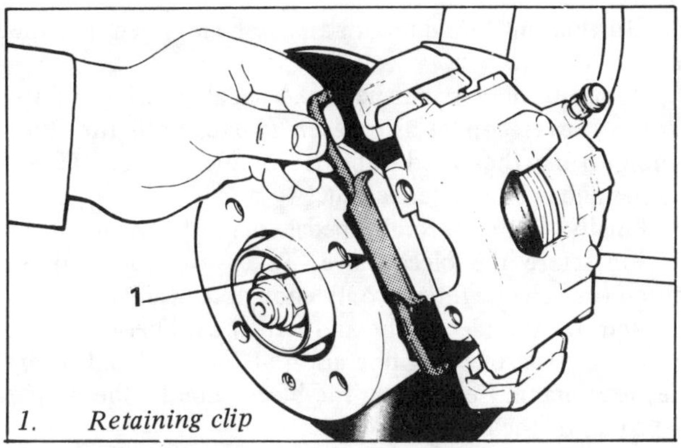

Fig. L:1 Removing retaining clip from caliper

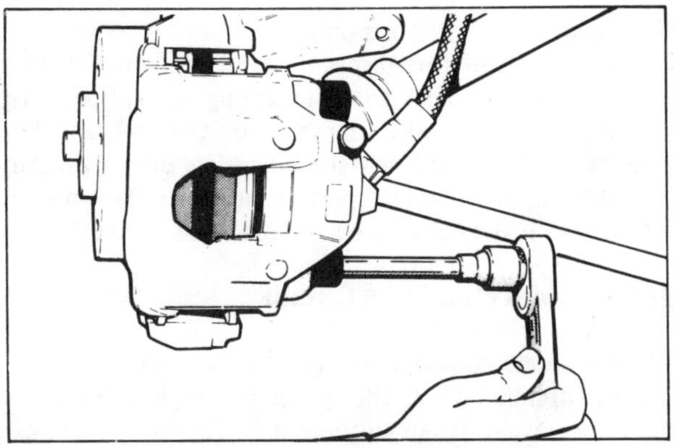

Fig. L:2 Removing caliper retaining bolts

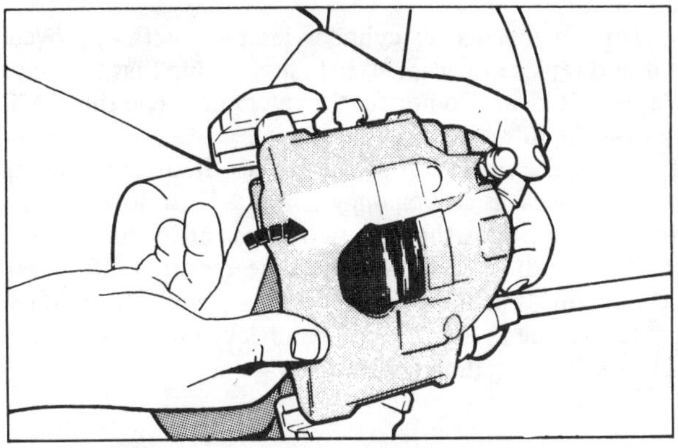

Fig. L:3 Removing caliper piston housing

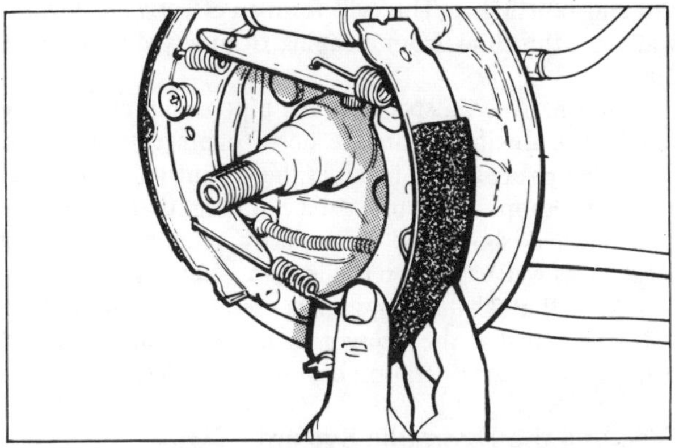

Fig. L:4 Removing leading shoe from back plate

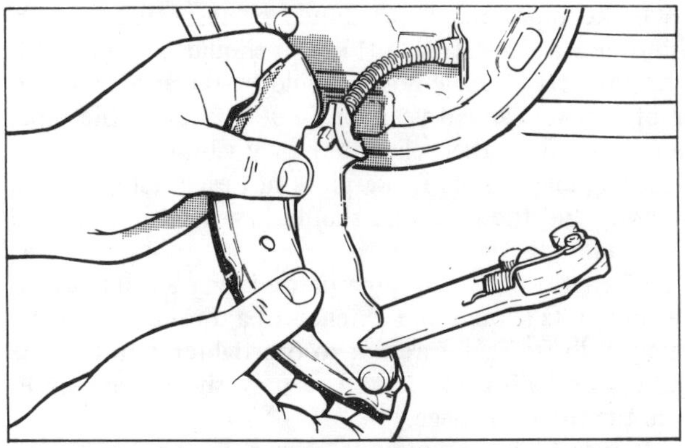

Fig. L:5 Detaching handbrake cable from lever

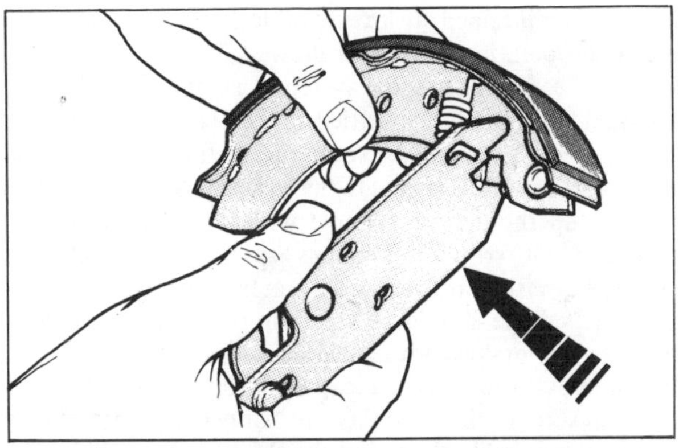

Fig. L:6 Removing spacer strut from shoe

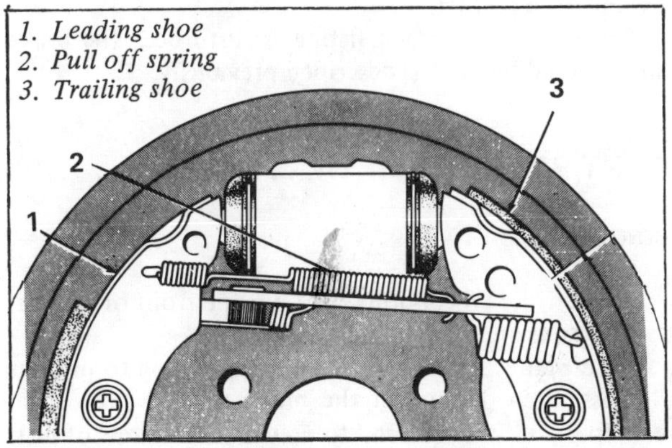

Fig. L:7 Later type brake shoe strut/spring arrangement

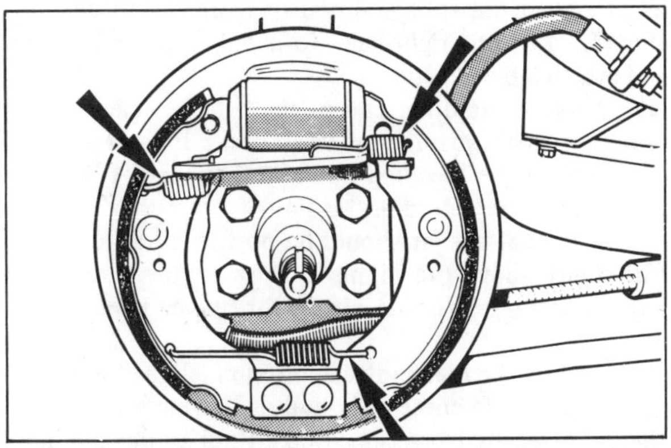

Fig. L:8 Position of shoe return springs

the overhaul of a particular unit.

All seals should be inspected carefully before fitting, even when new. Check that the sealing lips are perfectly formed, concentric with the bore of the seal, and free from 'knife edge' cuts, surface blemishes or marks. Any seal which does not appear perfect, no matter how minute the blemish may appear to be, should be discarded.

Blocking the Hydraulic Circuit

Whenever components are disconnected from the brake hydraulic circuit, the system will rapidly empty unless measures are taken to prevent it. An effective way of doing this is to block the top of the brake fluid reservoir by wrapping a piece of plastic sheet around the reservoir cap and refitting it. This will form an effective seal which will stop the fluid from escaping from any disconnected union.

Alternatively, a special brake pipe clamp can be used to pinch a flexible hydraulic hose to prevent fluid loss. Such clamps are available under the Girling name from car accessory shops. Any brake pipe unions that have been disconnected should be temporarily plugged to prevent seepage, as well as to stop the ingress of any dirt.

Note that whatever system is used to stop the system, it is essential that the reservoir is unblocked and pipes unclamped after the operation has been completed.

Bleeding the Hydraulic System

The fluid level in the brake master cylinder reservoir must be maintained at a reasonable level throughout the bleeding operation as, if it is allowed to drop excessively, air may be drawn into the system. Use only fresh hydraulic fluid of the type specified in TECHNICAL DATA at the end of this manual. Never re-use fluid which has already been passed through the system. Take care when topping up the master cylinder with fluid since it is an effective paint stripper, and any spilt on the bodywork should be washed off with cold water immediately.

1. The vehicle should be standing on level ground when bleeding the brakes, not jacked up at one side or end.
2. Remove the reservoir cap and top up the reservoir, making sure both chambers are topped up with the recommended fluid. Note that the fluid level should be topped up regularly during bleeding operations.
3. Remove the rubber dust cap from the bleed nipple (Fig. L:9) on the right-hand (offside) front brake caliper and attach a bleed tube.
4. Immerse the other end of the tube in a clean jar containing a small quantity of clean brake fluid (Fig. L:10). Throughout the bleeding operation, the end of the tube must remain immersed in this fluid.

   Note that the jar should be positioned at least 12 in (300 mm) above the nipple as shown. This is to prevent any air leaking past the nipple threads and into the hydraulic system.
5. Slacken the bleed valve approximately half a turn and then depress the brake pedal fully, allowing it to return to the fully released position. Brake fluid and/or air should have been pumped into the jar. If not, slacken the bleed valve further until fluid and/or air can be pumped into the jar.
6. Continue depressing the brake pedal, pausing briefly (about three seconds) after each stroke, until the fluid coming from the bleed tube is completely free of air bubbles and is completely clean.
7. Finally, with the brake pedal in the fully depressed position, close the bleed valve. Take care not to overtighten the valve; tighten it only enough to seal it.
8. Remove the bleed tube and refit the rubber dust cap.
9. Repeat the bleeding operation at the left-hand (nearside) rear brake. Connecting the bleed tube to the nipple at the rear of the brake backplate.
10. Repeat the bleeding operation at the left-hand (nearside) front brake, followed by the right-hand (offside) rear brake.
11. Top up the master cylinder reservoir with approved fluid and replace the cap, after first checking that the vent hole in it is clear. Do not fill the reservoir above the MAX mark on the side.
12. Check the operation of the brakes If, after bleeding, the brake pedal is still 'spongy', or goes right down to the floor, this indicates that air is still present in the system, and the bleeding operation must be repeated. If subsequent attempts at bleeding still fail to produce a satisfactory result, the system should be checked for leaks, as air is obviously being drawn into the system.

BRAKE PIPES & HOSES . . . . [5]

Brake pipes and hoses should be inspected for corrosion and leakage (Fig. L:11) and should be replaced if at all suspect. Flexible hoses should be replaced as a matter of course every 3 years or 36,000 miles. If the pipes come loose from their clips they may vibrate and fatigue fracturing may result. Brake pipes rust especially between the body and the pipe, and should have any mud washed off to aid inspection.

When fitting a new pipe or replacing an old fitting, the tube nuts should be tightened to a torque of approximately 8 lb ft. Take care not to overtighten a nut or the fitting will distort and leak. Use only short spanners to avoid the risk of damage.

Before fitting a tube nut, ensure that both sides of the connection are clean, free from grit which will prevent the joint from seating.

Whenever a hydraulic fitting is disturbed, the brake system should be bled as described previously.

FRONT BRAKE CALIPER. . . . . [6]

Removal

1. Slacken the wheel bolts, jack up the front of the car, fit axle stands and remove the road wheel.
2. Fit a brake pipe clamp to the flexible hose to prevent fluid loss, then disconnect the brake flexible hose from the caliper and plug the end to prevent the ingress of dirt.
3. Release the two caliper retaining bolts (Fig. L:12)

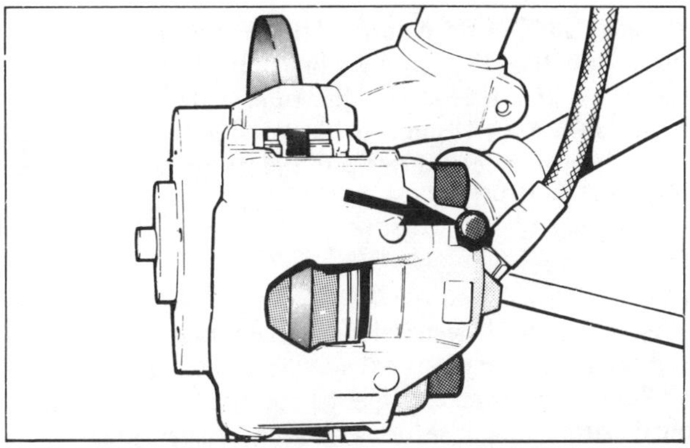

Fig. L:9 Location of front brake bleed nipple

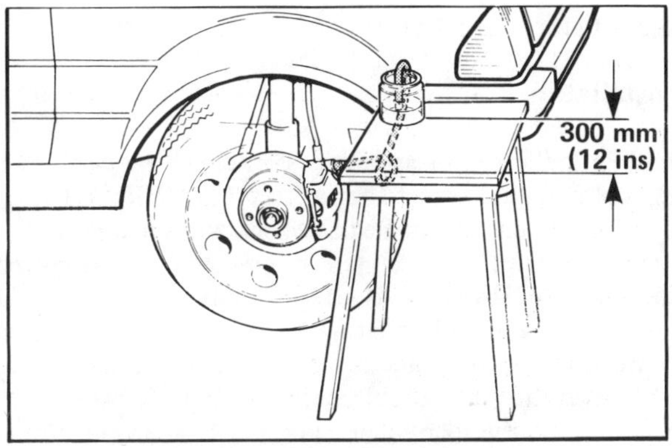

Fig. L:10 Method of bleeding the brakes

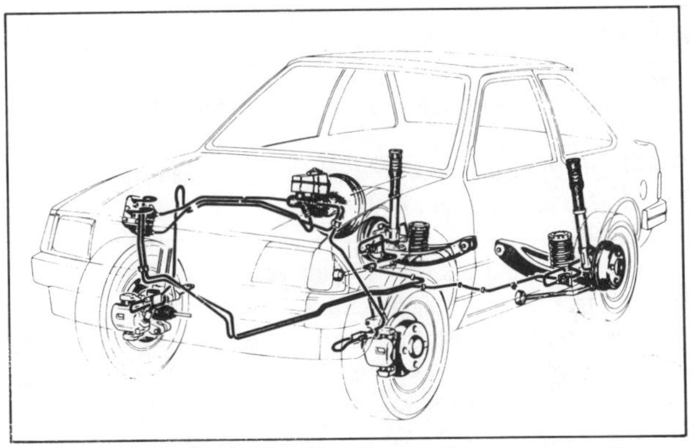

Fig. L:11 Brake hydraulic pipes and hose connections

Fig. L:12 Caliper retaining bolts arrowed

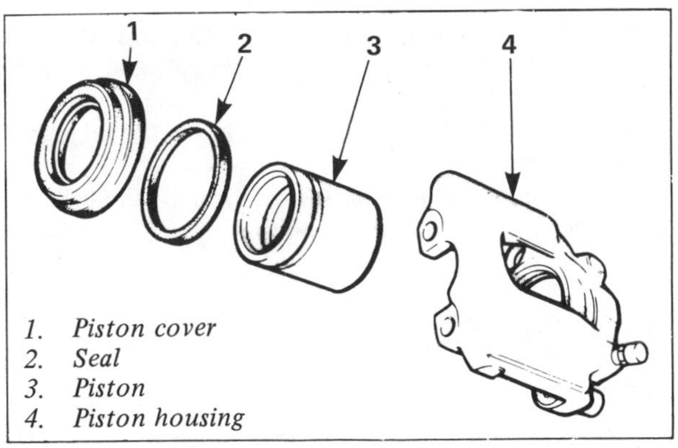

Fig. L:13 Exploded view of caliper piston and seals

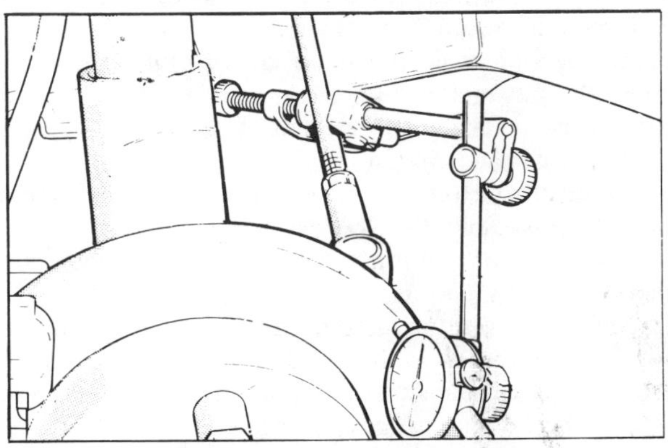

Fig. L:14 Method of measuring disc run-out

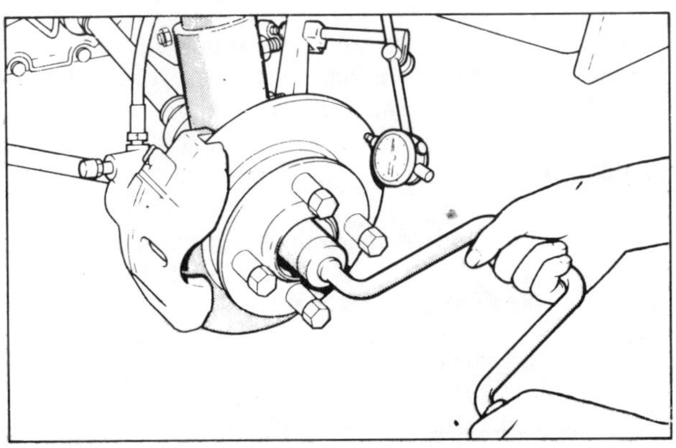

Fig. L:15 Turning disc when checking run-out

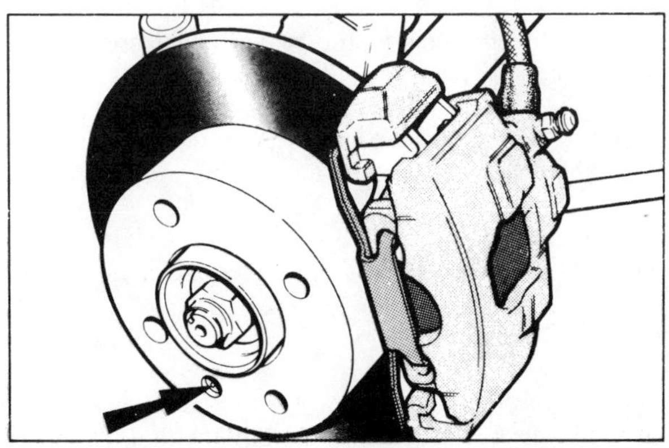

Fig. L:16 Disc retaining screw arrowed

and lift away the caliper.

Installation

1. Refit the caliper assembly, tightening the two retaining bolts to the correct torque - see TECHNICAL DATA.
2. Unplug the brake flexible hose and reconnect it to the caliper inlet port. Tighten the connector to the correct torque - see TECHNICAL DATA. Make sure the hose is not twisted and will be clear of the body and suspension components during steering and suspension movement.
3. Bleed the brake circuit as described previously.
4. Replace the road wheel, lower the car to the ground and fully tighten the road wheel bolts.

NOTE: If the caliper is being removed to allow removal of the front hub/disc assembly to gain access to the suspension components, it will not be necessary to disconnect the flexible brake hose. All that will be necessary will be to support the caliper from the body with a piece of wire, making sure the hose is not strained in any way.

Caliper Overhaul (Fig. L:13).

1. Remove the caliper piston housing as detailed previously under 'Front Brake Pad, Replacement', disconnecting the flexible brake hose and plugging it to allow the piston housing to be removed from the car.
2. Make sure the working conditions are spotlessly clean.
3. Remove the rubber cover over the end of the piston.
4. Apply an air line to the brake fluid inlet port. This will force the piston from its bore.
5. Using a blunt instrument, such as a plastic needle remove the piston seal from the annular groove in the piston housing. Take great care not to scratch the cylinder bore during this operation.
6. Wash the piston and piston bore in methylated spirits or clean brake fluid. Do NOT use petrol or paraffin.
7. Make sure the piston and its bore are free from score marks. This is important.
8. Fit a new piston seal to the annular groove in the piston housing.
9. Lubricate the piston with clean brake fluid and push it into the housing as far as it will go. Take care not to damage the piston seal.
10. Fit new rubber bellows between the piston and housing.
11. Refit the housing to the caliper and reconnect the brake hose.
12. Bleed the brake circuit as described previously.
13. Refit the road wheel, lower the car to the ground and fully tighten the road wheel bolts.

FRONT BRAKE DISC .... [7]

Runout Check

To check the disc runout, jack up the car and remove the road wheel. Refit the wheel bolts with 0.75 in (19 mm) spacers between the bolt heads and discs. Mount a dial gauge so that its pointer rests on the disc face and then rotate the disc through one complete revolution. (Figs. L:14 and L:15). The disc runout figure should be within 0.006 in (0,15 mm) of the total indicator reading. If a reading in excess of this is found the cause should be sought and rectified.

Disc Removal

1. Slacken the wheel bolts, jack up the car, fit axle stands and then remove the road wheel.
2. Remove the brake caliper assembly as described previously, suspending it securely from the suspension.
3. Remove the disc retaining screw from the hub (Fig. L:16) and lift away the disc (Fig. L:17).

Inspection

Inspect the disc friction surfaces. Moderate scoring is permissible, but heavy scoring or grooving, cracking or pitting of the surface, or excessive corrosion build up necessitates replacement of the disc.

The disc may be trued up by grinding down to within the minimum permissible thickness specified in TECHNICAL DATA, but replacement is preferable.

Make sure all disc/hub mating surfaces are free from dirt and dust. Also make sure the friction surface of the disc is free from dirt, oil and brake fluid.

Installation

1. Make sure the mating face of the hub and disc are spotlessly clean then refit the disc assembly and replace the retaining screw.
2. Replace the caliper assembly, tightening the retaining bolts to the specified torque.
3. Replace the road wheel, lower the car to the ground and fully tighten the road wheel bolts.

REAR BRAKE CYLINDERS .... [8]

Removal

1. Remove the rear brake shoe assembly as described previously.
2. Disconnect the brake pipe from the cylinder inlet port and plug the pipe to prevent loss of brake fluid.
3. Release the two retaining bolts at the back of the carrier plate (Fig. L:18) and remove the wheel cylinder.

Installation

This is a reversal of the removal procedure, noting that the brake circuit should be bled as described previously at the beginning of this chapter.

Overhaul (Fig. L:19)

1. Remove the wheel cylinder as described previously.
2. Pull the rubber boots from the ends of the cylinder

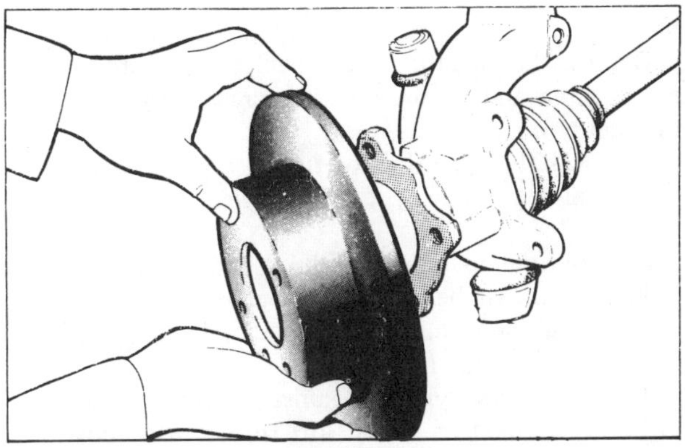

Fig. L:17 Removing disc assembly from hub

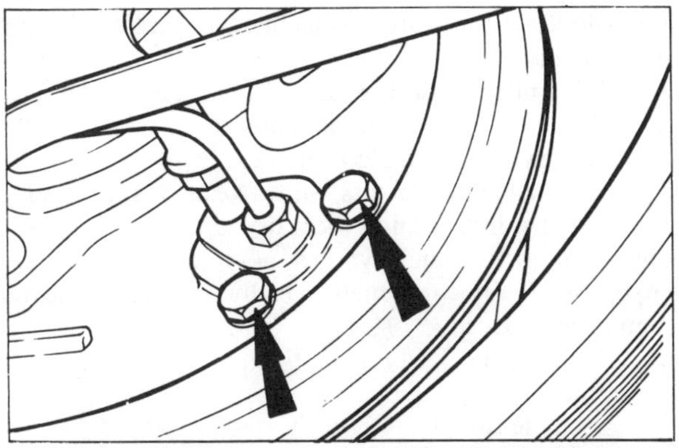

Fig. L:18 Rear wheel cylinder retaining bolts

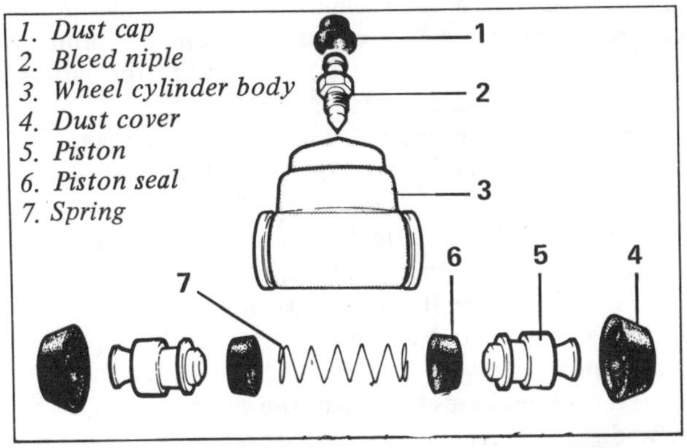

Fig. L:19 Exploded view of rear wheel cylinder

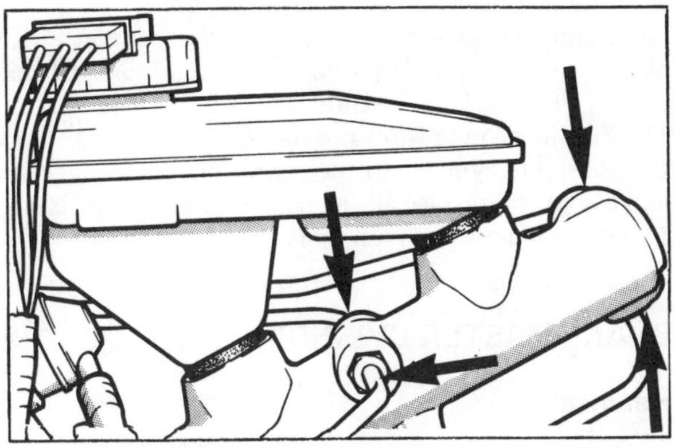

Fig. L:20 Master cylinder brake pipe connections

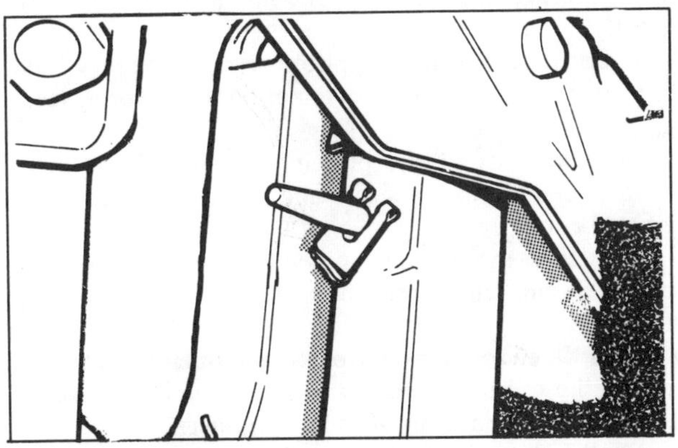

Fig. L:21 Master cylinder push rod spring clip

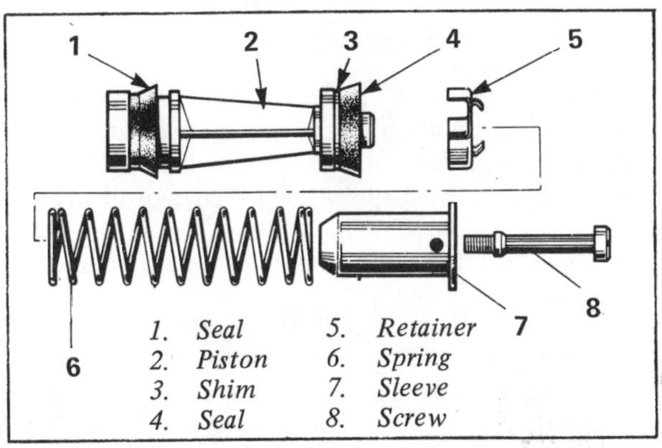

Fig. L:22 Master cylinder primary piston assembly components

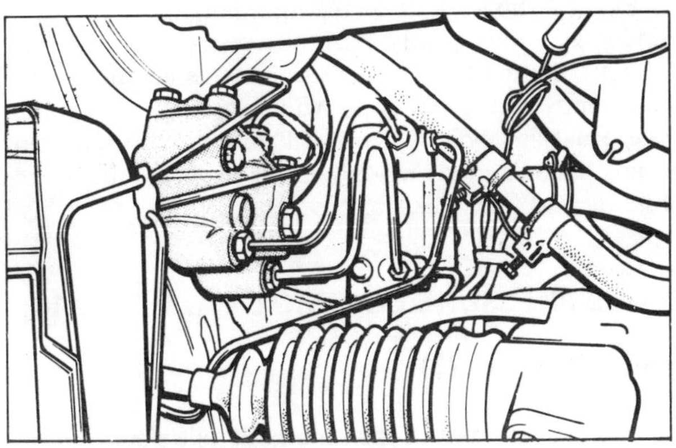

Fig. L: 23 Brake pressure control valve bolts

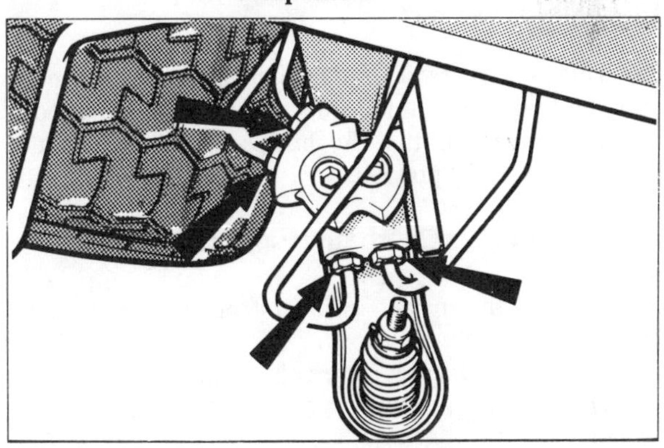

Fig. L:24 Light laden valve pipe connections - Van

and withdraw the piston assemblies. Separate the boots from the pistons.
3. Slide out the piston seals and then remove the spring from the centre of the cylinder bore.
4. Unscrew the brake bleed nipple.
5. Clean all components with clean brake fluid or methylated spirits, wiping them dry with a lint free cloth. Check the pistons and cylinder bore for scoring - this is important. If any component is badly scratched or scored, it must be replaced.
6. Screw in the bleed nipple, tightening it sufficiently to prevent leaks.
7. Assemble the rubber boots to the pistons.
8. Lubricate the piston, seal and cylinder bore with clean brake fluid.
9. Fit one piston and boot into the cylinder.
10. From the open end of the cylinder, slide in a NEW piston seal, spring and second seal, making sure the open ends of the seals face the spring.
11. Fit the second piston and rubber boot.
12. Refit the cylinder to the carrier plate, reconnect the brake hose and replace all the remaining components.
13. Bleed the brake circuit as described previously.

BRAKE MASTER CYLINDER............[9]

Removal

1. Remove the reservoir cap and drain all the fluid from the reservoir chambers by opening the bleed nipple on both front brake calipers. Close the bleed nipple when the reservoir is empty.
2. Place a large thick cloth under the master cylinder to catch any spilt brake fluid to protect the paint. Disconnect the fluid pipes from the master cylinder, plugging the ends to prevent fluid spillage and ingress of dirt (Fig. L: 20).
3. Disconnect the wiring from the fluid level indicator switch.
4. Prise off the spring clip securing the master cylinder pushrod (Fig. L:21).
5. Release the two nuts and spring washers holding the master cylinder to the bulkhead or servo cylinder and lift away the master cylinder.

Installation

This is a reversal of the removal procedure, noting that the brake pipe union nuts should be tightened to the correct torque and that the brakes should be bled afterwards as described previously.

Master Cylinder Overhaul (Fig. L:25)

1. Remove the master cylinder from the car as described previously.
2. Carefully prise the reservoir from the body of the master cylinder and remove the rubber seals from the inlet ports.
3. Clamp the master cylinder in a vice fitted with 'soft' jaws.
4. Depress the operating rod, to release any pressure on the stop screw and remove the stop screw through the forward inlet port.
5. Pull back the operating rod rubber boot to expose the piston retaining circlip. Use circlip pliers to remove the clip.
6. Slide out the operating rod, boot and washer.
7. Remove the rubber boot from the operating rod.
8. Withdraw the primary piston assembly from the master cylinder bore (Fig. L:22).
9. Tap the master cylinder against the palm of the hand or a block of wood or similar soft surface to remove the secondary piston assembly.
10. Unscrew the primary piston spring retaining screw and remove the screw and retaining sleeve from the spring. Remove the spring, retainer and seal from the primary piston and, taking great care not to damage the piston, gently lever the remaining seal from the piston.
11. Remove the secondary piston spring, retainer and seal and then gently lever the remaining seal from the piston.
12. Clean the master cylinder and pistons with methylated spirits or clean brake fluid. Do NOT use mineral based oils such as petrol or paraffin. Dry the components with a lint free cloth or blow them dry with an air line.
13. Inspect the pistons and master cylinder bore for visible score marks, ridges or corrosion. This is important. If any of these conditions is evident, the component must be renewed.
14. Fit new seals to the secondary piston and secure them with the spring and retainer. Make sure the seals are fitted correctly.
15. Lubricate the secondary piston with clean brake fluid and gently insert it into the cylinder bore, the spring leading.
16. Fit new seals to the primary piston securing them with the retainer, spring, sleeve and screw. Make sure they are fitted correctly.
17. Lubricate the primary piston assembly with clean brake fluid and gently insert it in the cylinder bore, spring leading.
18. Slide a new rubber boot on to the operating rod and then fit the rod in the master cylinder securing it with a new circlip fitted in front of the rod washer.
19. Fit the operating rod rubber boot to the end of the master cylinder.
20. Fit new rubber seals to the inlet ports of the master cylinder and then gently push the reservoir into place.
21. Refit the master cylinder assembly as described previously and bleed the brakes. After bleeding, depress the brake pedal and hold it depressed for approximately ten seconds. Examine the master cylinder to ensure that there are no signs of fluid leakage.

BRAKE PRESSURE CONTROL VALVE...[10]

Removal (Fig. L:23)

1. Jack up the front of the car and fit axle stands. Refer to the jacking information contained under ROUTINE

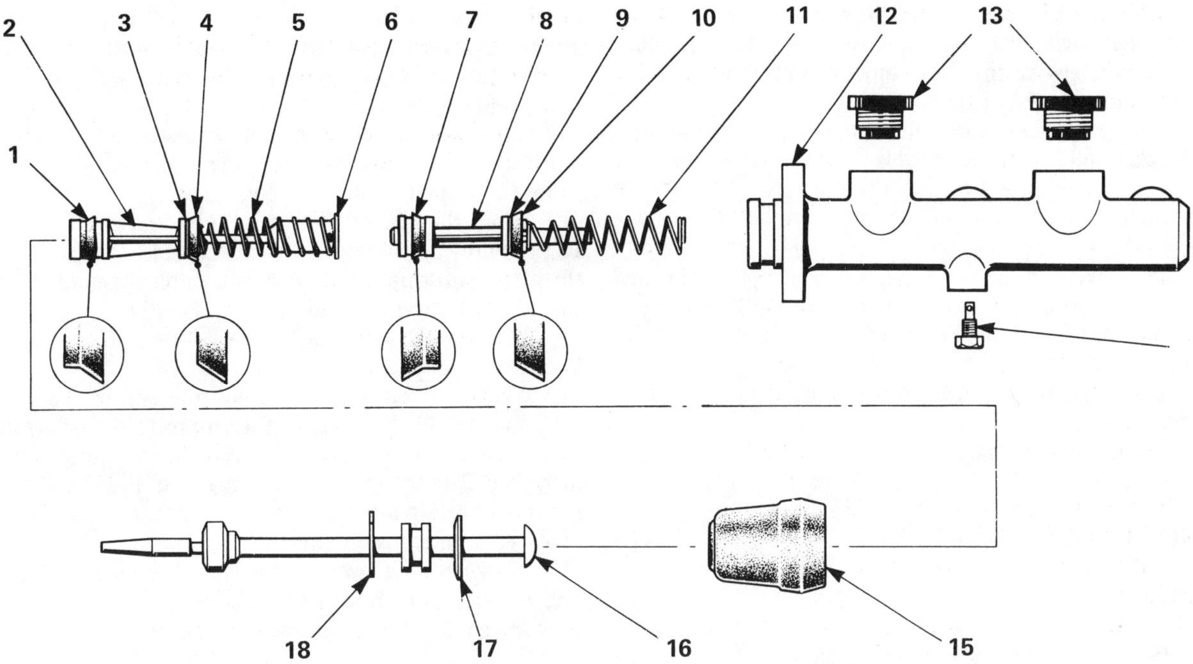

1. *Seal*
2. *Primary piston*
3. *Shim*
4. *Seal*
5. *Spring*
6. *Retainer*
7. *Seal*
8. *Secondary piston*
9. *Shim*
10. *Seal*
11. *Spring*
12. *Master cylinder*
13. *Reservoir seals*
14. *Piston stop pin*
15. *Boot*
16. *Operating rod*
17. *Washer*
18. *Circlip*

Fig. L:25 Exploded view of brake master cylinder primary and secondary piston components

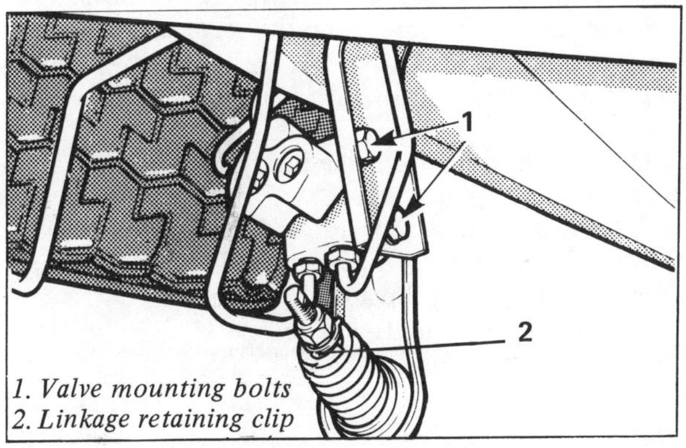

Fig. L:26 Light laden valve mounting connections

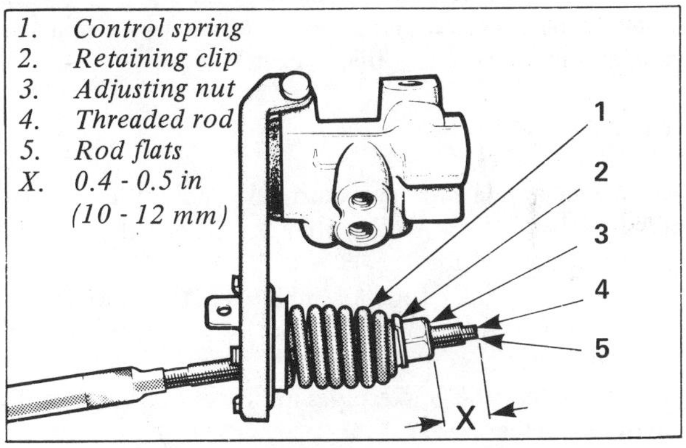

Fig. L:27 Light laden valve nut adjustment

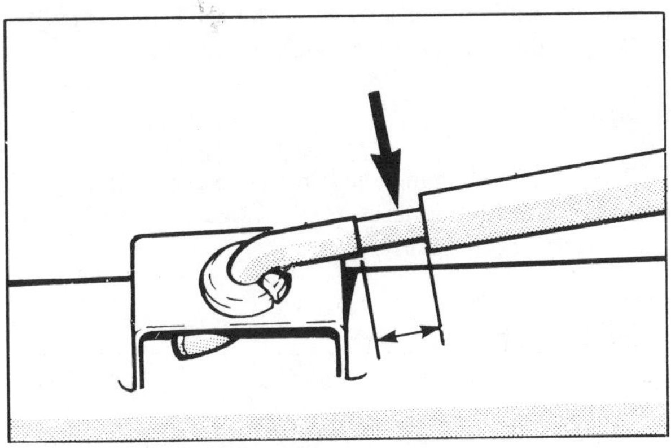

Fig. L:28 Linkage adjustment measurement point

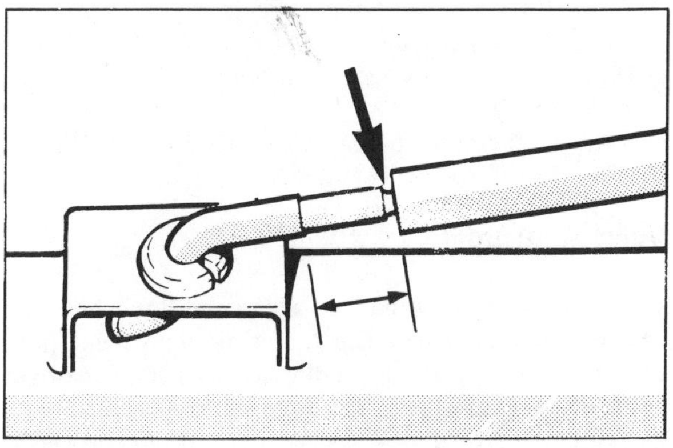

Fig. L:29 Linkage adjustment point with new springs

MAINTENANCE at the front of the book.
2. Place a thick cloth under the valve to catch any spilt brake fluid. Remove the brake pipes from the valve, plugging the ends to prevent fluid loss.
3. Remove the two bolts holding the valve to its mounting bracket and lift away the valve.

Installation

1. Fit the valve to its mounting bracket, tightening the bolts to the correct torque - see TECHNICAL DATA.
2. Reconnect the brake pipes and tighten their unions to the correct torque.
3. Bleed the brakes as described at the beginning of this chapter.
4. Lower the car to the ground.

LIGHT LADEN VALVE - VAN . . . . [11]

Removal

1. Disconnect the brake pipes from the valve by undoing the four unions (Fig. L:25), and drain the fluid into a container.
2. Remove the two bolts securing the valve to the body mounting bracket as shown in Fig. L:26.
3. Lower the valve and remove it by sliding the spacer tube assembly off the link rod in the axle bracket.
4. Detach the link rod from the axle bracket.
5. Check the axle bracket bush condition. If it is worn, it should be removed and a new one fitted by coiling the bush and pressing it into the bracket hole.

Installation

1. Apply a smear of grease to axle end of the link rod and a blob of grease to the other end. Refit the link rod into the bracket bush.
2. Slide the spacer tube assembly over the link rod and refit the valve to the body bracket in the reverse order of removal (Fig. L:26).
3. Reconnect the brake pipes in the reverse order of disconnection (Fig. L:25).
4. Refill the hydraulic system and bleed the brakes as detailed previously.
NOTE: Make sure that the valve is kept 'open' (valve actuating lever hard against the valve body) during the bleeding operation, by using a suitable clip or wire attached to the axle tube.
5. Adjust the valve linkage as detailed in the following section.

Valve Adjustment

The van must be in an unladen condition when adjusting the valve linkage. The following adjustment procedure is in three parts depending whether new or original parts are being used.
1. If new valve or linkage has been fitted, proceed as follows:
a) Measure the distance from the end of the threaded adjustment rod to the end face of the adjusting nut (Fig. L:27). The measurement should be 0.4 - 0.5 in (10 - 12 mm). If necessary, adjust by turning the nut.
b) Rotate the spacer tube until the distance between end of the spacer tube and the stepped shoulder of the link rod is positioned as shown in Fig. L:28.
c) Crimp the end of the spacer tube at a point adjacent to the knurled section of the tube, against the flats of the threaded adjustment rod. This will prevent the tube from turning and altering the adjustment.
2. Adjustment of existing valve linkage, proceed as follows:
a) Measure the distance between the end of the spacer tube and stepped shoulder of the link rod (Fig. L:28).
b) If necessary adjust by holding the threaded adjustment rod flats to prevent it turning (Fig. L:27) and rotating the adjusting nut, one way or the other until the correct measurement is obtained.
NOTE: When new rear springs have been fitted, the setting position for the end of the spacer tube is to align it with the groove in the link rod (Fig. L:29).

BRAKE SERVO . . . . [12]

Removal

1. Remove the brake pedal to push rod spring clip.
2. Remove the brake master cylinder from the servo as detailed previously under 'Master Cylinder - Removal'.
3. Undo the four nuts securing the servo to the mounting bracket (Fig. L:30).
4. Detach the non-return valve and vacuum hose from the servo.
5. Detach the spring from the linkage arm at the rear of the servo.
6. Pull the servo unit forward off the mounting bracket and disconnect the servo operating rod from the linkage by removing the spring clip and clevis pin.
7. Remove the servo unit.

Overhaul

As detailed previously in the 'Introduction' special tools are needed to overhaul the servo unit which the DIY mechanic is unlikely to possess. If the servo unit is faulty or is leaking fluid, a new unit should be fitted.

Installation

The servo unit is installed in the reverse order of removal and the hydraulic system refilled and bled afterwards as detailed previously.

HANDBRAKE CABLE . . . . [13]

Adjustment

1. Raise and support the rear of the vehicle as detailed

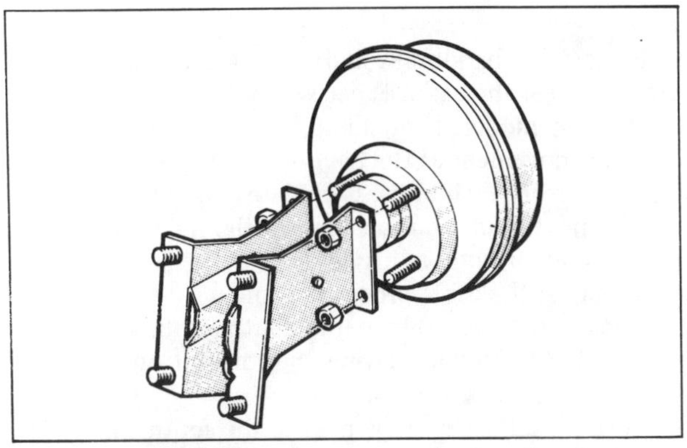

Fig. L:30 Brake servo to mounting bracket fitting

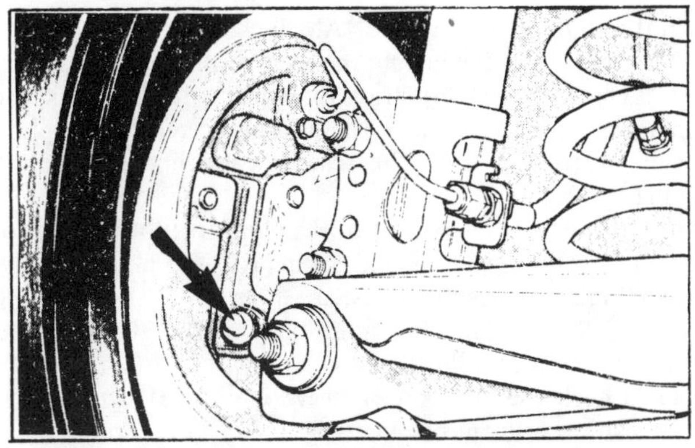

Fig. L:31 Location of handbrake backplate plunger

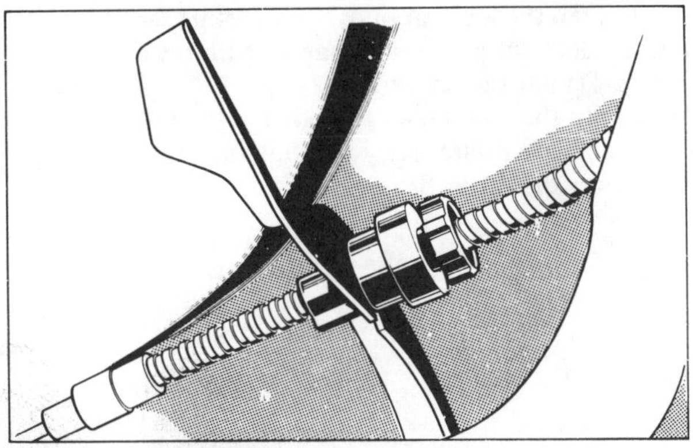

Fig. L:32 Handbrake cable adjuster - early type

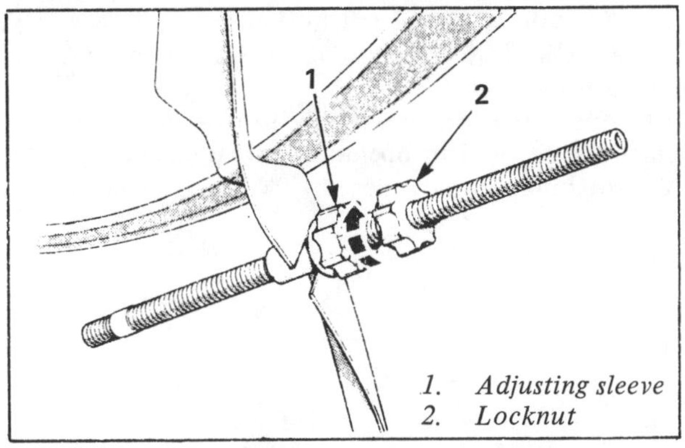

Fig. L:33 Handbrake cable adjuster - later type

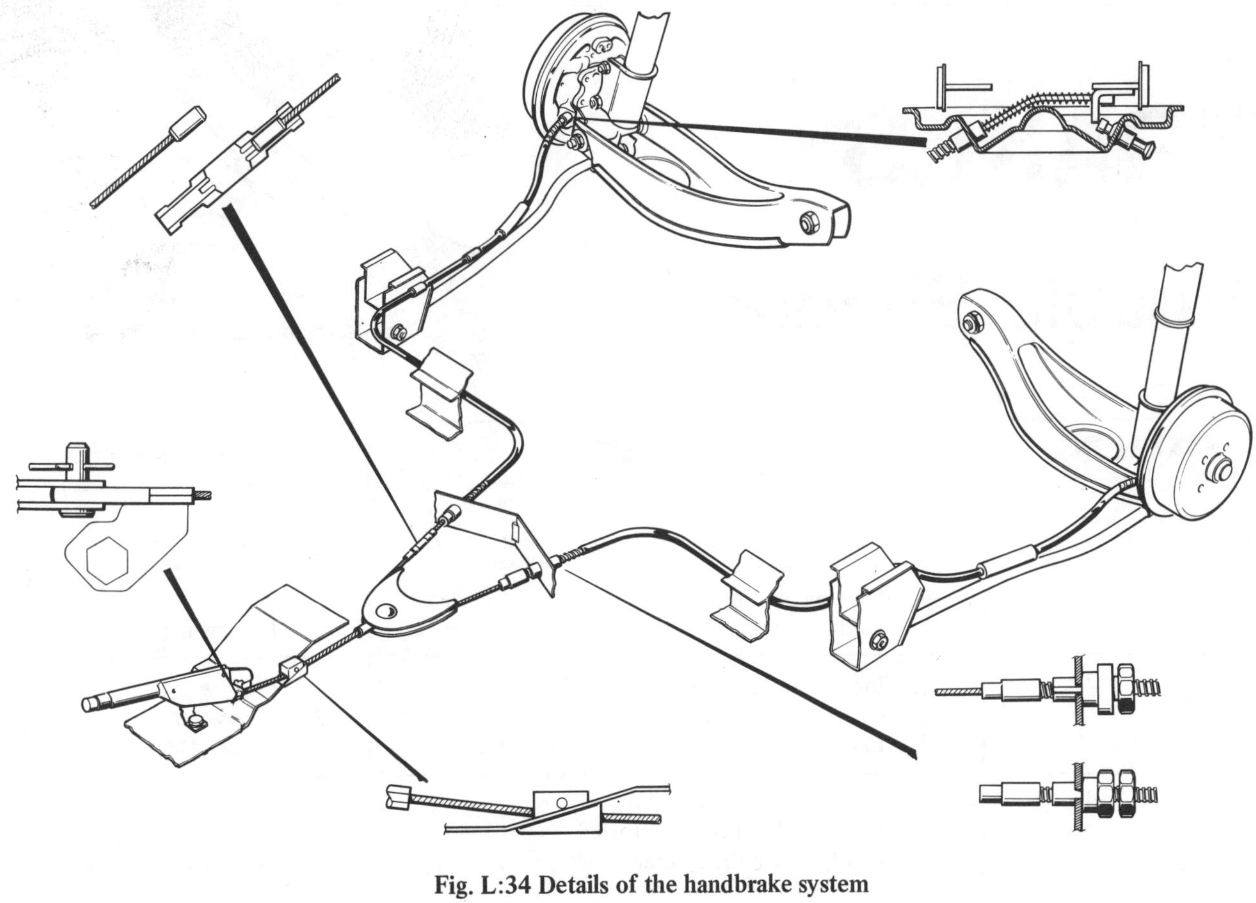

Fig. L:34 Details of the handbrake system

in the ROUTINE MAINTENANCE chapter.

2. Check that the handbrake lever is fully released.

3. On early models prior to January 1982, check the amount of movement at the handbrake relay lever plungers protruding from the rear of each brake backplate (Fig. L:31).

4. If the total movement of the plungers (both added together) is between 0.04 and 0.2 in (1,0 - 2,5 mm) then the handbrake is correctly adjusted.

If the movement is more or less than the specified figure, adjust as follows:

a) Check that the keyed sleeve is correctly engaged in the cable abutment bracket (Fig. L:32).

b) Unlock and separate the adjusting nut using a screwdriver between the nut and adjacent sleeve.

c) Turn the adjusting nut to eliminate any slack in the cable until it is just possible to rotate the backplate plungers by hand.

d) Apply the handbrake fully to seat the adjuster nut into the sleeve. This operation will automatically reduce the total plunger movement by 0.02 to 0.04 in (0,5 - 1,0 mm).

e) Release the handbrake and recheck the plunger movement. Readjust the cable as necessary.

5. On later models from January 1982 (Build Code CL) check for movement at the handbrake relay lever plungers protruding from each brake backplate (Fig. L:31). No perceptible movement of the plungers should be felt if the handbrake is correctly adjusted.

To adjust the cable proceed as follows:

a) Slacken off the cable at the abutment bracket adjuster (Fig. L:32) by unscrewing the locknut and adjuster. Do not use a screwdriver for this.

b) Turn the adjuster sleeve next to the abutment bracket until the total movement of both plungers added together equals 0.02 - 0.08 in (0,5 - 2,0 mm).

c) Tighten the locknut onto the adjusting sleeve by hand until the locknut protrusions engage with the sleeve indentations. Do not tighten further.

d) Apply the handbrake fully then release it. Check that the rear wheels rotate freely without any brake 'bind'.

6. Lower the car to the ground.

BRAKES

Trouble Shooter

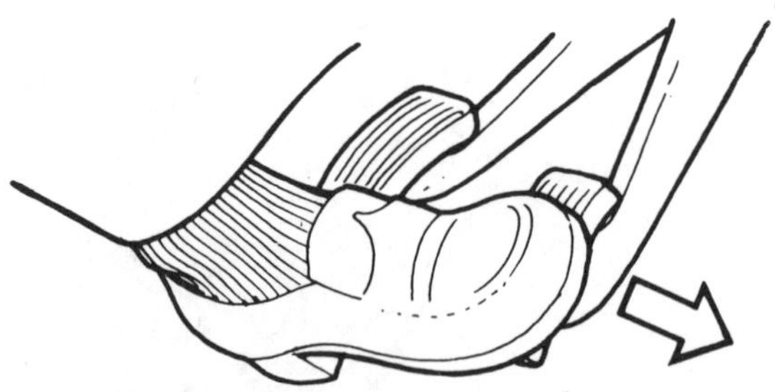

FAULT	CAUSE	CURE
Excessive brake pedal travel	1. Brakes need adjusting or replacement. 2. Air in system. 3. Leaking or contaminated fluid. 4. Faulty master cylinder.	1. Adjust or renew brake shoes. 2/3. Bleed hydraulic system. 4. Fit new master cylinder.
Brake fade	1. Incorrect pad or lining material. 2. Old or contaminated fluid. 3. Excessive use of brakes or car overloaded.	1. Fit new pads or shoes. 2. Renew brake fluid. 3. Check vehicle load.
Spongy brake pedal	1. Air in hydraulic system. 2. Shoes badly lined or distorted. 3. Faulty hydraulic cylinder.	1. Bleed system. 2. Fit new pads or shoes. 3. Check hydraulic circuit.

FAULT	CAUSE	CURE
Brake pedal too hard	1. Seized wheel cylinder or caliper piston. 2. Glazed friction material. 3. No clearance on master cylinder operating rod.	1. Replace seized component. 2. Fit new shoes/pads. 3. Adjust push rod if possible.
Brake pedal requires pumping or pedal sinks to floot	1. Brakes wrongly adjusted. 2. Air in hydraulic system. 3. Fluid leak from component or brake pipe. 4. Loss of fluid from master cylinder. 5. Seized caliper piston.	1. Adjust brakes. 2. Bleed system. 3/4. Check hydraulic circuit and replace parts as necessary. 5. Replace caliper.
Brakes grab when applied	1. Contaminated friction material. 2. Wrong linings fitted. 3. Scored drums or discs. 4. Rusty drums or discs.	1/2. Replace (don't clean) pads or shoes. 3. Fit new drum or disc. 4. Clean or replace drums or discs.
Brake squeal	1. Worn retaining pins (disc). 2. Faulty damping shims or shoe retaining clips. 3. Dust in drum. 4. Loose backplate or caliper. 5. Linings contaminated with brake fluid or hub grease.	1. Fit new pins. 2. Fit new shims or clips. 3. Remove dust from drums/shoe. 4. Tighten caliper or backplate. 5. Replace leaking cylinder and contaminated shoes or pads.
Brake judder	1. Distorted discs or drums. 2. No clearance at master cylinder operating rod. 3. Shoe tension springs either broken or weak. 4. Wheel cylinder or caliper piston seizing. 5. Faulty self-adjusting mechanism. 6. Seized handbrake mechanism.	1. Replace discs or drums. 2. Adjust rod if possible. 3. Replace tension springs. 4. Fit new caliper or cylinder. 5. Check mechanism. 6. Check handbrake operation.
Brake pull to one side	1. Contaminated friction material on one side (grease, oil or brake fluid). 2. Loose backplate. 3. Seized cylinder. 4. Faulty suspension or steering.	1. Replace shoes/pads in axlesets. 2. Tighten backplate. 3. Replace seized cylinder. 4. Check suspension and steering.
Handbrake ineffective	1. Worn rear shoes or pads. 2. Brakes require adjusting. 3. Faulty handbrake linkage. 4. Cable or rod requires adjustment.	1. Fit new pads/shoes. 2. Adjust brakes. 3. Check linkage and operating mechanism. 4. Adjust cable or rod.
Servo (where fitted) late in operation	1. Blocked filter. 2. Bad vacuum sealing or restricted air inlet.	1. Clean or replace filter. 2. Tighten vacuum hose connections and check hoses.
Loss of servo action	1. Air leak in servo - vacuum low.	1. Either overhaul servo or replace.
Loss of fluid (Servo only)	1. Servo seal failure. 2. Scored servo bore.	1. Replace or overhaul servo.
Loss of fluid - general	1. Wheel cylinder or caliper seal failure. 2. Damaged or corroded fluid pipes.	1. Replace wheel cylinder/caliper. 2. Inspect and fit new pipes.

General Electrics

INTRODUCTION .... [1]

The Escort has a Negative-earth electrical system, and all components are of conventional design and fitting. Certain components such as the instrument panel, incorporate printed circuit boards and extra care must be taken when removing and refitting bulbs or wires, also when dismantling the component, to avoid damaging the circuit board.

The direction indicator and flasher units are housed in the central fuse box in the engine compartment along with additional relays according to the model and level of equipment.

All of the electrical components are straightforward for the DIY mechanic to replace, and no special tools are needed.

When working on any part of the Escort electrical system, it is very important that the battery earth terminal is disconnected first to avoid any chance of damage due to accidental short circuits. Note that some of the circuits are 'live' even with the ignition switched off.

CHARGING THE BATTERY .... [2]

A high-speed battery charger must not be used as a starting aid. A high-speed charger may only be used if the battery has been completely disconnected from the car's electrical system. Certain types of 'maintenance-free' batteries, for example, the lead-calcium type, can be seriously damaged by high-speed chargers.

It is preferable to slow or trickle-charge a battery as then there is little chance of damage to the plates, provided that the electrolyte is at the correct level in all cells.

When charging a battery from a charger, make sure that the following precautions are observed:

1. Check that the charger output voltage is the same as the nominal voltage of the battery.
2. The charger Positive (+) lead is connected to the Positive (+) battery terminal.
3. The charger Negative (-) lead is connected to the Negative (-) battery terminal.

When boost starting a car using another battery, the following precautions must be observed to avoid the possibility of serious damage to the charging system or electrical components of the car:

1. The booster battery must be of the same nominal voltage as the car's battery.
2. The interconnecting cables (jump leads) are of sufficient capacity to carry the starting current.
3. The cables are connected, one at a time, to the booster battery first, not the other way round.
4. The cables are connected between the battery terminals in the following order, Positive (+) to Positive (+), and then Negative (-) to an earth point on the car body (Fig. M:1).
5. The engine speed must be reduced to 1000 rpm, or below, before disconnecting the booster battery.
6. The car's battery MUST NEVER be disconnected while the engine is running.

BULB & LAMP UNIT REPLACEMENT ..... [3]

It is good practice to carry a stock of spare bulbs in a padded container, and know how to replace them as well. You can buy packs of light bulbs suitable for your car, or make up your own kit. Since you are most likely to notice that a bulb has gone when it is dark, carry a torch in the car.

If any of the lenses are cracked or the reflectors corroded, they should be replaced.

Headlamps/Parking Lamp

The headlamp bulb can be easily replaced from within the engine compartment. Pull the multi-plug from the back of the bulb. Remove the rubber gaiter (Fig. M:2). Rotate the spring clip or push the spring clip prongs together, dependant upon type. Pull the bulb from the lamp housing.

Do not touch the glass part of the bulb because the finger marks will darken with the heat of the bulb in use and markedly reduce the light output.

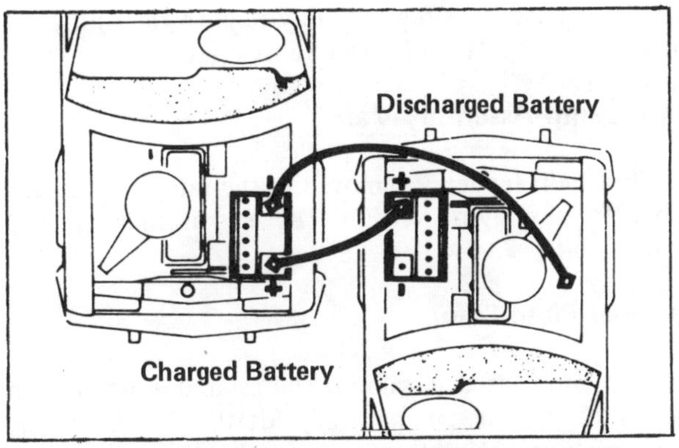

Fig. M:1 Battery jump lead connections

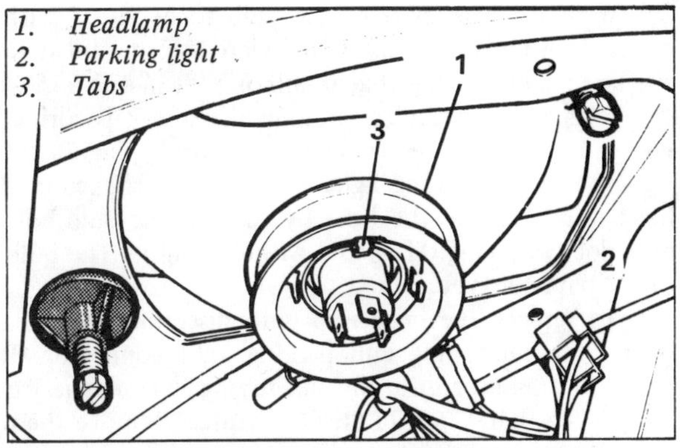

Fig. M:2 Headlamp bulb socket

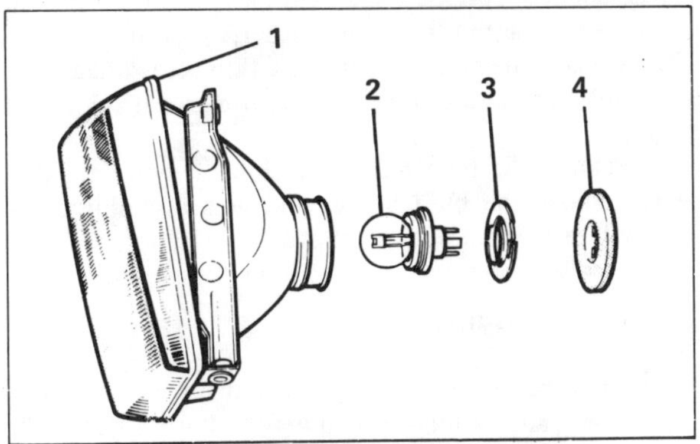

1. Lens 3. Locking clip
2. Bulb 4. Gaiter

Fig. M:3 Tungsten type headlamp

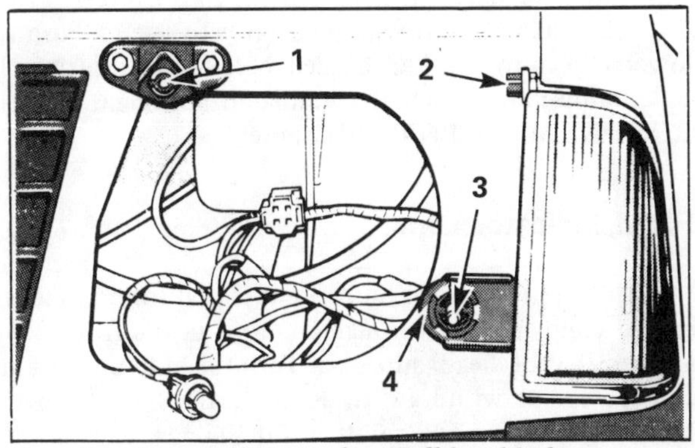

1. Upper adjuster 3. Ball head bolt
2. Upper clip 4. Indicator support

Fig. M:4 Headlamp mounting

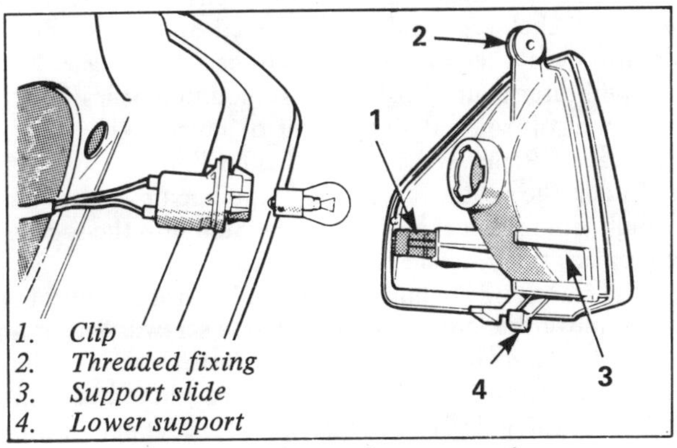

Fig. M:5 Indicator lamp mounting

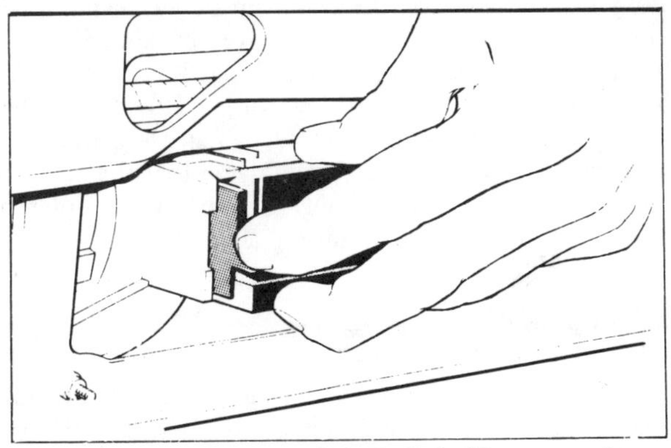
Fig. M:6 Saloon rear light bulb holder removal

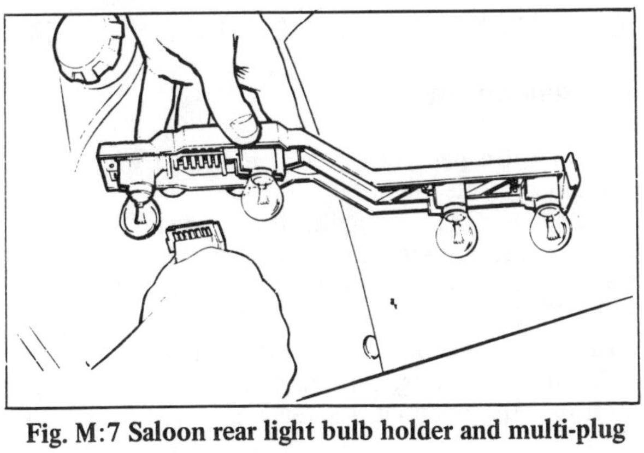
Fig. M:7 Saloon rear light bulb holder and multi-plug

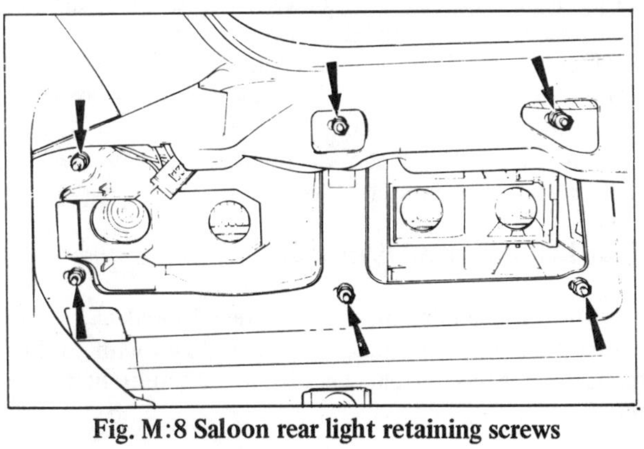
Fig. M:8 Saloon rear light retaining screws

Wipe the glass part of the bulb to be fitted with a clean cloth moistened with methylated spirits. Fit it into the lamp shell, noting that it will only fit in one position.

Engage the retaining clip, making sure it is correctly engaged. Fit the rubber cover and the wiring plug.

The front parking light is fitted to a socket in the headlamp. Remove the bulb by twisting the bulb holder anti-clockwise (Fig. M:3) and withdrawing it. The bulb is of the bayonet type.

To replace the headlamp unit, disconnect the headlamp bulb multi-plug and parking light connector. Remove the radiator grille by unclipping it from the front panel - see BODY & FITTINGS chapter. Remove the upper adjuster retaining bolts (Fig. M:4) and the side clip by rotating 90 degrees. Pull the assembly forward to release from the lower clip. Fit the headlamp in the reverse order. Twist the side clip through 90 degrees to secure. Push the lower clip onto the ball headed bolt. After refitting the bulb multi-plug, check the headlamp alignment, see the ROUTINE MAINTENANCE Chapter.

Front Indicator Lamp

The front indicator bulbs are replaced from inside the engine compartment through a cutout in the front panel adjacent to the headlamp unit. Turn the bulb holder anti-clockwise and withdraw it. Remove the bulb from the holder by pushing and rotating. Fit the new bulb and the holder in the reverse order.

To replace the front indicator lamp unit, first remove the bulb. Remove the headlamp as described in the previous section. Remove the clip from the headlamp lower mounting ball headed bolt. Remove the lower adjuster by twisting the plastic mounting collar. Release the lamp by levering the latches with a screwdriver (Fig. M:5).

If a new lens is to be fitted, mount the lens on the support bracket. Refit the assembly so that the clips engage. Reassemble the assembly in the reverse order or removal, then adjust the headlamp alignment.

Rear Lamp Assembly-Saloon

Access to the rear panel bulbs is through the rear luggage compartment. Push the printed circuit retaining latches inwards (Fig. M:6) and pull the bulb holder assembly away from the lamp body. Bulbs are of the bayonet fitting type. Release the holder, guiding the bulbs into their reflectors. Clip the latch into place.

Replace the lamp assembly bulb holder by disconnecting the multi-plug (Fig. M:7). Remove the lens assembly by removing the six retaining nuts (Fig. M:8). If fitting a new lens, ensure that the sealing ring is fitted correctly. Refit in the reverse order of removal.

Rear Lamp Assembly-Estate

The bulb holder on estate models is held by two retaining tabs. Push the holder up then down while pulling forward. The holder can be withdrawn. Refit in the reverse sequence.

To remove the lamp assembly remove the four cross head screws (Fig. M:9). Fit using a new gasket.

Rear Lamp Assembly-Van

The bulb holder is removed in the same way as described for the saloon version. The lens assembly is held by four bolts.

Number Plate Lamp

The lamp fitted to saloons and estated is mounted in the middle of the bumper (Fig. M:10). To replace the bulb, reach up behind the bumper and pull free the multi-plug. Twist the bulb holder base to remove. The bulb is of the glass socket type. Pull the bulb glass from the plastic base. Refit the base in the reverse order.

To replace the lamp unit insert a thin screwdriver between lamp body and bumper, then lever free. Disconnect the plug.

To replace the bulb on van versions, it will be necessary to remove the bulb holder and lens assembly complete, by levering as described above.

Interior Lamp Bulbs

Replace the glove compartment bulb by opening the compartment and pulling the bulb free. Push a new bulb into the socket.

The hazard warning light bulb is mounted beneath a red lens on top of the steering column cowl. Pull the cover upward, taking care not to damage the lens. The bulb is of the glass socket type and can be pulled out. Push a new bulb into the socket and push the lens back into place.

The heater control light bulb is mounted beneath the heater control panel. Push the control levers to the top of their travel. Remove the motor switch knob by pulling off. Unclip the heater fascia trim by pulling upwards. Pull the bulb from the socket. Push a new bulb into the socket then replace the trim and heater controls.

The interior light bulb is mounted beneath a pushfit lens. Remove the lens by levering with a screwdriver, protecting the paintwork using a suitable block or cloth. The lamp assembly can be removed after the wiring has been unclipped. The bulb is of the bayonet type, clipped within the lamp.

FUSES, RELAYS & FLASHER UNIT ......[4]

Fuse Replacement

Failure of a particular fuse is indicated when all of the circuits protected by it fail.

If a new fuse fails, establish the cause and rectify the fault before fitting a second replacement, which should be of the correct value. Do not fit a fuse of a higher rating then specified.

The fuse box is mounted in the engine bay on the driver's side inner wing. As well as providing easy access to the fuses, the box holds the relays and flasher unit. Re-

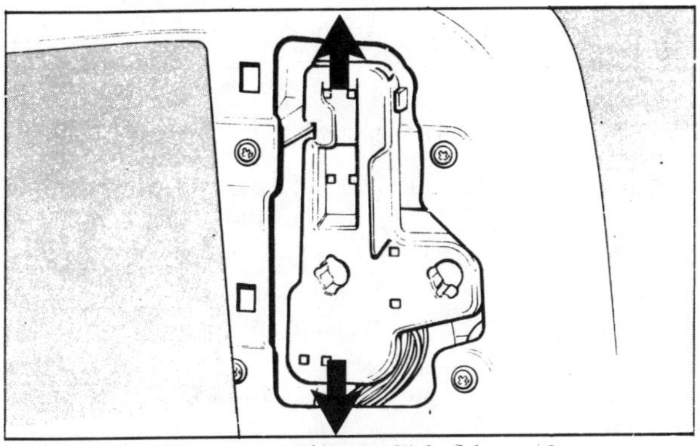

Fig. M:9 Estate rear light bulb holder and screws

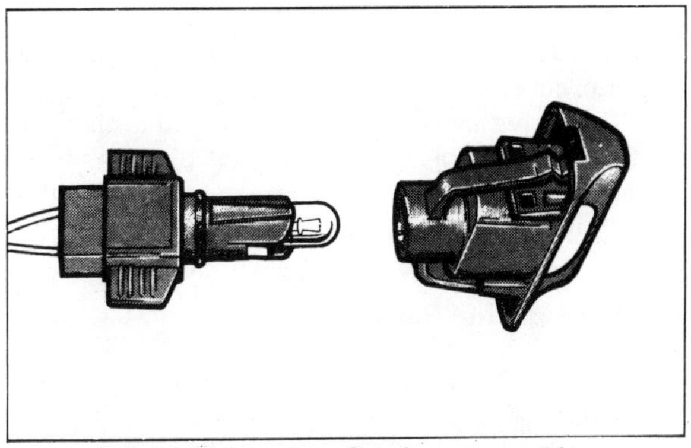

Fig. M:10 Number plate lamp assembly

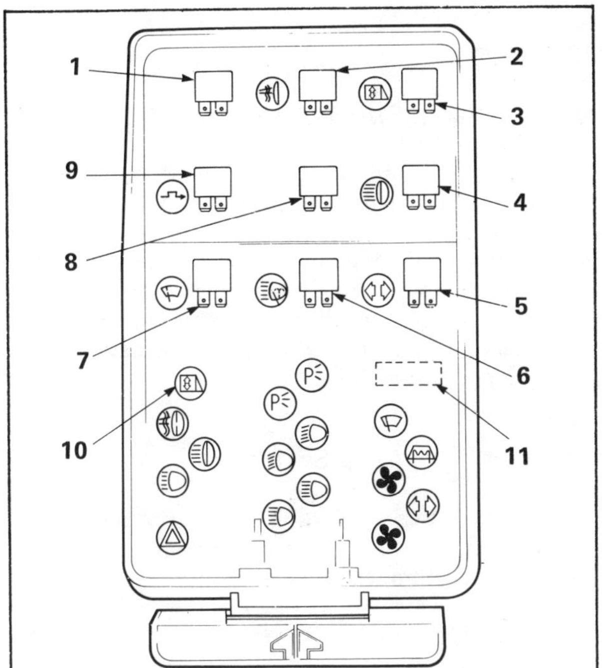

1. *Spare*
2. *Auxiliary fog lamp relay*
3. *Electric window relay*
4. *Auxiliary driving lamp relay*
5. *Direction indicator relay*
6. *Headlamp wash relay*
7. *Wiper delay relay*
8. *Spare*
9. *Ingition relay*
10. *Overload circuit breaker (electric window system)*
11. *Overload circuit breaker (central locking system)*

Fig. M:11 Fuse box showing fuses and relays

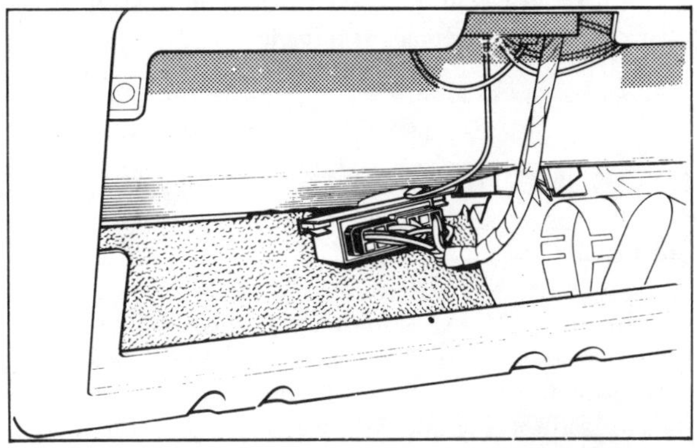

Fig. M:12 Central locking relay mounting

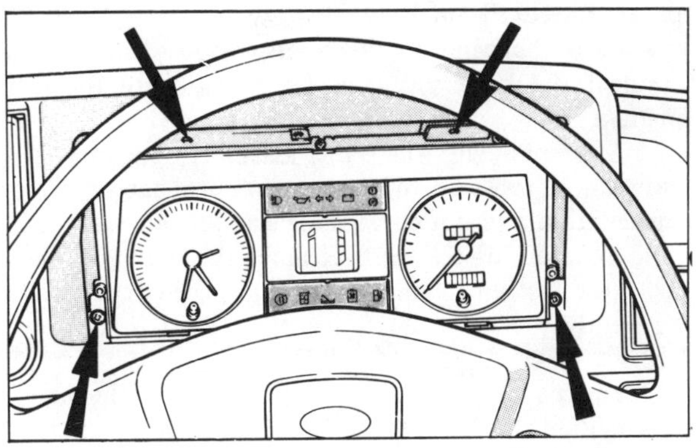

Fig. M:13 Instrument cluster screws

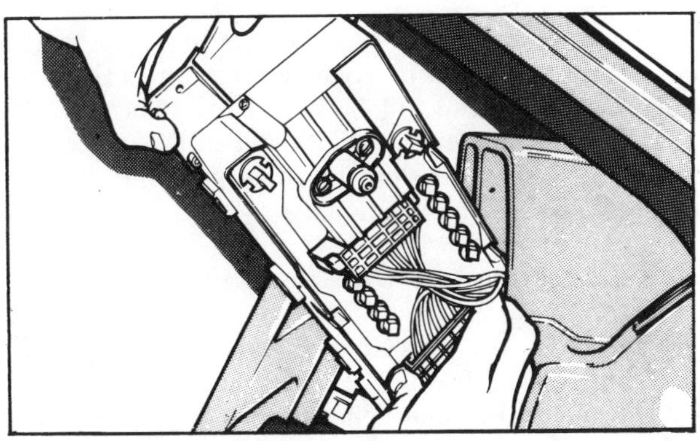

Fig. M:14 Instrument cluster removal showing plugs and bulbs

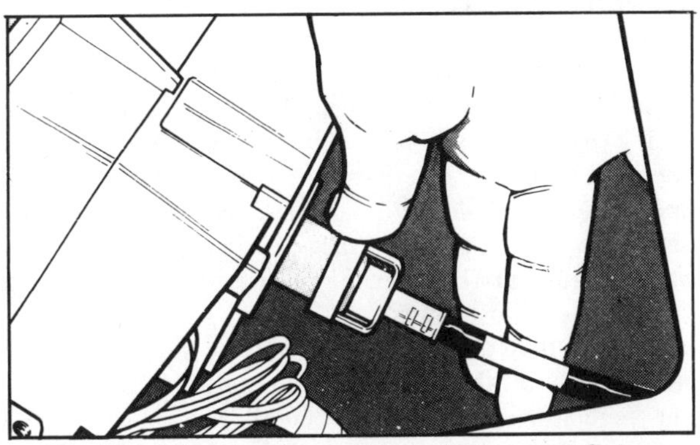

Fig. M:15 Speedometer cable to instrument fixing

lays and fuses can be identified by the symbols on the fuse box cover (Fig. M:11).

The relays and fuses are a pushfit within the housing. Remove by withdrawing vertically. Fit the component ensuring that the pins engage. Check operation of the relay and component.

Central Locking System Relay

The central locking operating relay is mounted beneath the passenger's side dashboard. Remove the dashboard trim panel and pull the relay from its spring clip (Fig. M:12). Disconnect the relay multi-plugs and withdraw. Refit in the reverse order. Note that the multi-plug connectors are colour coded for refitting. After refitting check operation. Replace trim panel.

INSTRUMENT PANEL.................. [5]

Removal

Remove the instrument panel surrounding bezel by removing the two upper screws. Pull the bezel back to free from the lower clips. The instrument cluster is held by two screws (Fig. M:13). Release the speedometer cable from the back of the panel, see the BODY & FITTINGS chapter. Pull the cluster gently away from the panel and disconnect the wiring multi-plugs. Remove the instrument panel cluster (Fig. M:14).

Refit the panel in the reverse order of removal. Reconnect the battery and check operation of the instruments.

After removing the panel assembly, the printed circuit can be replaced. With the panel held face down on a surface that will not damage the glass, remove all the bulb holders and plug retainers. Remove the nuts and washers from the instrument mounting studs. The circuit can then be gently pressed from fixing pegs. Refit in the reverse order.

The panel glass can be replaced with the panel in place. Remove the bezel as described above. Remove the exposed retaining screws and pull the glass from the panel. Refit in the reverse order.

Instrument Replacement

The instruments on the panel can be replaced after removing the instrument cluster from the dashboard, then taking off the glass cover. Each instrument unit is held by the nuts holding the terminals to the printed circuit. The speedometer is held by a pair of screws on either side of the cable mount. Remove the screws and remove the relevant instrument. Refit in the reverse order. Do not overtighten the securing nuts.

Note that the instrument voltage regulator unit is mounted with the fuel and temperature gauge units. If the regulator requires replacement, the gauges must be replaced at the same time.

IGNITION SWITCH .... [6]

Replacement

The ignition switch key barrel can be removed separately from the steering lock body as follows:

1. Remove the steering column lower shroud.
2. Fit the ignition key and turn it to position 1.
3. Use a flat bladed screwdriver to depress the retaining clip.
4. Holding the clip depressed, carefully pull the key and barrel out.
5. Fit a new barrel with the key in place, pushing it into the housing until the rim touches the lock housing.

SWITCHES .... [7]

Steering Column Switch Assembly

The direction indicator, horn, warning flasher, lights, windscreen wipers and washers are controlled by the multi-switch at the top of the steering column. Remove the steering column shrouds. Disconnect the multi-plugs and remove the two screws holding the switches together (Fig. M:17). Fit the switch in the reverse order. Do not overtighten the securing screws. Test the operation of the switch before refitting the shrouds.

Instrument Panel Light Switch

Remove the two screws holding the trinket tray. Pull the knob from the switch. Remove the securing nut. Withdraw the switch through the tray aperture. Pull the connector from the switch.

Refit the switch in the reverse order of removal.

Luggage Space Lamp Switch

Open the tailgate and remove the trim panel covering the switch. Disconnect the wire. Remove the retaining screw (Fig. M:18). Withdraw the switch. Refit the switch in the reverse order.

Electric Window Lift Switch

The electric window switches are mounted in the door arm rest panel. Remove by levering with a flat bladed screwdriver. Place a suitable block or cloth beneath the screwdriver to protect the trim material. Disconnect the multi-plug. Refit by pushing into the space until it engages. Check the switch operation.

Central Locking Door Switch

Fully raise the driver's door window. Remove the door trim as described in the BODY & FITTINGS chapter. Disconnect the multi-plug to the switch, then the single lead and the cable securing clips (Fig. M:19). Free

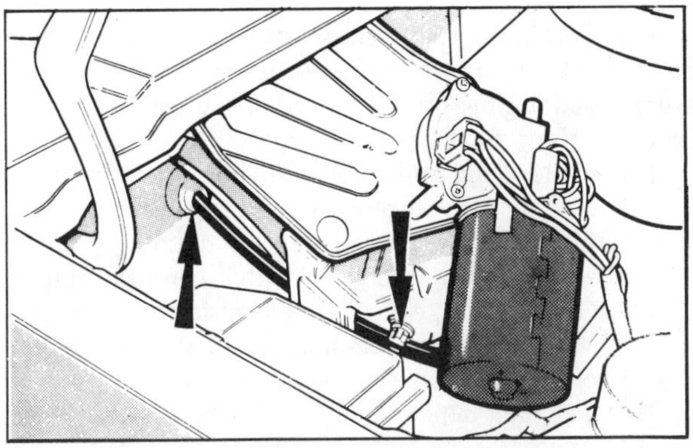

Fig. M:16 Speedometer cable routing

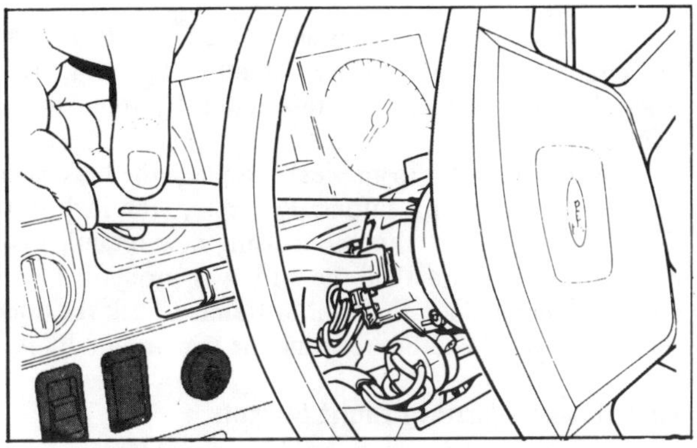

Fig. M:17 Steering column switch removal

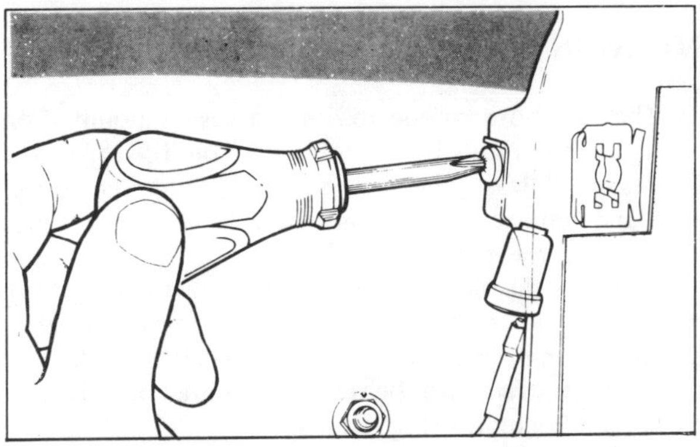

Fig. M:18 Luggage space lamp retaining screw

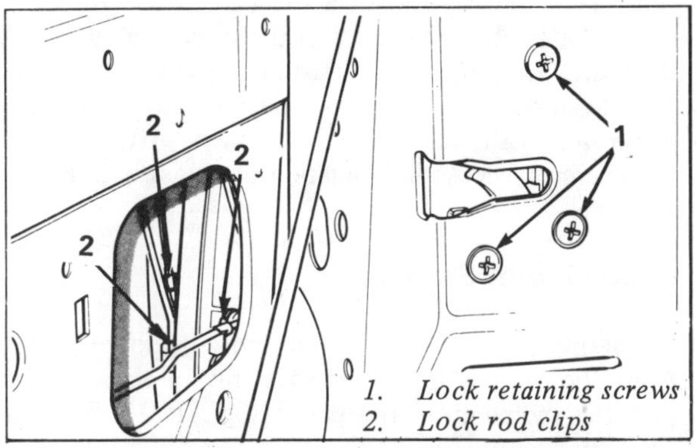

Fig. M:19 Central locking switch location

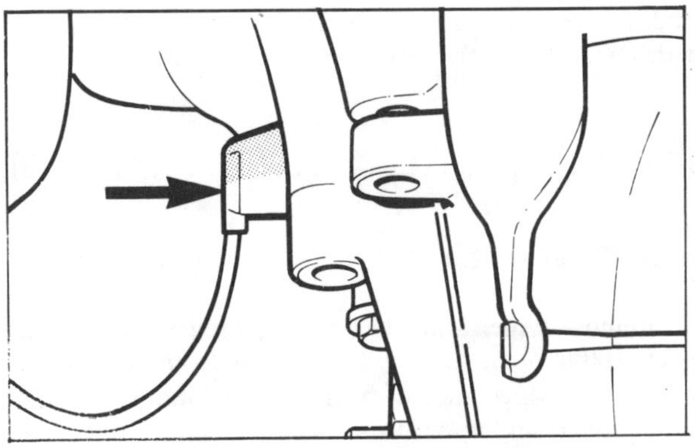

Fig. M:20 Reversing light switch

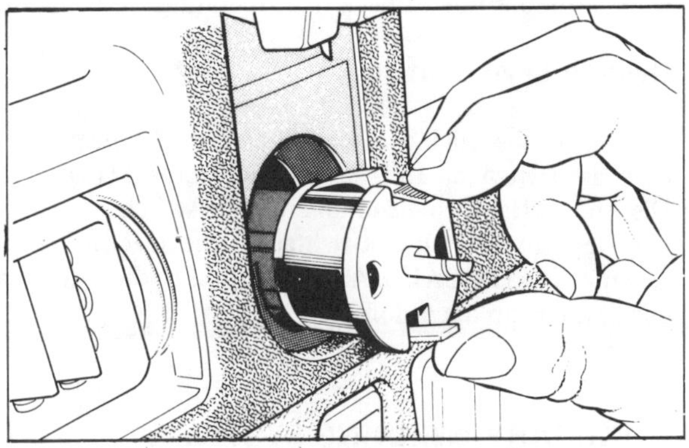

Fig. M:21 Heater motor switch

Fig. M:22 Auxiliary warning assembly control

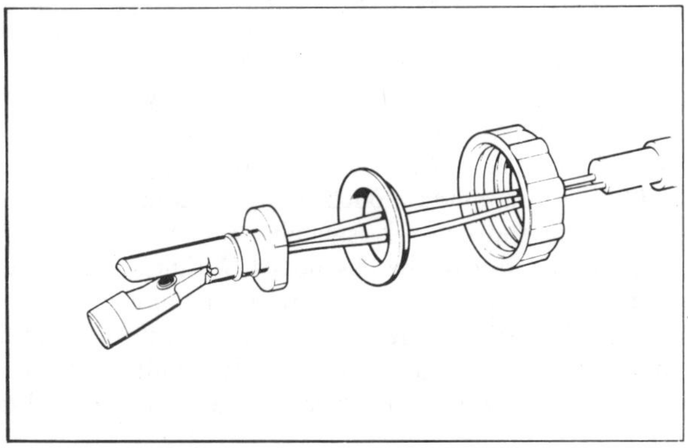

Fig. M:23 Expansion tank float unit

the rod locks from the clips and remove the lock retaining screws. Remove the lock assembly by withdrawing through door cutout. Free the switch by removing the two securing screws.

Refit the switch to the lock. Replace the lock assembly through the door cutout. Reconnect the lock in the reverse order. Secure the wiring using new retaining clips if necessary. Ensure that the wiring is properly located and not rubbing upon the door or frame. Check the operation of the switch before refitting the trim assembly.

Handbrake Warning Light Switch

Expose the handbrake lever. If a console is fitted, remove the retaining screws and withdraw. Prise the switch cover from the lever with a flat bladed screwdriver. Remove the lower screw, slacken the upper screw and slide the switch from the lever. Disconnect the lead.

Reconnect the lead and position the switch. The upper screw should locate in the slot. After tightening both screws, check the operation of the switch. Refit the cover.

Heater Motor Switch

Remove the heater motor switch knob by pulling from the shaft. Push the latches toward the centre and pull the switch from the panel (Fig. M:21). Disconnect the wiring multi-plug.

Reconnect the plug and push the switch into the housing. The latches should fully engage. Push the knob back into position. Check the operation of the switch.

Reverse Light Switch

Inside the engine compartment on the near side locate the switch on the side of the gearbox (Fig. M:20). Disconnect the wiring and unscrew the switch.

Refit the switch and tighten to the specified torque, see the TECHNICAL DATA. Reconnect the wiring and check the lamp operation.

AUXILIARY WARNING SYSTEM.........[8]

The auxiliary warning system monitors the coolant, washer fluid, fuel levels as well as the brake pad wear. All models apart from the 1.1 OHV engine cars, are also equipped with a low oil warning system.

The warning system is controlled by an electronic unit mounted behind the dashboard. To remove it, disconnect the battery, remove the radio speaker and grille (dash mounted models only). Disconnect the multi-plug from the control assembly. Remove the two nuts holding the unit to the dashboard. Withdraw, taking care not to knock it or the electronic components may be damaged. Refit in the reverse order, ensuring that the nylon nuts are not overtightened. The multi-plug connector must make a good connection with the control unit (Fig. M:22).

The warning system sensors can be easily replaced. The oil level warning sensor is on the end of the dipstick. Ensure that the plug on the end of the dipstick is firmly in place, with no trace of oil contamination.

The washer fluid and coolant are monitored by floats in the side of the washer reservoir and expansion or degas tanks. Replace the switch assembly by draining the relevant container, then levering the switch from the side of the tank (Fig. M:23). Note that expansion tank models are fitted with a threaded retaining ring that should be unscrewed before attempting to lever the float free.

The fuel tank sender unit is described in the FUEL SYSTEM chapter. The brake pad wear system is described in the BRAKES chapter.

INTERIOR HEATER FAN MOTOR........[9]

The fan motor is located in the casing mounted on the engine bulkhead. Remove the seal from the top of the air intake. Release the five springs holding the intake cover and remove at the front. Detach the motor multi-plug and unscrew the earth lead.

Unscrew the fan housing retaining nuts (Fig. M:24), and lift out the fan housing. Detach the fan casings halves by levering apart the plastic latches (Fig. M:25). Remove the retaining strap with the resistor mounted on top, then remove the fan motor (Fig. M:26).

Refit the fan motor to the housing. Refit the casing and push together. The casings are retained by metal clips which should be obtained from a Ford dealer. Check the condition of the sealing strip, replacing if necessary. Refit the assembly in the reverse order of removal.

WIPERS ....[10]

Front Wiper Motor

Remove wiper arm and blade assemblies. Remove nut covers fixing nuts, washers and spacers from pivot shafts. Disconnect wiper motor from wiring loom by pulling multi-plug from connectors. Remove two fixing bolts from wiper motor mounting and withdraw motor complete with linkage and mounting (Fig. M:27). Remove spacers from pivot shafts and place motor (linkage uppermost), on a suitable workbench.

Remove securing nut and withdraw motor arm from motor spindle.

Remove three bolts securing motor to mounting and withdraw motor (Fig. M:28).

Align motor spindle to park position. Position motor spindle through cut-out in mounting panel and secure motor with three retaining bolts (Fig. M:28). Fit motor arm in straight alignment to link assembly (Fig. M:29). Secure motor arm to spindle with retaining nut. Place spacers onto pivot shafts and position wiper motor and linkage into body ensuring that pivot shafts protrude through body fixing holes. Fit upper spacers and washers

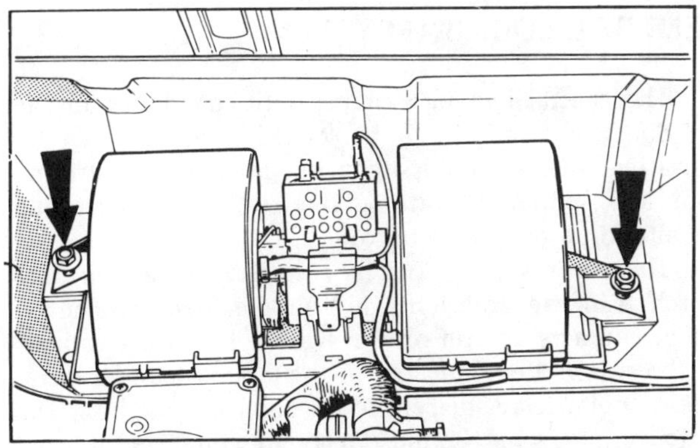

Fig. M:24 Fan motor retaining bolts

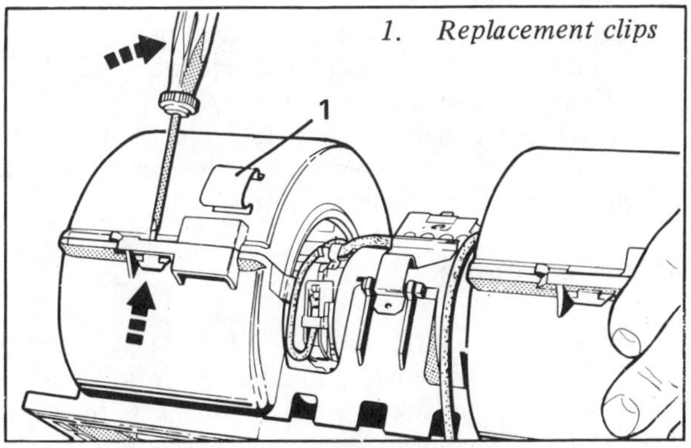

Fig. M:25 Fan cover removal

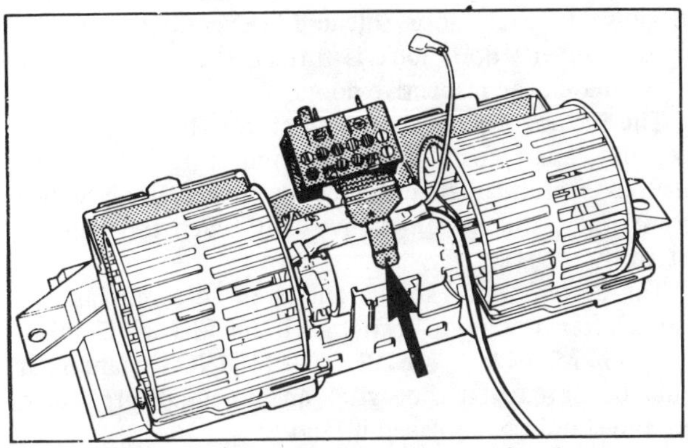

Fig. M:26 Motor retaining strap and resistor

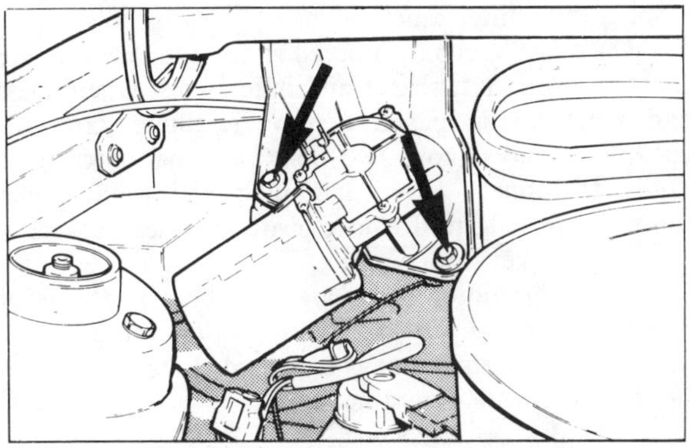

Fig. M:27 Front wiper motor mounting

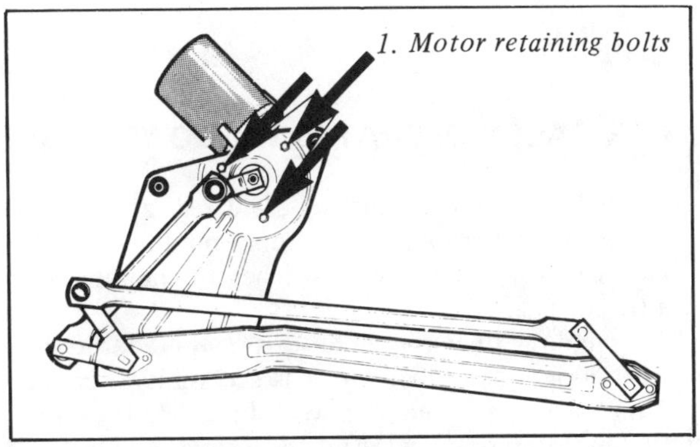

Fig. M:28 Dismantling front wiper assembly

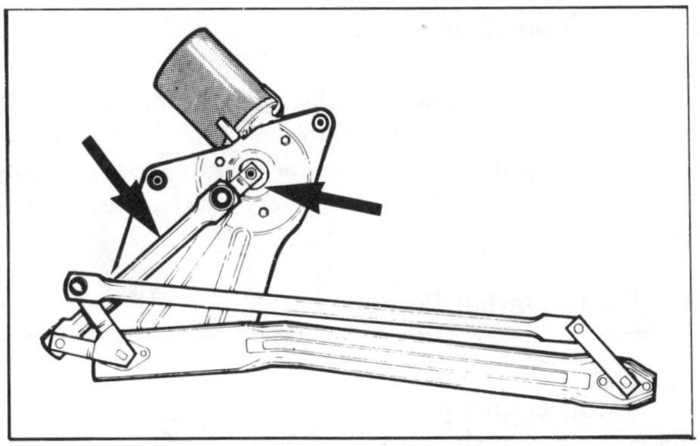

Fig. M:29 Motor arm in alignment

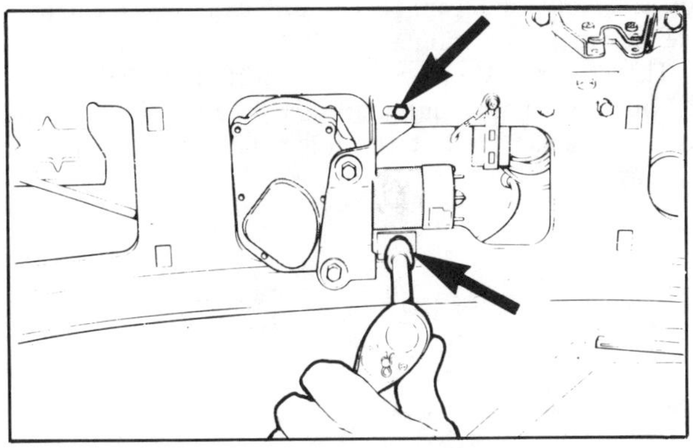

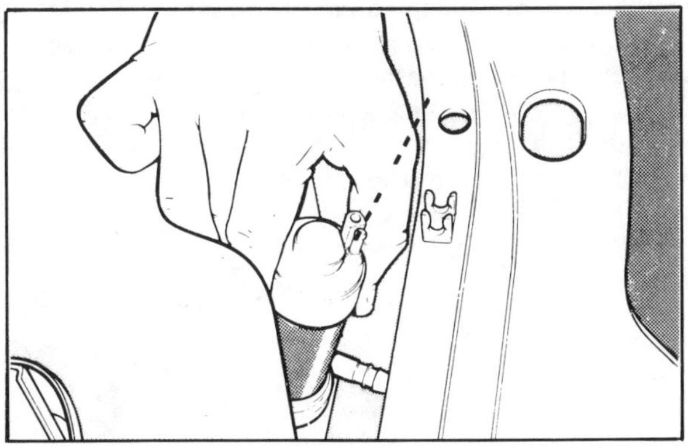

Fig. M:31 Headlamp washer pump removal

to pivot shafts and fit securing nuts finger tight.

Align bolt holes in wiper motor mounting panel with body fixings and secure using two retaining bolts. Fully tighten the pivot shaft securing nuts and fit nut covers. Reconnect loom to wiper motor. Fit wiper arm assemblies and check operation of wiper motor.

Rear Wiper Motor

Remove wiper arm assembly. Open tailgate and remove trim panel. Remove retaining screw from earth wire tag and remove bolts securing fixing brackets to tailgate (Fig. M:30). Disconnect loom multi-plug and remove wiper motor and bracket assembly from tailgate. Remove securing brackets from motor by removing two bolts and one nut. If necessary, remove grommet and pivot shaft from tailgate outer panel.

On estate versions, proceed as follows. Remove pivot shaft securing nut, spacer and outer seals. Open tailgate and remove trim panel. Remove retaining screw from earth wire and remove two bolts securing motor to rear door. Disconnect loom multi-plug and remove wiper motor from tailgate. Remove pivot shaft seal, spacer and securing bracket from motor.

Refit the motor and shaft assemblies in the reverse order of removal. Loosely fit the motor retaining bolts until the pivot shaft nut is tightened. Lubricate the rubber grommet with soap if necessary.

WASHERS .... [11]

Front Screen Washer Pump

Drain washer reservoir. Disconnect loom connector and washer pipe. Ease top of washer pump from reservoir and pull pump upwards to remove. Ensure grommet seats correctly in reservoir refit and reconnect pump. Refill reservoir. Check operation of pump.

Headlamp Washer Pump

Disconnect headlamp washer pump from loom multi-plug and drain washer fluid from reservoir. Slacken clamping screw from reservoir bracket and disconnect pipe from headlamp washer pump. Ease top of pump from reservoir and pull pump upwards to withdraw (Fig. M:31).

Ensure rubber seal is located correctly in reservoir and refit headlamp washer pump. Secure reservoir with clamping screw and reconnect headlamp washer pipe. Reconnect wiring, refill reservoir and check headlamp washer pump for correct operation.

CENTRAL LOCKING SYSTEM .... [12]

The central locking system (if fitted) allows all the the doors and tailgate to be locked simultaneously by activating their solenoids upon operation of the driver's door lock, either by turning the key or by depressing the interior door locking button.

The driver's door lock is connected to an earthing switch using an additional lug on the lock mechanism which contacts an arm on the switch. The two terminals on the switch are connected to the input side of a central locking pulse relay adjacent to the auxiliary fuse box. The outlet terminals on the pulse relay are connected to each door lock solenoid. When the driver's door is locked or unlocked, one of the two terminals on the switch is earthed, and the resulting current reversal releases a short current pulse to each door solenoid. Depending on which way the driver's door lock is turned, the solenoids will lock or unlock the respective doors.

The locking system carries no current for most of the time, and, apart from the wire connecting the driver's door lock switch to the pulse relay which carries Negative current only, the system is live only when the pulse is emitted by the relay.

The driver's door switch and door solenoids are accessible after the door trim panel and lock mechanism have been removed - see BODY FITTINGS chapter. It should be noted that the switch and solenoids are sealed units and should be replaced if faulty.

An overload circuit breaker is mounted in the fuse box, see the 'Fuses' section earlier for details of the units' location.

ELECTRICALLY OPERATED WINDOWS. . [13]

Electrically operated windows are a regular production option on the front windows only on L, GL, Ghia and XR3 saloons. The window switches are mounted on the centre console.

The window operating motor is mounted on the respective door inner panel which is accessible after the door trim panel has been removed. Full details of the operating motor and winder regulator removal and installation are given in the BODY FITTINGS chapter. The motor can be replaced independently of the winder regulator assembly and can be engaged with the regulator cable rack in any position.

The motor has an overload protection device which cuts out the motor automatically when resistance to movement is detected at the fully open or fully closed positions.

Explanation of Wiring Diagrams

The diagrams reproduced on the following pages have been taken from the new series of Autodata Wiring Diagrams which are now being supplied to the motor trade and auto electricians.

Autodata Wiring Diagrams are terminal diagrams without the interconnecting wires. The cable connected to each terminal is shown as a short line followed by an address code. The code identifies the destination component and terminal with the colour of the cable.

Each component on the diagram is identified by its graphical symbol and code and is positioned as far as possible according to its location on the car. The key for the component codes is shown overleaf.

COMPONENT CODE

GRAPHICAL SYMBOL

TERMINAL CODE

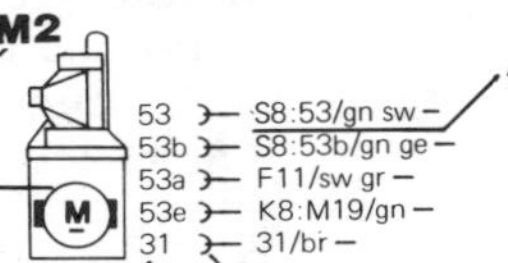

ADDRESS CODE

CONNECTION SYMBOL

COMPONENT CODE

The Component Code consists of a letter and number. The letter identifies the component group and the number identifies the specific component.

- A - Assemblies
- B - Transducers
- C - Capacitors
- E - Miscellaneous
- F - Protective Devices
- G - Generators
- H - Warning Lamps
- K - Relays
- M - Motors
- N - Regulators
- P - Meters/Gauges
- R - Resistors
- S - Switches
- T - Transformer
- V - Semi conductors
- W - Aerials
- X - Connectors
- Y - Solenoids

GRAPHICAL SYMBOL

The Graphical Symbol consists of an outline representation of the component enclosing a standardised symbol. The standardised symbols are based on those used in DIN Standards.

CONNECTION SYMBOL

The Connection Symbol identifies the type of connection used.

- Disconnectible (e.g. Lucar)
- Fixed (e.g. soldered joint)
- Plug and socket
- Multi-pole connector

TERMINAL CODE

The Terminal Code identifies the actual terminal marking to be found on the component. These markings normally conform to DIN Standard and are specific to particular circuits.

Many multi-pole connectors are identified with a numerical sequence.

Terminals are not always marked, particularly on simple components with a limited number of connections.

ADDRESS CODE

The Address Code identifies the destination and colour of a cable to another component terminal within the electrical system.

S8:53/gn sw

Component Code

Terminal Code

Colour Code

The Component Code shows the component to which the cable is connected at its other end.

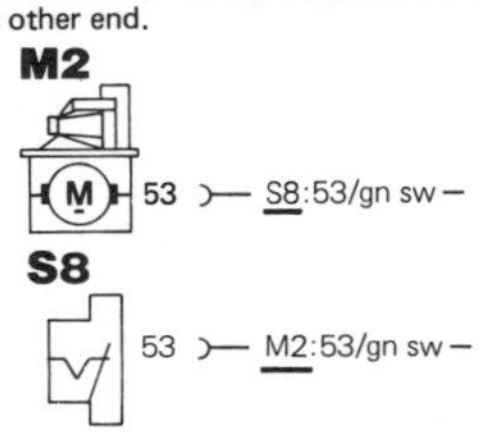

The Terminal Code shows the markings at the destination terminal.

53 S8:53/gn sw –

53 M2:53/gn sw –

The Colour Code shows the cable colour(s) according to the key shown below. If code appears as 'z' then connection is made by means other than cable e.g. printed circuit or internal connection.

53 S8:53/gn sw –

53 M2:53/gn sw –

bl = blue		rt = red	
br = brown		sw = black	
el = cream		vi = violet	
ge = yellow		ws = white	
gn = green			
gr = grey		z = printed circuit or internal connection	
nf = neutral			
rs = pink		y = high tension cable	

SYMBOLS & CODES

- line fuse
- resistor cable
- sw ws ⟩ sw ge — colour change
- sw ws (sw ge) — alternative colour
- sw — earth return cable
- direct earth (no cable) e.g. earthed through component body
- resistor
- **S3(R4)** — Two components combined as a unit.
- cable leading to optional or alternative equipment
- alternative method of wiring according to model/year etc (marked)

TERMINAL MARKINGS

The following terminal markings have been extracted from the DIN Standard and represent some of the most commonly used circuits.

30 - Input direct from battery positive.

31 - Return circuit direct to battery negative (earth).

15 - Switch in line with battery positive (e.g. circuit live when ignition switch ON).

ALTERNATIVE OR OPTIONAL EQUIPMENT DIAGRAM

When the main wiring diagram indicates that alternative or optional equipment is fitted, i.e. the cable through the address code is indicated with a dotted line, reference should be made to the Alternative or Optional Equipment Diagram.

GRAPHICAL SYMBOL
For example: Alternator

COMPONENT CODE

Motor — M2
Wiper

COLOUR CHANGE
Colour change on cable between components and destination: the component code is not repeated

ADDRESS CODE
Address code shows destination of wire leaving Fuse 5.

S30:86/ws gn
Heated Rear Window Switch
Terminal number on switch
White cable with green tracer

NON CABLE CONNECTION
Non Cable Connection indicated by z in Colour Code position i.e. by printed circuit or internal connection.

IN LINE FUSE
In Line Fuse between component and destination

EARTH CONNECTION
Earth Connection directly through component body (or printed circuit) to chassis

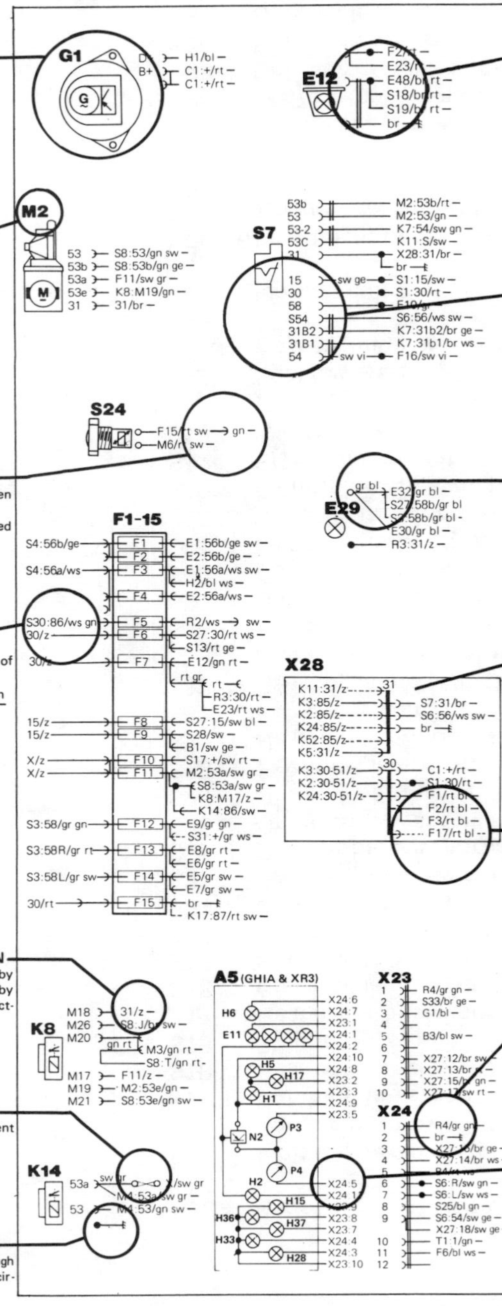

SOLDERED JOINT IN LOOM
Soldered joint where cable divides for several destinations. Normally used when joint is remote from component i.e. hidden in loom.

TERMINAL CODES
Terminal Codes with the numbers/letters which are on the actual components.

CABLE COLOUR
Cable Colour leaving terminal with subsequent division via connector to several component destinations.

RELAY PLATE
Relay plate simplified to show connections made through the internal bus bars. The bus bars are identified by the appropriate DIN codes e.g. 30 and 31.

OPTIONAL EQUIPMENT
Optional Equipment connection indicated by broken line.

EARTH CONNECTION
Earth Connection via cable.

TERMINAL IDENTIFICATION
Terminal identification for multi-pin plugs show the number of the pins in numerical order. The connections to the component are identified with the connector number with its corresponding pin number e.g. X24:5.

Key to Wiring Diagrams

A1	Radio/stereo unit
A5	Instrument panel
A6	Digital clock
A8	Rear lamp assembly LH
A10	Rear lamp assembly RH
B1	Horn (single or LH)
B3	Fuel gauge tank transmitter
B4	Coolant temperature transmitter
B12	Brake pad wear sensor LHF
B13	Brake pad wear sensor RHF
B28	EGR system control
C1	Battery
E1	Headlamp LH
E2	Headlamp RH
E5	Side lamp LH
E6	Side lamp RH
E7	Tail lamp LH
E8	Tail lamp RH
E9	Number plate lamp (single or LH)
E10	Number plate lamp RH
E11	Panel illumination lamp/s
E12	Interior lamp/s front
E13	Distributor
E14	Fog lamp LH
E15	Fog lamp RH
E16	Driving lamp/spot lamp LH
E17	Driving lamp/spot lamp RH
E18	Rear fog lamp (single or LH)
E19	Rear fog lamp RH
E20	Reversing lamp (single or LH)
E21	Reversing lamp RH
E23	Clock
E24	Glove box lamp
E26	Boot lamp
E28	Heater control lamp
E48	Interior lamp/s rear
F1-18	Fuse box
G1	Alternator
H1	Ignition warning lamp
H2	Main beam warning lamp
H3	Stop lamp LH
H4	Stop lamp RH
H5	Oil pressure warning lamp
H6	Indicator warning lamp (single or LH)
H8	Indicator lamp LH front
H9	Indicator lamp RH front
H10	Indicator lamp LH rear
H11	Indicator lamp RH rear
H13	Indicator repeater lamp LH
H14	Indicator repeater lamp RH
H15	Low fuel warning lamp
H17	Handbrake on warning lamp
H28	Brake pad wear warning lamp
H33	Low washer fluid warning lamp
H36	Low coolant warning lamp
H37	Low oil warning lamp
K2	Driving lamps/fog lamp relay
K3	Ignition warning lamp relay
K5	Indicator relay
K7	Intermittent wiper relay
K11	Headlamp wipe/wash relay
K24	Window lift motor relay
K52	Driving/foglamps relay
M1	Starter motor
M2	Wiper motor
M3	Washer motor
M4	Rear wiper motor
M5	Rear washer motor
M6	Coolant blower motor
M7	Heater blower motor
M10	Headlamp wash (single or LH)
M14	Window lift motor LH front
M15	Window lift motor RH front
N2	Voltage stabiliser
P3	Fuel gauge
P4	Coolant temperature gauge
P7	Tachometer
R1	Ignition ballast resistor
R2	Heated rear window
R3	Cigarette lighter
R4	Panel lamps dimmer rheostat
R33	Wiper intermittent resistor
S1	Ignition/starter switch
S6	Combination switch I
S7	Combination switch II
S10	Rear wipe/wash switch
S13	Stop lamps switch
S17	Heater motor switch
S18	Courtesy lamp switch LH front
S19	Courtesy lamp switch RH front
S20	Courtesy lamp switch LH rear
S21	Courtesy lamp switch RH rear
S22	Courtesy lamp switch tailgate/boot
S24	Coolant blower motor thermoswitch
S25	Oil pressure warning lamp switch
S26	Oil level warning lamp switch
S28	Reversing lamps switch
S30	Heated rear window switch
S31	Fog lamps switch
S32	Rear fog lamps switch
S33	Brake fluid level warning lamp switch
S35	Handbrake warning lamp switch
S36	Glovebox lamp switch
S50	Coolant low warning lamp switch
S64	Window lift switch LH front door
S65	Window lift switch RH front door
S100	Central locking switch
S124	EGR vacuum switch
S126	Window lift dual switch RH
S127	Driving lamp switch
S135	Fuel low warning switch
S136	Washer fluid low warning switch
T1	Ignition coil
X20-25	Instrument panel multi-plug
X27	Module box connector
X28	Relay plate
Y17	Idle fuel cut-off valve
Y28	EGR system valve
Y29	Door lock impulse unit
Y30	Door lock actuator - passenger front
Y31	Door lock actuator - LH rear
Y32	Door lock actuator - RH rear
Y33	Tailgate actuator

Wiring Diagram

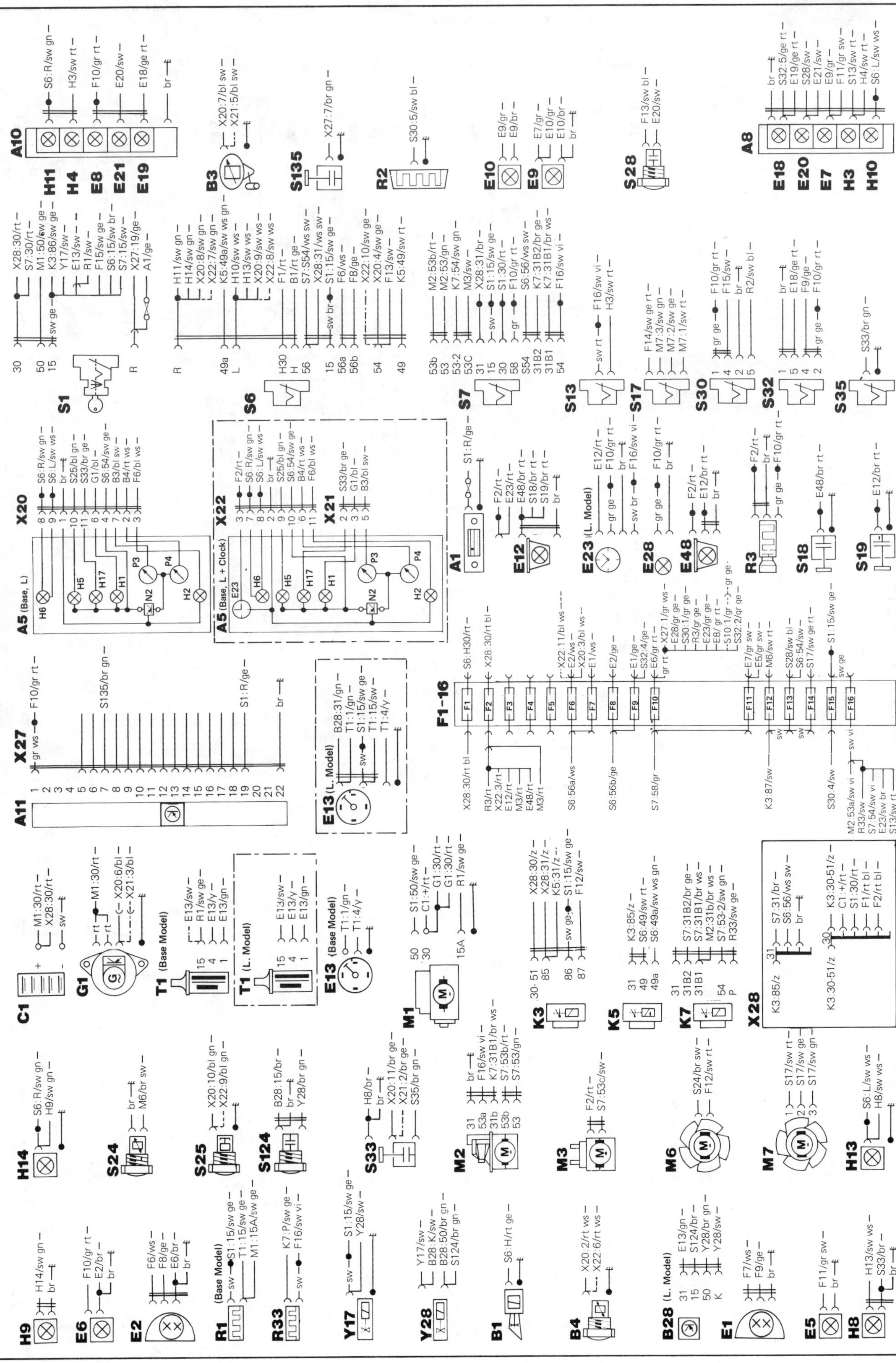

Base, L, and Estate versions.

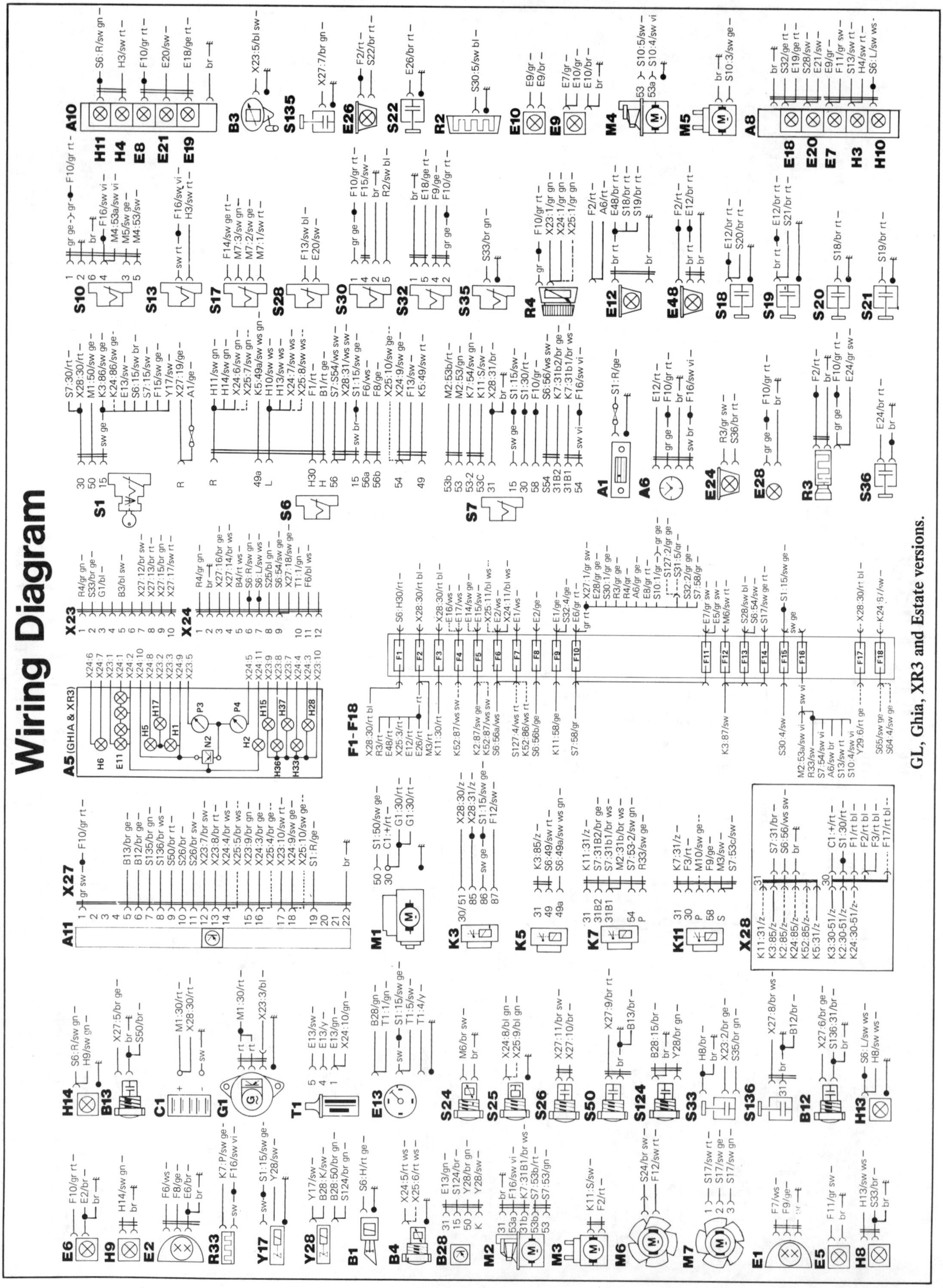
Wiring Diagram
GL, Ghia, XR3 and Estate versions.

Wiring Diagrams

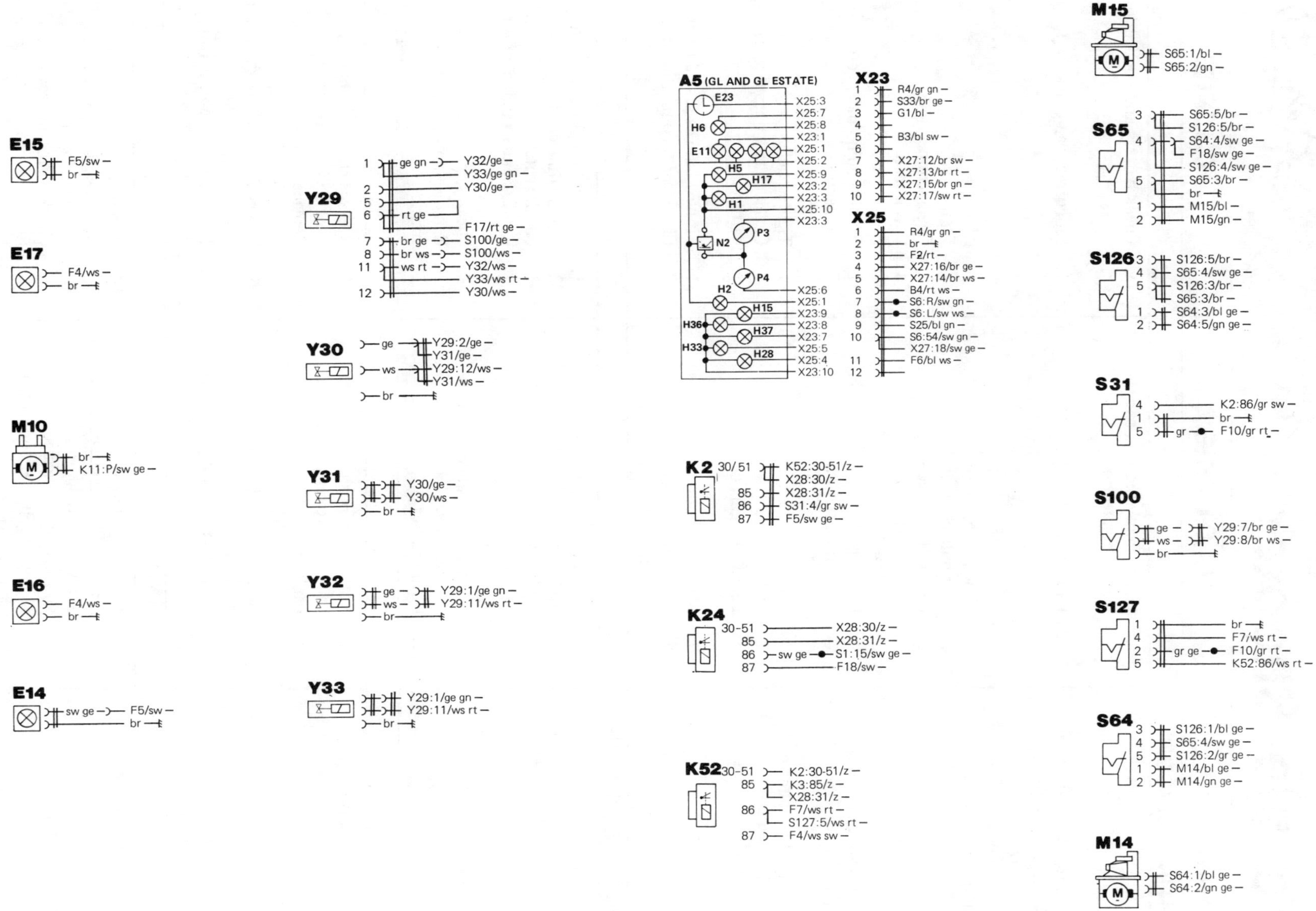

Optional Equipment.

ELECTRICAL
Trouble Shooter

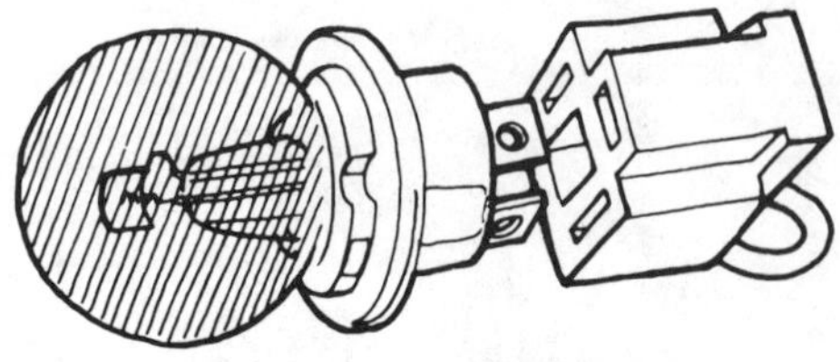

FAULT	CAUSE	CURE
LIGHTING CIRCUIT		
No lights (or very dim)	1. Flat or faulty battery, bad battery connections.	1. Check battery and connections.
Side and rear lights inoperative although stoplights and flashers work	1. Fuse blown.	1. Fit correct value fuse.
One lamp fails	1. Blown bulb. 2. Poor bulb contact. 3. Bad earth connection. 4. Broken feed.	1. Fit new bulb. 2/3. Check connections. 4. Check feed.
Flasher warning bulb stays on or flashers twice as fast	1. Faulty bulb or connection on front or rear of offending side.	1. Fit new bulb, make good connection.
Lights dim when idling or at low speed	1. Loose drive belt. 2. Flat battery. 3. Faulty charging circuit.	1. Tighten belt. 2/3. Check charge output and battery.
One dim light	1. Blackened bulb. 2. Bad earth. 3. Tarnished reflector.	1/3. Fit new bulb or sealed-beam. 2. Check earth connections.
WINDSCREEN WIPERS		
Wipers do not work	1. Blown fuse. 2. Poor connection. 3. Faulty switch. 4. Faulty motor.	1. Fit new fuse of correct rating. 2. Check connections. 3. Check switch. 4. Remove and examine motor.
Motor operates slowly	1. Excessive resistance in circuit or wiper drive. 2. Worn brushes.	1. Check wiper circuit. 2. Remove motor and check brushes.
HORN		
Horn(s) do not work	1. Faulty horn push. 2. Faulty or broken connection. 3. Faulty horn(s).	1. Replace steering column switch. 2. Check wiring continuity. 3. Replace horn(s).
Horn operates continuously	1. Horn push earthed or stuck down. 2. Horn cable to horn push earthed.	1. Remedy cause of earthing out or free off horn push. 2. Trace cause of cable earthing and re-insulate.
FUEL GAUGE		
Fuel gauge not reading	1. No fuel in tank. 2. Sender unit to gauge cable broken or disconnected. 3. No earth on fuel gauge. 4. Supply to gauge disconnected. 5. Gauge faulty.	1. Add fuel. 2. Reconnector repair cable. 3. Reconnect earth. 4. Check wiring and reconnect or repair. 5. Replace gauge.
Fuel gauge registering full	1. Sender unit to gauge cable earthed.	1. Trace cause of earthing out and re-insulate.

Body & Fittings

INTRODUCTION . . . . [1]

All the body fittings are easy to get at and to remove by the DIY mechanic. Some of the fittings however, are secured with Torx type screws which have socket heads requiring a special Torx adaptor when unscrewing or tightening them. These adaptors can usually be purchased from a local Auto Engineering supply shop.

On models with central locking system and electric front window lifts, the removal and installation procedures are similar to the manual version described although it is important to disconnect the battery earth lead when working on the components to avoid accidental short-circuits.

FRONT & REAR BUMPERS . . . . [2]

Removal

1. Open the bonnet or the tailgate and remove the two nuts on each side securing the bumper mountings to the brackets (Figs. N:1 & N:2).
2. Remove the bumper assembly complete with outer quarter bumpers from the front of the car.
3. The outer quarter bumpers, once fitted are permanent and need to be destroyed to remove them from the main bumper bar. The locking tangs should be pulled off, as shown in Fig. N:3 to detach.

NOTE: If damaged, the retaining clip securing the quarter bumper to the body can be removed by twisting it through 90° and withdrawing it.

Installation

1. Align quarter bumper with bumper bar and snap it into position.
2. Position the bumper assembly so that bracket studs locate in the body and the quarter bumper clips engage.
3. Holding the bumper in position, refit and secure the four retaining nuts.

Overrider - Remove & Install

1. Locate overrider retaining screw on underside of overrider, slacken clamp screw and lift overrider off bumper, (Fig. N:4).
2. Position overrider on bumper, ensure clamps are correctly located over bumper rim then tighten clamp screw. On variants with headlamp wash facility the overriders must locate into a cut out on the bumper.

BONNET PANEL . . . . [3]

Removal

1. Open the bonnet and support with the stay.
2. If the original bonnet is to be refitted, it will facilitate alignment of the bonnet on installation if the positions of the side hinges are marked by scribing round them at this stage.
3. An assistant will be needed to lift off the bonnet, and the area at the rear corners of the bonnet should be padded with cloth to prevent damage to the paintwork as the hinge bolts are released.
4. Undo the two bolts securing each side hinge and carefully lift off the bonnet after disconnecting the support stay.

Installation

When installing the bonnet, take care to avoid damaging the paintwork with the rear corners of the bonnet. Refit the hinge bolts loosely at first, then align thc hinges with the marks previously made and tighten the bolts.

Lower the bonnet and check that the alignment is correct in relation to the wings before pressing it into the locked position. If necessary, realign the bonnet by slackening the hinge bolts.

BONNET RELEASE CABLE & LOCK . . . . . . [4]

Cable Removal

1. Remove the three screws and detach the lower steering column shroud.
2. Remove the screw and detach the cable/bracket assembly from steering column (Fig. N:5).
3. From inside the engine compartment, pull the cable grommet from latch bracket and disengage the cable nipple from latch lever (Fig. N:5). Unclip cable from the sides of the engine compartment.
4. Pull the cable through the bulkhead into the passeger compartment.

Installation

Install the cable in the reverse order of removal.

Bonnet Lock

The bonnet lock can be removed by disconnecting the cable from the latch then undoing the three screws. Withdraw the lock from under the front panel.

Make sure the lock sliding plate is well greased before installing the lock in the reverse order of removal.

Bonnet Lock - Adjustment

Bonnet lock adjustment is done by screwing the spring-loaded striker in or out with a screwdriver.

The bonnet should fully latch under its own weight when released from 12 in (300 mm) above its pop-up position.

When released from a height of 3 in (75 mm), the secondary catch only should engage.

TAILGATE . . . . [5]

Removal

1. With the tailgate open, remove the tailgate aperture weatherstrip from the rear header bar. Prise the headlining clips from weatherstrip flange.
2. Remove two screws securing 'C' pillar trim panel on each side. Pull the headlining down around the area of the hinge bolts.
3. Separate the ball and socket connecting the tailgate damper to 'C' pillar, using a small screwdriver to withdraw the nylon locking pin.
4. Support the tailgate, then remove the two nuts from each tailgate hinge bracket and remove the tailgate, (Fig. N:6).

Installation

1. Position the tailgate hinges on the bolts and secure with two nuts.
2. Reconnect the tailgate damper socket to the ball on the 'C' pillar by pushing the socket over the ball and refitting the nylon locking pin.
3. Refit the headlining over the weatherstrip flange and secure with clips.
4. Refit the upper 'C' pillar trim panels.
5. If necessary, move the tailgate on the hinge brackets to equalise gap.
6. Check the height of the tailgate at rear end relative to quarter panels. If necessary slacken bolts securing lock striker and raise or lower striker to achieve desired setting. Check height of the tailgate relative to quarter panels at front end. If necessary slacken bolts securing hinges to body mounting brackets and move tailgate up or down as required.

DOOR TRIM PANELS . . . . [6]

Door Trim Panel - Removal

1. On Ghia variants only, carefully detach the door trim panel capping first.
2. Remove the door window regulator handle and the door pull/armrest. On variants with electrically operated windows pull out the switches and remove door pocket finisher to gain access to the arm rest securing screws.
3. Carefully prise the remote control handle bezel off its retaining lugs by pushing it rearwards.
4. Locate each panel clip by feeling the trim panel surface around the periphery of the panel. Carefully lever out each clip using a wide blade tool or screwdriver and pull off trim panel. Where fitted, remove the door pocket.

Installation

1. Make sure the panel retaining clips are not damaged and are correctly located in the trim panel.
2. Refit the panel to the door frame and make sure all the retaining clips are engaged.
3. Refit the regulator, armrest or door pull and the trim panel capping on Ghia variants, in the reverse order of removal.

Tailgate Trim Panel - Removal

1. On cars with rear wash wiper, remove the motor cover section by turning the retaining clips through 90°.
2. Using a wide blade tool or screwdriver, carefully prise the panel trim clips from their locations and detach the panel (Fig. N:7).

Installation

1. Align the panel clips with the holes in the tailgate inner panel and clip the panel into position.

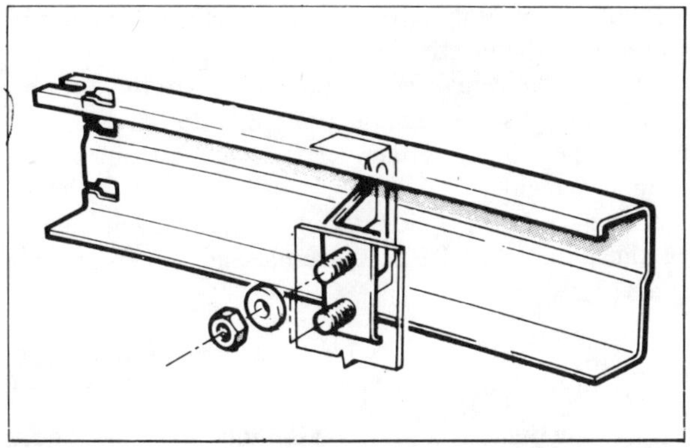

Fig. N:1 Front bumper bracket fitting arrangement

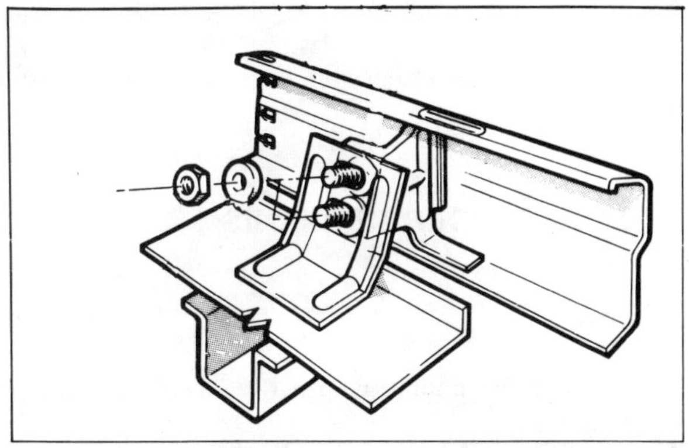

Fig. N:2 Rear bumper bracket fitting arrangement

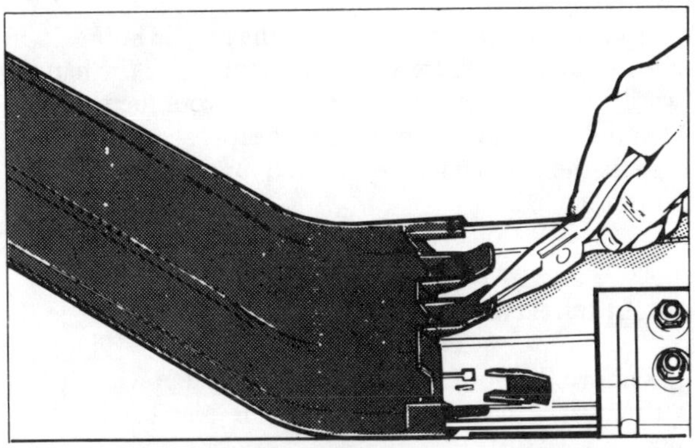

Fig. N:3 Removing outer quarter bumper tangs

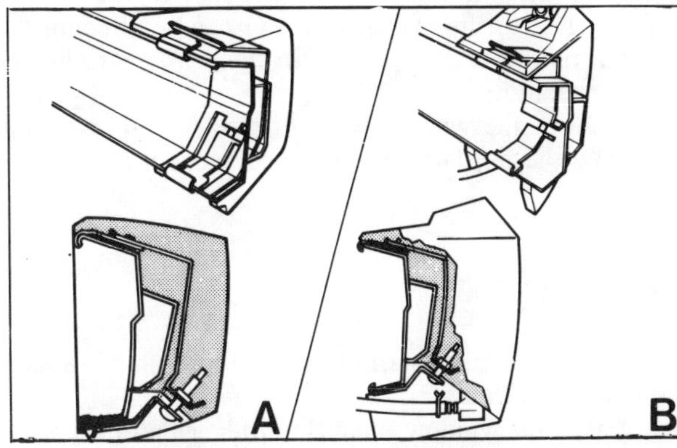

Fig. N:4 Overrider fitting, (A) with headlamp wash (B)

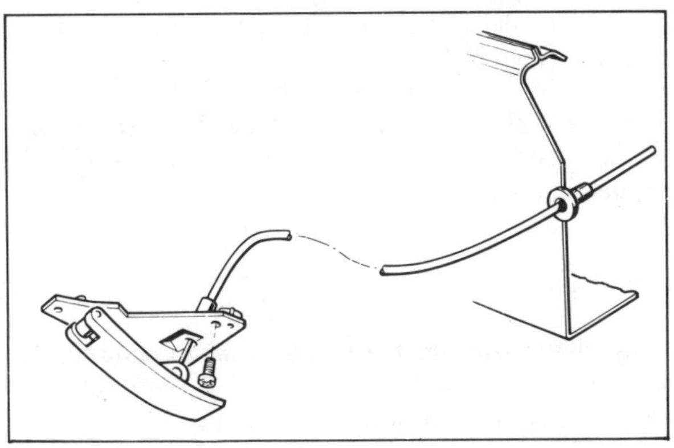

Fig. N:5 Details of bonnet release cable fitting

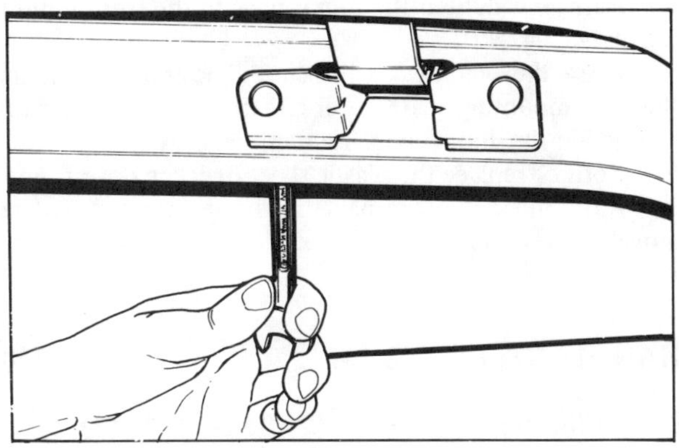

Fig. N:6 Undoing tailgate hinge bracket nuts

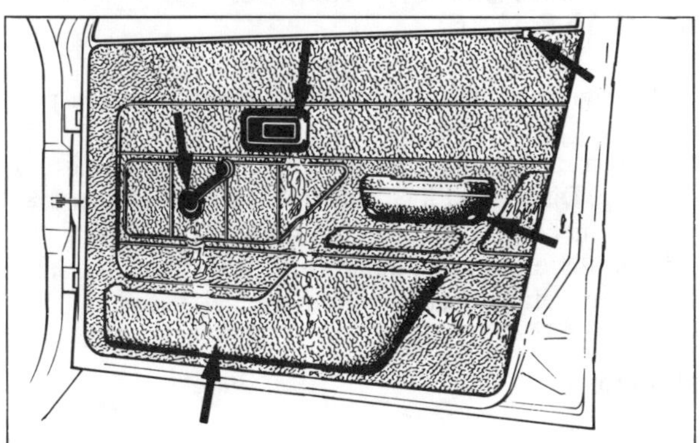

Fig. N:7 Door trim panel removal, fittings arrowed

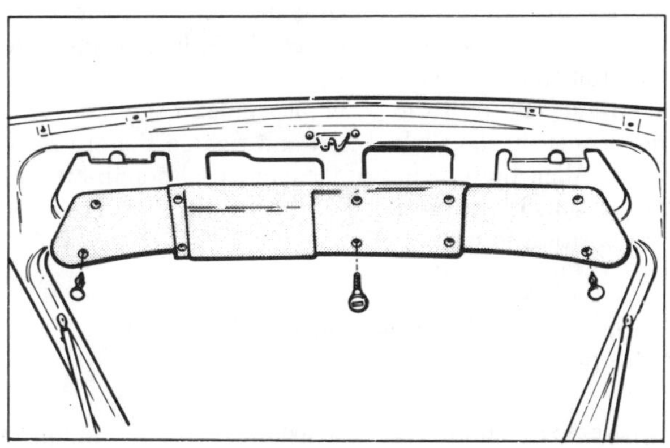

Fig. N:8 Tailgate trim panel retaining clips

2. If applicable, align the holes in the motor cover section of the panel with the holes in the tailgate inner panel and push a clip into each hole then turn through 90° to secure.

WINDOW WINDER MECHANISM . . . . [7]

Removal

1. Remove the door trim panel as detailed in the previous section.
2. Carefully peel back the plastic sheet to expose the inside of the door.
3. Wind the window up or down so that the window bracket slide to the regulator runner can be disconnected through the lower cut out. Lower the window to the bottom of the door.
4. Remove the winder regulator screws, (Fig. N:9) and withdraw the regulator through the lower cut out (Fig. N: 10).

Installation

1. Refit the regulator through the door frame lower cut out (Fig. N:10).
2. Attach the window channel to the regulator sliders.
3. Loosely refit the three screws to the rear triangular mounting plate (Fig. N:9).
4. Refit and tighten the four screws to the front mounting plate (Fig. N:9).
5. Raise the window to the closed position and adjust the rear mounting plate to ensure that the glass is fully engaged in the door frame. Tighten the screws.
6. Refit or replace the plastic sheet over the door frame.
7. Refit the door trim panel in the reverse order of removal.

HANDLES & LOCKS . . . . [8]

Door Window Regulator Handle (Fig. N:11)

To remove the handle, carefully prise the circular plastic insert from the regulator shaft end of handle. Remove the crosshead screw and detach the handle and bezel from the regulator shaft.

To install, refit the bezel on regulator shaft and align the handle on the shaft. Insert and tighten the crosshead screw. Align the notches of the circular plastic insert with the cut outs in the shaft end of the handle. Press the insert into position to secure.

Door Remote Control Handle - Removal (Fig. N:12)

1. Remove remote control handle bezel by sliding it rearwards.
2. Remove the door trim panel as detailed previously and peel back the plastic sheet around handle.
3. Remove the screw securing the handle to the door inner panel and push the handle assembly into the door. Detach the operating rod damper from its cut out in the door inner panel.
4. Push the clip off the operating rod and detach the rod from the door lock. Remove handle and rod from door (Fig. N:12).

Installation

1. Place the handle and rod assembly into the door. Turn the lock lever bushing clip parallel to the door base then refit the rod end. Secure it by snapping the clip over the handle.
2. From inside the door, position the handle in the door frame and loosely secure with the screw. Slide the handle forward to take up any slack in the rod then finally tighten the screw. Refit the plastic sheet.
3. Refit the trim panel, bezel and door fittings in reverse order of removal.

Door Latch - Removal

1. Remove the door trim panel as detailed previously.
2. Peel back the plastic sheet at the rear of the door frame.
3. On central locking system models, disconnect the battery earth strap.
4. On all models, disconnect the operating rods at the latch levers (Figs. N:13 & N:14).
5. Undo the screws on the outside of the door frame and withdraw the latch from inside the door. Disconnect the lock solenoid wiring at the connector on central locking system models.

Installation

Install the latch in the reverse order of removal. Reconnect the wiring at the connectors on central locking system models, and reconnect the battery..

Tailgate Lock Assembly - Removal (Figs. N:15 & N:16)

1. Open tailgate and remove trim panel as detailed previously.
2. On central locking system models, disconnect battery earth strap. Disconnect lock solenoid wiring at connector.
3; Disconnect operating rod from lock.
4. Undo three bolts and detach lock from tailgate frame.
5. Remove retaining clip and remove lock barrel.

Installation

1. Slide the pad over the lock cylinder housing so that the pad cut outs align with those on underside of housing

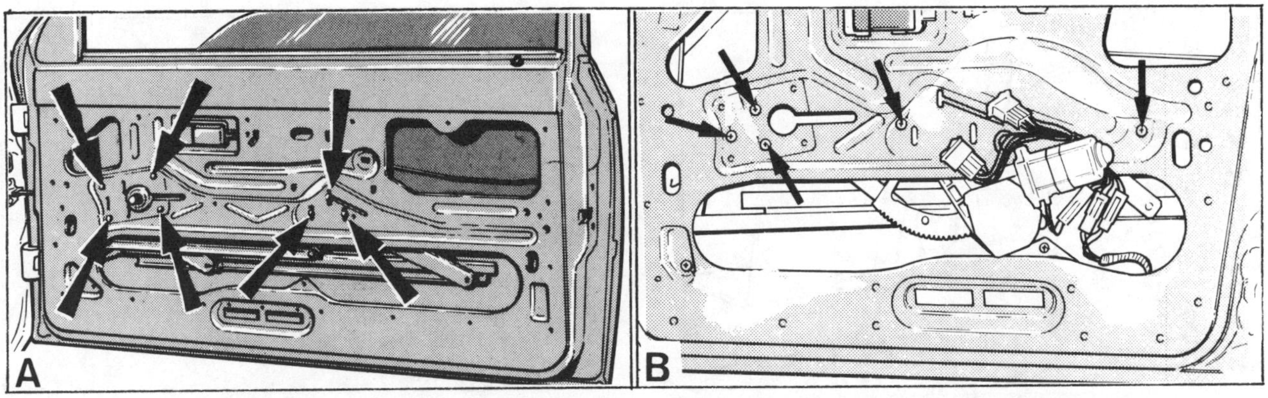

Fig. N:9 Location of regulator securing screws, manual type (A) electric window type (B)

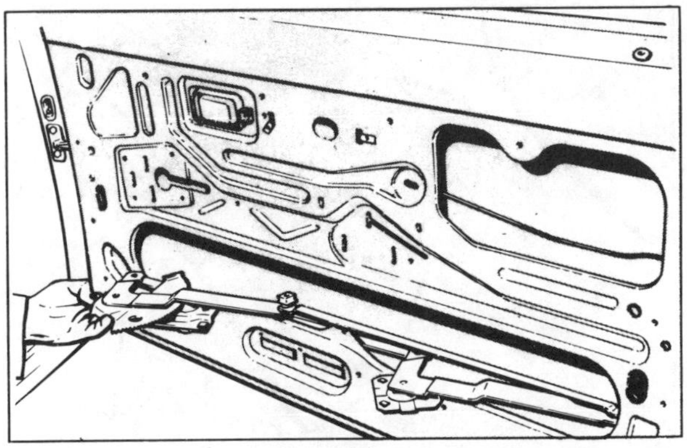

Fig. N:10 Removing manual regulator from door

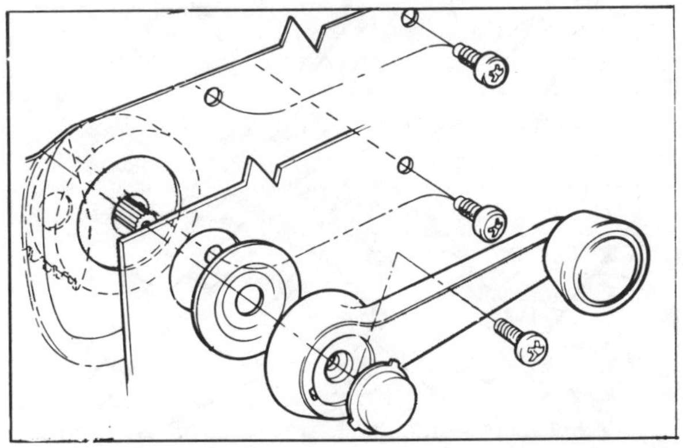

Fig. N:11 Exploded view of window regulator handle

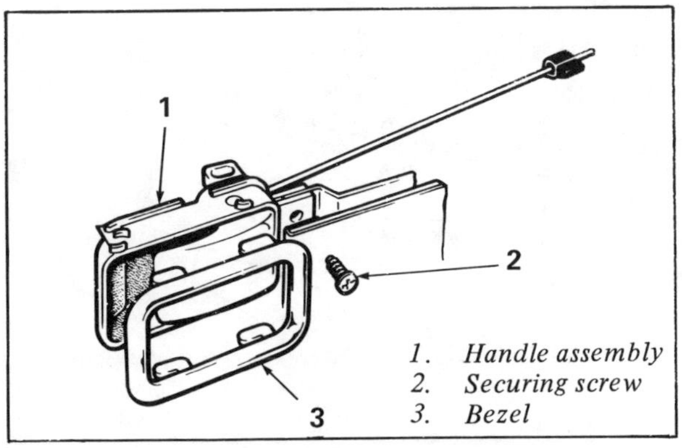

Fig. N:12 Door remote control handle fitting

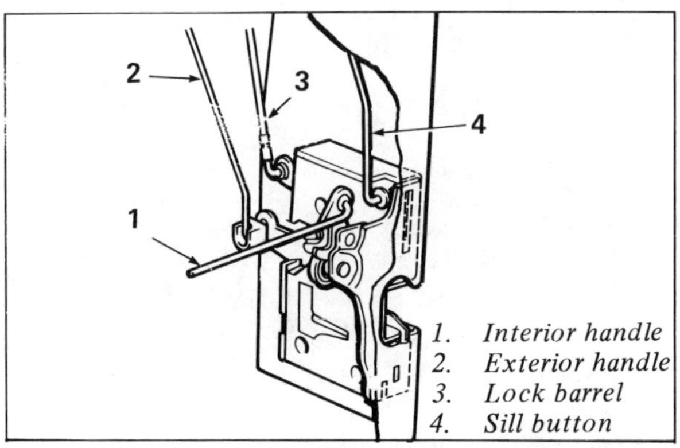

Fig. N:13 Door latch fitting and link rods - manual

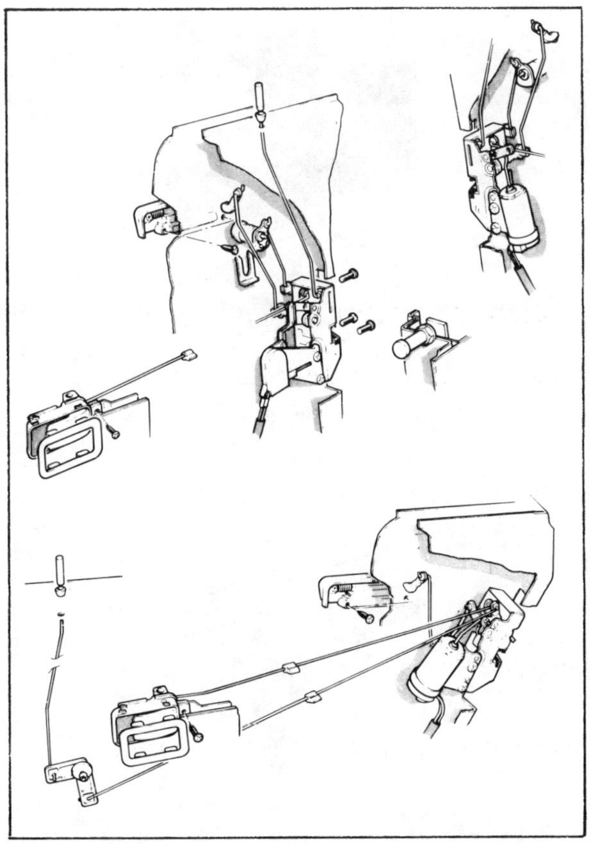

Fig. N:14 Door latch fitting - central locking system

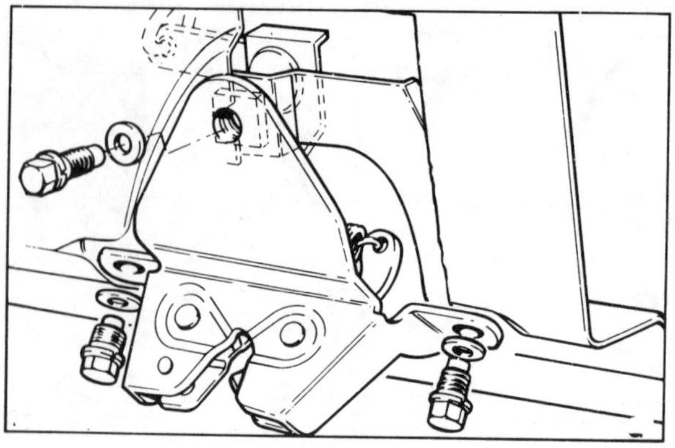

Fig. N:15 Tailgate lock fitting

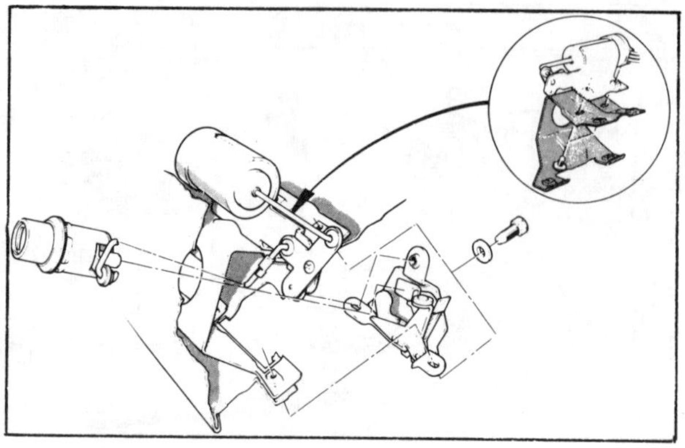

Fig. N:16 Tailgate lock-central locking system

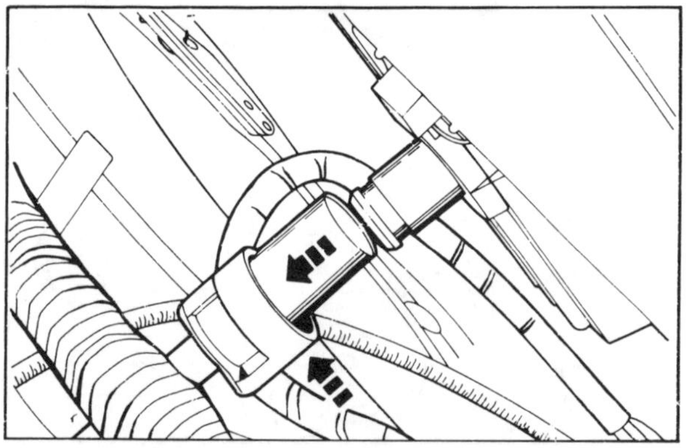

Fig. N:17 Releasing cable from speedometer

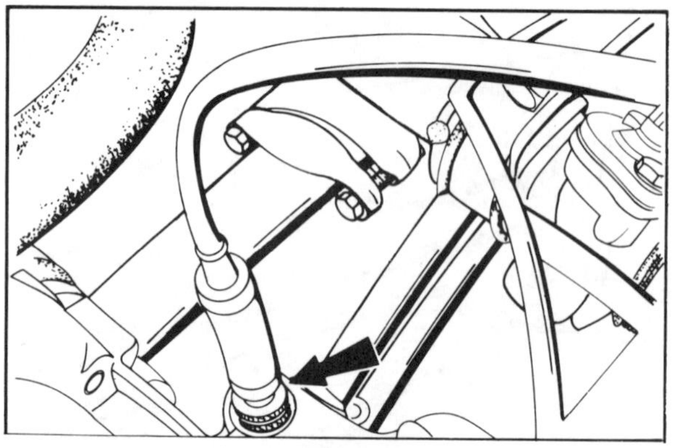

Fig. N:18 Disconnecting speedo cable from gearbox

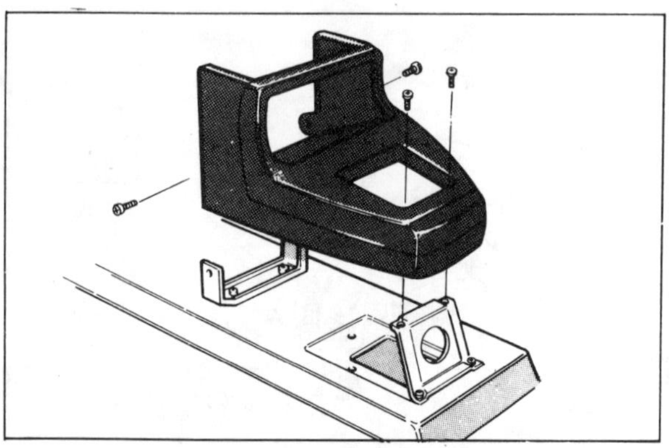

Fig. N:19 Centre console securing screws

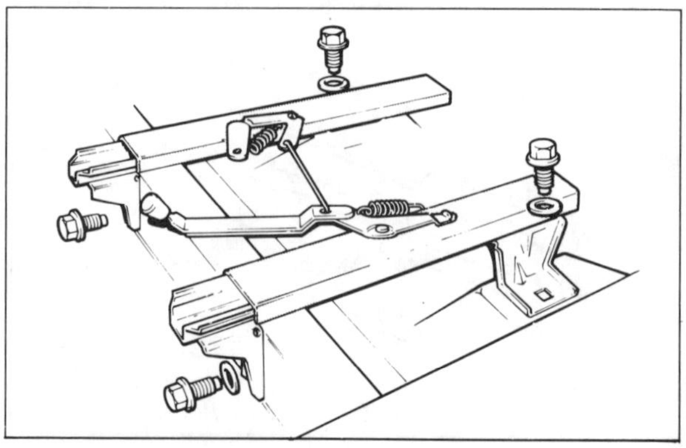

Fig. N:20 Front seat securing bolt positions

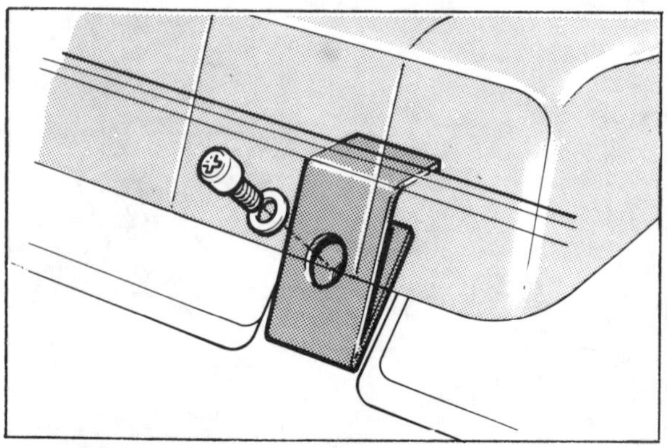

Fig. N:21 Rear seat cushion securing screw

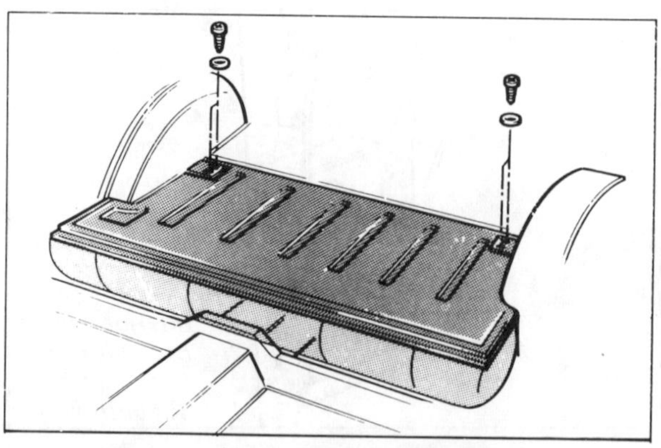

Fig. N:22 Rear seat backrest hinge screws

flange.
2. Insert the cylinder barrel through its cut out in the tailgate and, from inside tailgate frame, slide the retainer over the cylinder barrel lightly tap retainer with mallet to secure.
3. Refit the lock and secure with the three bolts.
4. Reconnect lock solenoid wiring on central locking models.
5. Refit tailgate trim panel.

Door Lock Barrel - Replacement (Fig. N:17)

1. Remove the door trim panel as detailed previously.
2. Pull down the plastic sheet to expose the door lock barrel.
3. Disconnect the link rod from the barrel lever.
4. Remove the retaining clip and withdraw the barrel from outside the door.
5. Fit the lock barrel and reconnect link rod in reverse order of removal. Refit plastic sheet and door trim panel.

SPEEDOMETER CABLE .... [9]

If worn or broken the speedometer cable should be replaced. Wear will be indicated by the speedometer needle wavering when travelling at a steady speed. Remove the instrument cluster as described in the 'Instrument Panel' section. The speedometer cable is held to the instrument by a retaining clip. Release the clip by pushing the latch prongs together (Fig. N:17).

Inside the engine compartment release the cable from the gearbox by unscrewing the knurled ring (Fig. N:18). Free the cable from the body clips and pull through the bulkhead grommet.

Fit the new cable through the bulkhead and secure to the body clips. The cable must be correctly routed or it may fail prematurely. Fit the cable ends to the connections, ensuring that the inner cable is properly fitted.

CENTRE CONSOLE.................. [10]

Removal

1. Unclip and remove the console end cover.
2. Remove the gear lever knob then pull the gear lever gaiter off the console and slide it off over the gear lever.
3. Remove the four screws shown in Fig. N:19, and remove the console.

Installation

1. Fit the console and secure it with the four screws.
2. Refit the gear lever gaiter and attach the gear knob in reverse order of removal.
3. Refit the console end cover by clipping into place.

SEATS .... [11]

Front Seat Removal (Fig. N:20)

1. Slide the seat forwards as far as the mechanism will allow to gain access to the bolts securing the rear of the slides to the floor pan. Remove the bolts, one each side.
2. Now, slide the seat rearwards and remove the bolts securing the front of the slides.
3. Detach the seat from the floor and remove it.

Installation

Position the seat on floor and align the holes in the slides. Fit and tighten the slide bolts, adjusting the seat position as necessary to gain access.

NOTE: Tighten the front bolts before the rear bolts and ensure that the seat slide latches are both fully engaged before tightening fully.

Rear Seat Cushion (Fig. N:21)

To remove the cushion, locate the hinges on either of the floorpan tunnel. Remove the torx screws, one each side, securing the seat to the floor. Lift the cushion out of the car.

Fit the cushion in the reverse order of removal and push it down to align the hinges with the holes in the floorpan. Fit and tighten the torx screws.

Rear Seat Backrest (Fig. N:22)

1. Lift up the rear seat cushion and fold down the rear seat back to expose the hinges.
2. Remove the six screws, three each side, attaching the backrest to the hinges.
3. Carefully pull the backrest from its location and remove it from the car.
4. Fit the backrest in position and align it with the holes in the hinges. Refit the securing screws, and refit the rear seat cushion.

CORROSION .... [12]

We all recognise rust when it starts to appear around parts of the car body. Leave it untreated and it will spread and eventually produce holes which will not only weaken the car body, but drastically reduce the car's value as well.

The only way to beat rust is to prevent it forming in the first place, or at the very least, slow it down by making sure that any parts where paint has been chipped off are touched-up as soon as possible.

All Escort models have extensive rustproofing measures applied at the factory which should last for several years. However, the complete underbody should be checked at least once a year to make sure that the protective layer has not been damaged. To do this, the car will have to be thoroughly cleaned underneath using a high pressure hose, or the car steam cleaned.

Examination will now show up any suspect areas where additional rustproofing protection is required. There are various under body rustproofing kits on the market designed specifically for the DIY motorist which are easy to apply.

Small stone chips on the paintwork can be carefully repaired using a touch-up paint kit obtainable from Ford dealers or car accessory shops - see HISTORY & IDENTIFICATION at the beginning of this manual for location of paint colour code.

Accessories

INTRODUCTION .... [1]

Various accessories are available specifically for the Escort, which can be added according to your needs. The most comprehensive accessory selection naturally comes from Ford who supply the original optional equipment through their dealer network for the Escort.

In this accessories chapter we give fitting instructions for a variety of accessories listed above. The instructions will also be useful if your car already has a particular accessory fitted, and you wish to change or remove it, when selling the car for example.

When removing or installing any electrical accessory, make sure that the battery earth strap is disconnected first. This will avoid the possibility of any short-circuits if another component has to be disturbed or disconnected.

RADIO FITTING .... [2]

All Escort models except the Base model 1.1 Saloon and Estate cars, have a radio as a standard fitting. The following instructions are for fitting a Ford radio/cassette unit to a basic model saloon or estate car.

1. Fit the speaker(s) and aerial as detailed in the following sections.
2. Disconnect the battery earth strap.
3. Remove the control knobs, escutcheons and trim cover from the front of the radio/cassette unit (Fig. O:6).
4. Fit the mounting plate to the front of the radio/cassette unit and secure with the two nuts on the control spindle sleeve (Fig. O:1).
5. Connect the aerial, speaker(s), supply and earth wires to the radio cassette unit connection points - see Fig. O:3. If an electric aerial is fitted, also connect the aerial relay feed. Make sure the aerial is fully extended and temporarily reconnect the battery.
6. Turn the ignition switch to the auxillary position and switch on the radio. Turn the volume control spindle fully clockwise.
7. Turn the tuning knob spindle to a weak signal on the Medium wave band around the 200 m mark.
8. Using a small screwdriver inserted through the small hole adjacent to the tuning knob spindle (Fig. O:1), adjust the aerial trimmer screw to obtain maximum volume. Note that for best results, the car should be away from tall buildings.
9. Switch off the radio and disconnect the battery earth strap. Disconnect the aerial, speaker, earth, supply and aerial relay wires from the rear of the radio.
10. Fit the plastic support bracket to the rear of the radio as shown in Fig. O:2.
11. Position all radio wires through the aperture then connect them to the rear of the radio at the positions shown in Fig. O:3.
12. Slide the radio/cassette unit into the aperture, making sure that the rear support bracket is engaged with its runner.
13. Secure the mounting plate to the fascia with four screws - see Fig. O:4.
14. Fit the trim cover panel making sure that it is fitted the correct way round - see Fig. O:5.
15. Refit the control knobs to the radio spindles in the order shown in Fig. O:6.
16. Reconnect the battery earth strap and test the operation of the radio and aerial.

SUPPRESSION .... [3]

Electrical interference is caused by the operation of the various electrical systems in the car. Changes in current flow are picked up by the radio and create noise. The amount of interference caused by different sources varies between cars. The suppression requirements should be checked after installation of the audio system. Steps can then be taken to counteract the specific problems of your car. However there are certain components that should be

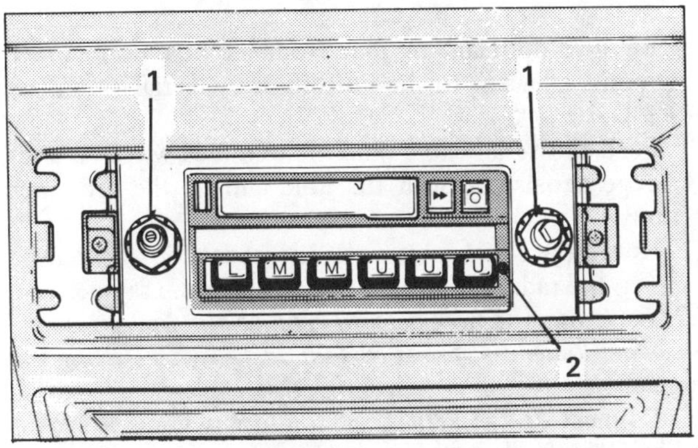

1. *Securing nuts*
2. *Aerial trimming screw*

Fig. O:1 Radio/cassette mounting plate and securing nuts

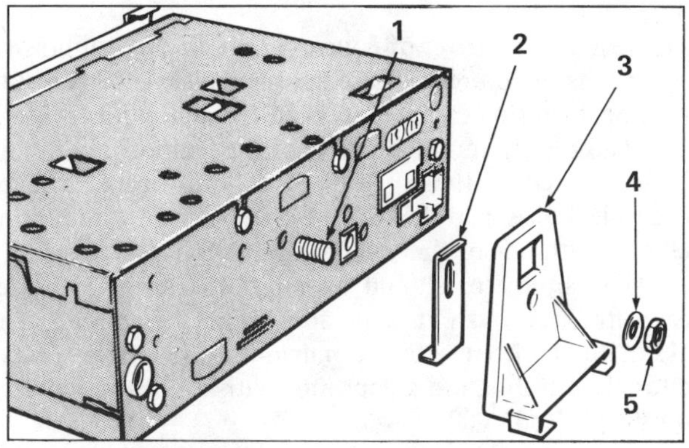

1. *Securing stud*
2. *Locating plate*
3. *Bracket*
4. *Washer*
5. *Securing nut*

Fig. O:2 Details of the rear support bracket fitting

1. *Aerial socket*
2. *Earth terminal*
3. *Speaker plug*
4. *Radio supply*
5. *Electric aerial relay supply*

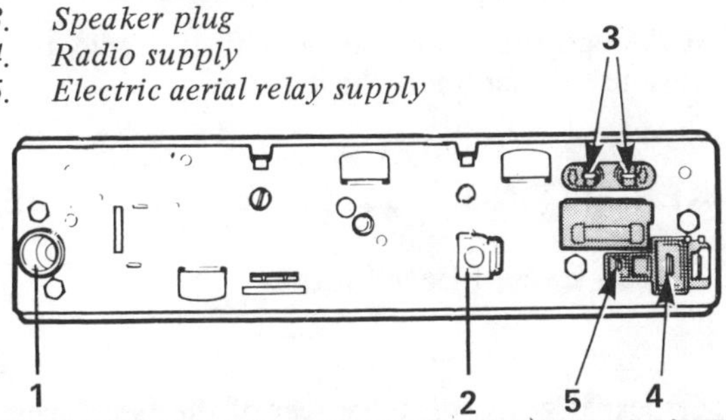

Fig. O:3 Details of wiring connections at rear of unit

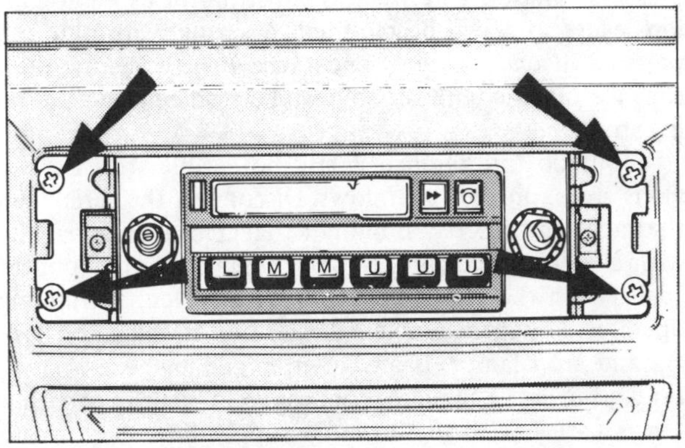

Fig. O:4 Mounting plate securing screws

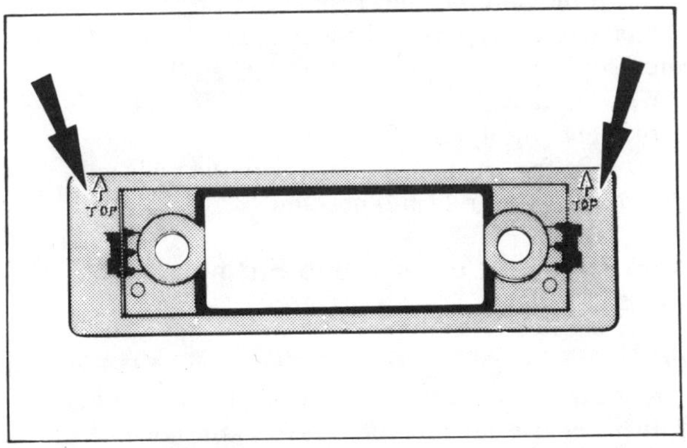

Fig. O:5 Radio/cassette trim cover location marks

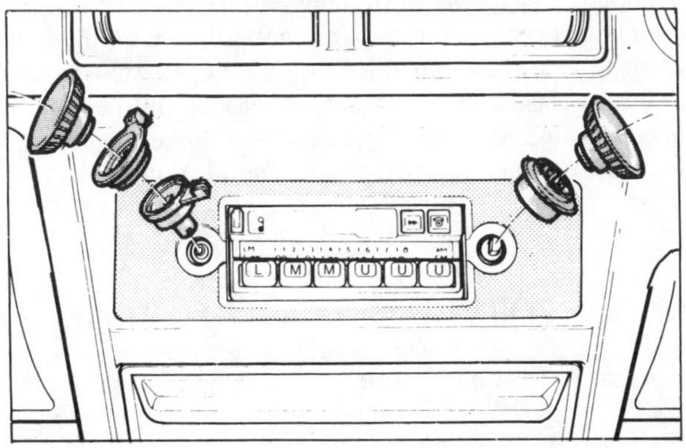

Fig. O:6 Correct fitting of radio/cassette control knobs

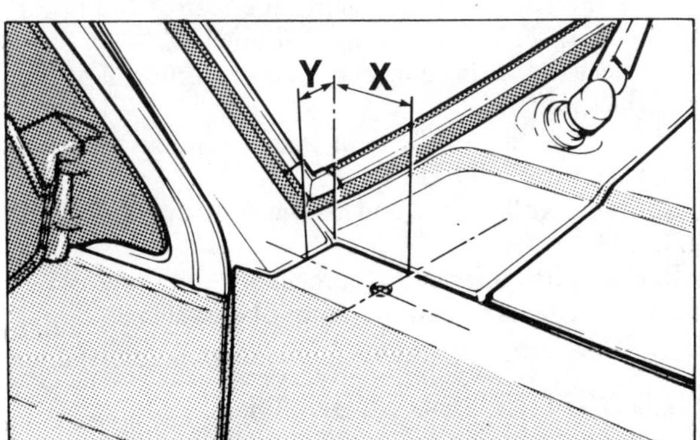

X = 1.97 in (50 mm)
Y = 0.73 in (18,5 mm)

Fig. O:7 Aerial hole position in front wing

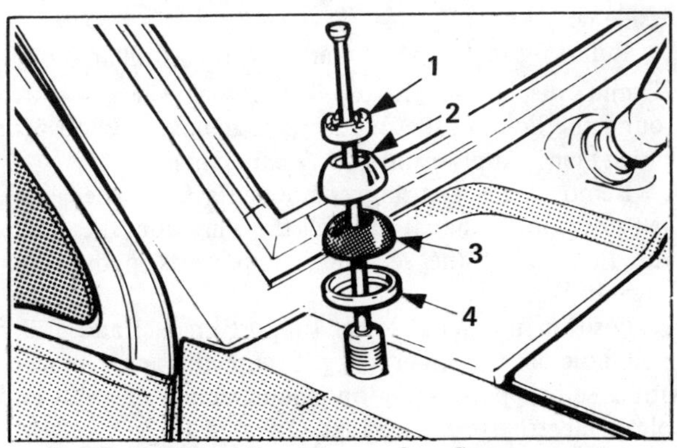

1. *Retaining collar nut*
2. *Bezel*
3. *Spacer*
4. *Sealing washer*

Fig. O:8 Correct assembly of aerial components

suppressed as a matter of course (if not already suppressed by the manufacturer). These are the ignition coil (on conventional ignition engine only) and the alternator.

The supply terminal (+) ve of the ignition coil should be connected to the fly lead of a 1 mfd. capacitor, for example Lucas part LS 627. Fit the capacitor mounting clip beneath one of the coil fixing clip bolts.

The alternator should be suppressed using a 3 mfd. capacitor, although it may already be so required. The alternator is fitted with a terminal for a suppression capacitor. If possible, use a capacitor with a Lucar connector on the fly lead.

If interference is still present after these initial measures, the following measures are recommended by Lucas. Change the high tension leads to a resistive high tension type. Suppress the distributor cap with in-line suppressors. These fit between the distributor cap and ignition leads, so 5 will be required. A set of spark plug suppressors fit over the plug ends, receiving the ignition lead cap. Finally the ignition coil can be earthed using a bonding strap.

If radio interference is still noticeable after the previous steps, proceed as follows. Disconnect the aerial from the radio and insert a dummy aerial. This is an aerial plug with a polystyrene capacitor from 68 to 82 pfd. connected between the pin and shroud. With the engine running, operate the wipers, washers, heater motor, indicator flashers and stop lamps. If any interference happens over the tuning range of the radio, connect an in-line choke such as Lucas part LS630 in the radio supply line. Any further interference from electrical components requires the connection of capacitors to the supply leads and earth. It may even be necessary to use in-line chokes in the supply leads. Check the earth connections of the engine transmission, indicator flasher circuit, wiper or washer motors, and the heater motor. If necessary, disconnect and clean the earth terminal to make good electrical ground contact.

AERIAL FITTING . . . . [4]

Manual & Electric Type

1. The aerial should be fitted to the rear of the front offside wing.
2. Referring to Fig. O:7, mark and then drill an 0.87 in (22 mm) dia hole in the top of the wing. Clean the area around the hole on the underside to remove any loose or flaking paint. Apply paint to any bare metal.
3. Remove the securing components from the aerial, then position the aerial through the wing from the underside. Loosely fit the securing components in the order shown in Fig. O:8.
4. Position the aerial lower support bracket and drill a small hole in the inner wing. Secure the aerial bracket with a self-tapping screw on manual aerials or a nut and bolt on electric types - see Figs. O:9 & O:10.
5. Fully tighten the aerial retaining collar nut (Fig. O: 8), but do not overtighten it.
6. Remove the blanking plug from the hole in the inner wing panel adjacent to the bottom of the aerial - see Figs. O:9 & O:10.
7. Feed the aerial cable (and the two feed wires on electric type aerials) through the hole and fit the grommet securely.
8. Route the aerial cable along the underside of the fasic fascia to the radio location and connect to the radio.

NOTE: It may be necessary to remove the lower insulation panel under the fascia to gain access when fitting the aerial cable.

9. Connect up the aerial feed and supply wires as shown in Fig. O:11, in the following order:
   1) Aerial connection to radio.
   2) White wire connection.
   3) Red wire connection.
   4) Ignition switch connector
   5) Line fuse, 8 amp.
   6) Aerial relay feed wire to radio.
10. Test the operation of the aerial with the radio as detailed previously under Radio "Installation".

SPEAKER FITTING . . . . [5]

Mono Type - Except Base Models

1. The speaker is fitted at the rear of the fascia panel. The grille can be removed by using a small screwdriver to lever it off the four retaining clips.
2. The speaker is attached by four screws (Fig. O:13). Remove the screws and detach the speaker.
3. When installing the speaker, make sure that the wires are connected firmly - see Fig. O:12.
4. To refit the grille, align the lugs on the grille with the clips and press the grille into position.

Stereo Type - Including Mono Fitting On Base Models

1. Remove the speaker grille button (arrowed in Fig. O: 14), then remove sufficient screws from the cowl panel to enable the speaker grille to be removed.
2. Feed the speaker wiring along the rear of the fascia and into the space behind the speaker bracket.
3. Connect the wiring connectors to the terminals at the rear of the speaker.
4. Fit the speaker to the door bracket and secure with the four screws (Fig. O:15).
5. Refit the grille cover and secure it with the button (Fig. O:14).
6. Repeat the installation procedure on the other speaker in the same way on stereo speaker systems.

CHILD SAFETY SEAT & BELTS. . . . . [6]

Only by fitting a safety seat or harness, secured firmly to the car body, can a child have a good chance of sur-

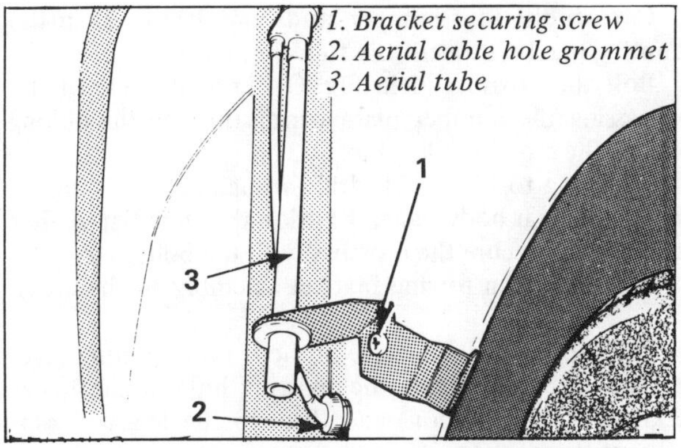

Fig. O:9 Manual aerial lower bracket fitting

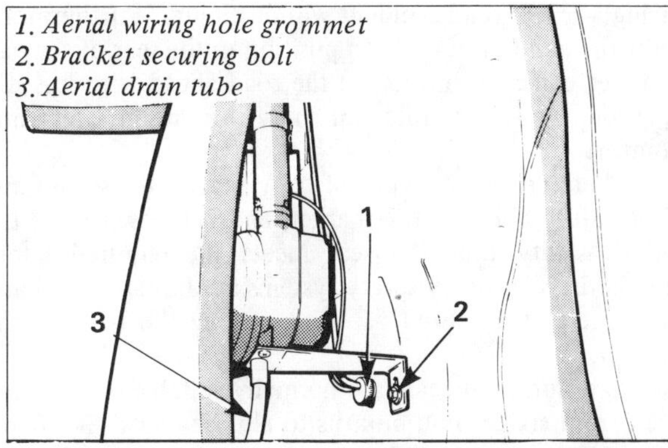

Fig. O:10 Electric aerial lower bracket fitting

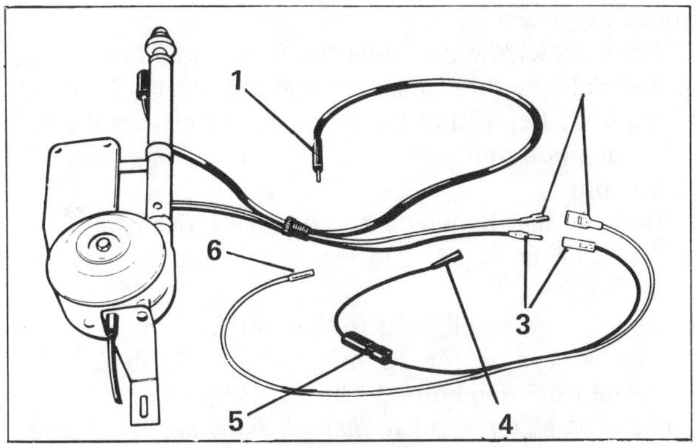

Fig. O:11 Electric aerial wire identification

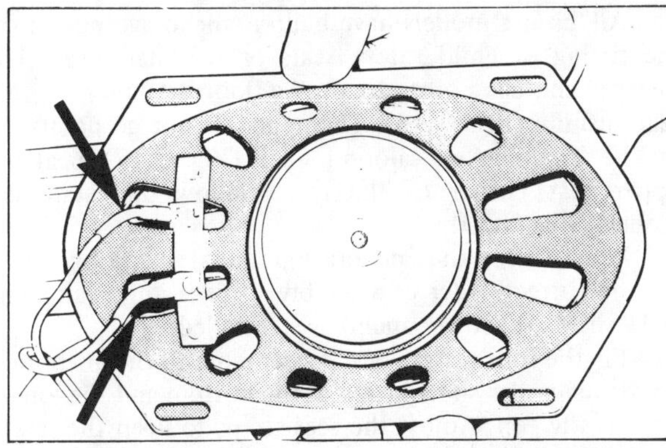
Fig. O:12 Mono speaker wire connections

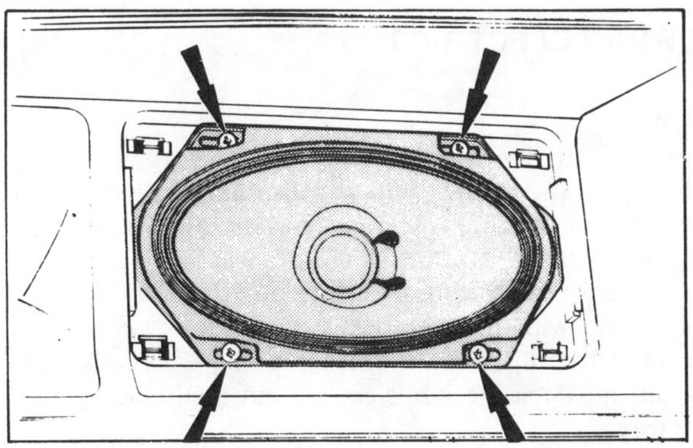
Fig. O:13 Mono speaker fitting in fascia panel

Fig. O:14 Stereo speaker cowl grille clip

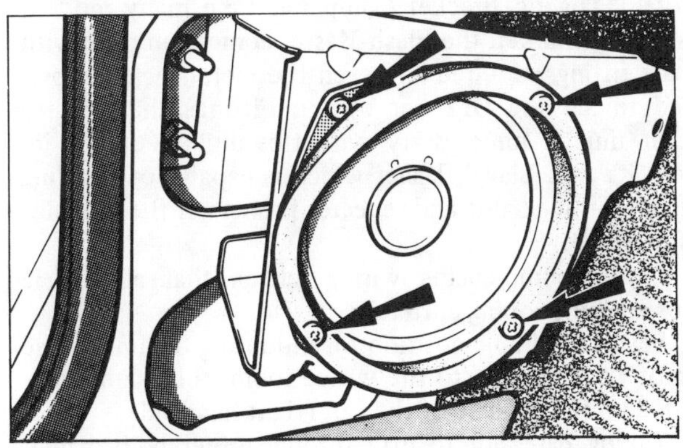
Fig. O:15 Stereo speaker securing screws

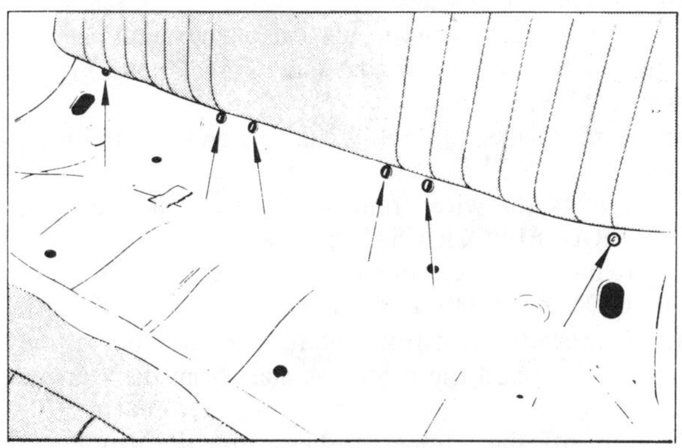
Fig. O:16 Seat belt mounting holes in floorpan

viving a severe road accident without injury. Furthermore, with the child safety seat in position in the rear of the car, a driver can concentrate on the road ahead with less distraction, while the child can enjoy the ride in safety and comfort.

There are many types of baby seat, cot restraint and child safety belts on the market, and it is essential to the child's safety that the seat meets the required safety standards. KL Jeenay safety systems are BSI approved and are generally accepted as being among the best on the market.

Baby or child seats are normally suitable for children aged from six to nine months to children aged five years. KL Jeenay safety seats or harnesses can, if necessary, be fitted two or three abreast, doubling up the anchor plates at anchorage points where necessary.

All Escort models have built-in anchorage points for the fitting of child safety seats or rear harnesses. The lower anchorage points are on the floorpan under the rear seat cushion (Fig. O:16). The upper anchorage points are on the 'D' pillar on saloon models (Fig. O:17) or at the upper rear corner of the rear window on estate car models (Fig. O:20).

When fitting the anchorage points, refer to Fig. O:19 for the correct order of assembly - this is important. Fig. O:18 shows the lower anchorages installed.

Fit the four restrain strap assemblies to the anchorage points using the short straps at the bottom and the longer ones at the top. Adjust the seat straps to keep the safety seat as high as possible. Attach the lower straps to the seat and adjust the webbing to suit, without overtightening. Adjust the harness straps to fit as tightly as comfortable. The lap straps should rest low over the bony part of the hip. Always ensure that the crotch strap is used.

To remove the seat, press the seat into the car seat cushion and disconnect the lower straps first.

TOWBAR FITTING .... [7]

Fitting a towbar to the Escort is quite simple and straightforward although the rear bumper must be removed first. The following fitting instructions apply to the Tanfield Towing bracket which is one-piece tubular steel design (with the AA seal of approval) and is ideal for the DIY owner to fit.

The bracket is supplied complete with all necessary fixing plates, nuts and bolts, and a socket mounting plate as standard (Fig. O:21).

1. Remove the nuts, bolts and plates from the towing bracket.
2. Unplug the wires from the number plate lamps - see GENERAL ELECTRICS chapter.
3. Remove the rear bumper from the car as detailed in the BODY & FITTINGS chapter.
4. Remove the mat from the rear luggage compartment floor and discard the rubber washers from the rear bumper studs.
5. Cut off the rubber washers from the rear bumper studs.
6. Cut off the four bumper studs and drill holes in the mounting brackets to suit the four 8 mm bolts.
7. Bolt the crossbar (1 in Fig. O:21) to the rear of the car, passing the number plate wiring through the oblong holes on the crossbar (9 in Fig. O:21).
8. Referring to Fig. O:21, drill through hole "2" and on through the rear body panel. Position the large flat washer inside the car, secure the crossbar with the bolt.
9. Bolt the main towing bracket assembly to the crossbar.
10. Drill up through the hole in the front mounting bracket, and on up through the existing hole in the lower thickness of box section, and through the luggage compartment floor.
11. Position the large reinforcing plate (5 in Fig. O:21) inside the car, and insert the long 10 mm bolt down through the box section.
12. From under the car, slide the front mounting to one side and slide the tubular spacer over the previously inserted long bolt. Reposition the front mounting over the end of the long bolt and secure the front mounting assembly with the nut.
13. Bolt the rear bumper into position with the reinforcing plates (7 in Fig. O:21) inside the luggage compartment.
14. Finally tighten all bolts to the correct torque. The recommended torque for 10 mm bolts is 35 lb ft (4,75 kg m), and for 8 mm bolts, 17 lb ft (2,3 kg m).

NOTE: It is recommended that all bolts are retightened after approximately 50 miles of towing.

TOWBAR ELECTRICS.................. [8]

Obviously if you are going to fit a towing bracket and ball to your Escort then you will also have to fit an electrical plug so that the trailer or caravan can be wired into the car's electrics. A lot of DIY types will tackle the most complicated of mechanical repairs but fight shy of anything to do with the electrical system. However, the fitting and wiring up of a tow bracket wiring kit is easily carried out provided the job is carried out methodically and with care.

The following text described the fitting of a complete Flash-Master wiring kit, which is also available from the Tanfield Towing Bracket Company. Like many modern kits on the market, the Flash-Master comes complete with all the fittings required and even the metal socket is prewired to aid the DIY car owner. Fitting a heavy duty flasher unit is unnecessary with this unit as the kit includes its own 'slave' flasher which is capable of handling hazard warning lights and repeater flashers, if fitted. Refer to Fig. O:23.

1. Before commencing wiring, check that all vehicle bulbs are functioning correctly.
2. Select a suitable position for mounting the unit in the rear of the car near to the wiring loom, this is normally located on the off-side of the car, (right-hand side).
3. Taking care not to drill through outer body panel, drill two fixing holes 2.75 in (70 mm) apart horizontally,

Fig. O:17 Upper mounting on 'D' pillar

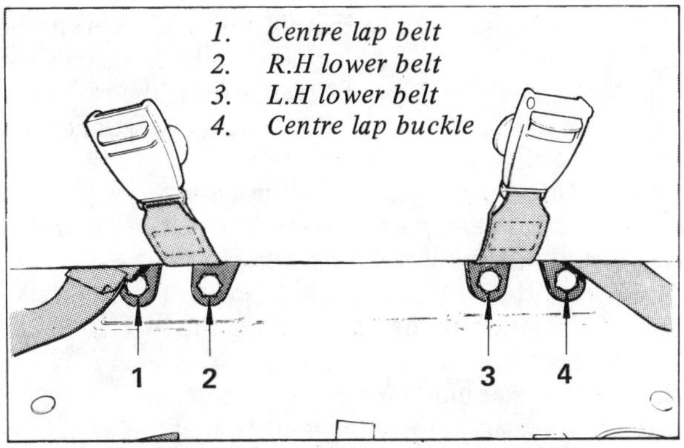

Fig. O:18 Lower belt anchorage installed

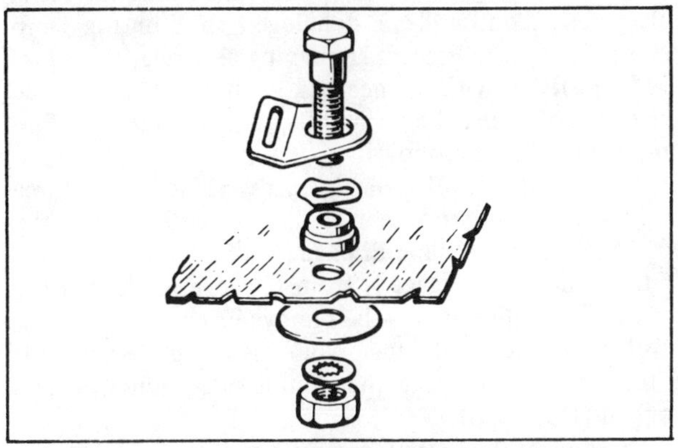

Fig. O:19 Correct assembly of KL anchorage

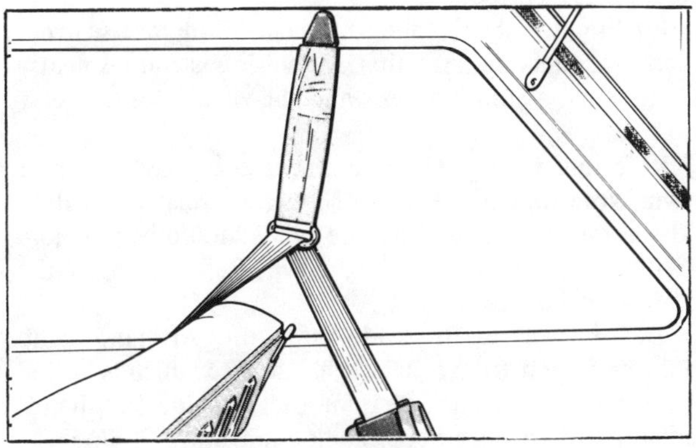

Fig. O:20 Upper anchorage point - estate car

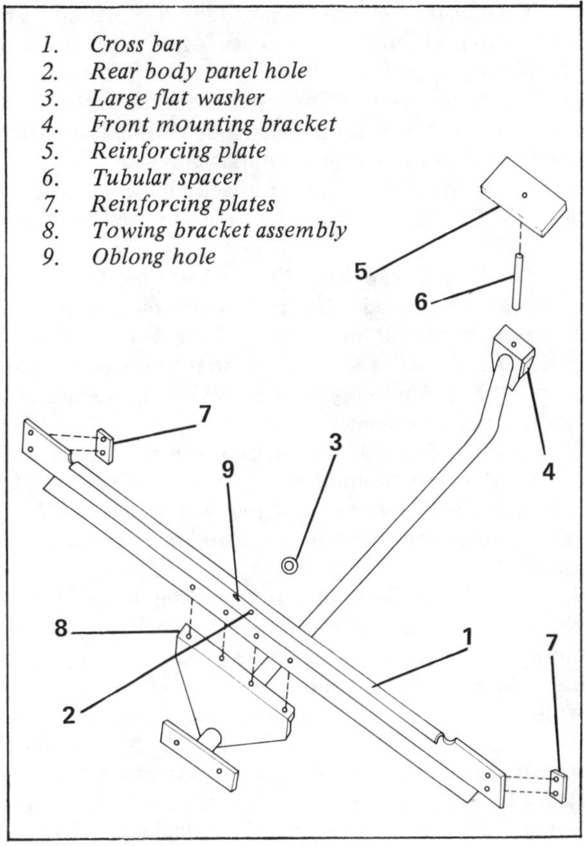

Fig. O:21 Details of the Tanfield towing bracket components

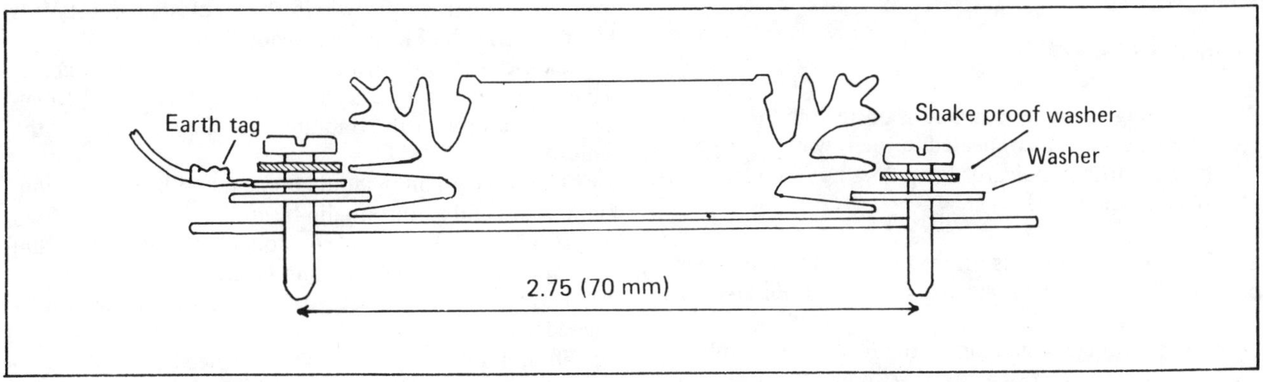

Fig. O:22 Flashmaster unit fitting arrangement

with a 1/8 in dia. drill, so that the unit will be positioned vertically with the connecting tags and leads facing downwards. Position and fix and the unit using the two screws and washers provided at this time also connect the BLACK earth lead tag (Fig. O:22).

NOTE: Make sure that the fixing with the self-tapping screws is to a metal part of the vehicle, otherwise proper earthing of the unit will not be possible.

4. Locate the wire in the vehicle loom that connects to the rear nearside flasher lamp. Using the special 'Scotchlok' connector, join this wire to the Green/Red wire from the Flash-Master unit. (When using 'Scotchlok' connectors it is not necessary to bare the insulation of the wire, as the connector is designed to pierce the insulation of the wires and so make good contact when clipped into place.

5. Now locate the wire leading to the rear offside flasher lamp. Connect this to the Green/White wire from the Flash-Master unit.

6. Uncoil the long BROWN wire from the unit and connect it to the POSITIVE feed from the vehicle battery, either using a ring terminal, routing the wire along to the battery or, if there is a positive feed already provided to the rear of the vehicle, then connect to that, using 'Scotchlok'.

7. The Purple wire from the Flash-Master unit is to be connected to the indicator lamp, which must be positioned so as to be within the sight of the driver, i.e. on the vehicle dashboard. The other tag on the warning indicator lamp is to be connected to the body of the vehicle as an earth, whre convenient.

8. Drill a 5/8 in dia. hole in boot floor in a convenient place and insert grommet. Affix socket mounting plate between towing bracket and tow ball and route 7 core cable through grommet to terminate at Flash-Master and loom.

9. Connect the 7 pin plug to the wiring loom as shown below using 'Scotchlok' connectors. Bear in mind that most sockets are pre-wired nowadays, but if the socket has to be wired beforehand refer to Fig. O:23.

White . . . . Earth
Black . . . . Left-hand tail light
Brown. . . . . Right-hand tail light
Red . . . . Stop lights
Green . . . . 'R' terminal on Flash-Master
Yellow . . . . 'L' terminal on Flash-Master

Fault Finding

Assuming that the wirng up is correct and the towed vehicle is hitched and connected to the towing vehicle:

1. If the warning indicator lamp does not flash when indicators are being used:

Check

a) That the 7 pin plug is firmly connected to the socket.

b) The live feed including the fuse to the Flash-Master unit.

c) The earth tag connection on the Flash-Master unit.

d) The indicator lamp for bulb failure and also the earthing tag.

2. If indicator warning lamp fails to flash when using right-hand flashers:

a) Check right-hand flasher bulb on towed vehicle, and the wiring from the 7 pin plug.

b) Check wiring connections to Flash-Master unit, right terminal tag. Conversely, check the left-hand flasher bulb and Flash-Master unit tag is warning lamp fails to flash when indicating to the left.

AUXILIARY LAMPS . . . . [9]

The following instructions apply to fitting the Hella replacement front grille and halogen spot or fog lamp accessories for the Escort. The lamps and relays are prewired and fitted with connectors, so they only have to be joined up following the wiring diagram - see Fig. O: 25 to complete the job, as follows:

1. Remove the front grille by pressing down the three securing clips at the top, then pulling the grille away at the top. Make sure the clips disengage completely and then lift the grille to disengage it from the locating lugs at the bottom. The grille can now be removed from the car.

2. Remove the Ford badge from the old grille and fit it to the Hella grille using the double-sided adhesive tape supplied (Fig. O:26).

3. Disconnect the battery earth strap.

4. Fit the spot or foglamps to the grille panel with the securing screws provided.

5. Fit the assembled grille to the car in the reverse order of removal, making sure that the wiring is routed neatly into the engine compartment and the vicinity of the relay position.

6. Referring to Fig. O:25, connect the relay wiring as shown. Note that the diagram shows the relative position of the component parts and the relay should be positioned with the terminals pointing down. Refer to Fig. O:24 for additional details.

7. The Brown earth lead from the left-hand lamp should be joined to the Brown earth wire which is bolted to the car body at the rear of the nearside (left-hand) headlamp unit, using a Scotchlock connector.

8. The Brown earth lead from the right-hand lamp has a small ring terminal on the free end and is joined to the front panel at the point where the air cleaner air intake is secured, using the PK screw provided.

9. Check that the wiring has been connected in accordance with the wiring diagram - see Fig. O:25, then reconnect the battery earth strap and test the operation of the lamps.

a) The driving lamps should come on with the headlamp main beam, and go out on dip beam.

b) Fog lamps should be controlled by the switch, and only operate while the dip beam is on.

10. Now check that the auxiliary lamps are correctly aligned.

11. If an additional on/off switch is needed for the driving lamps, connect from terminal 85 on the relay to the switch, and from the switch terminal to a good earth

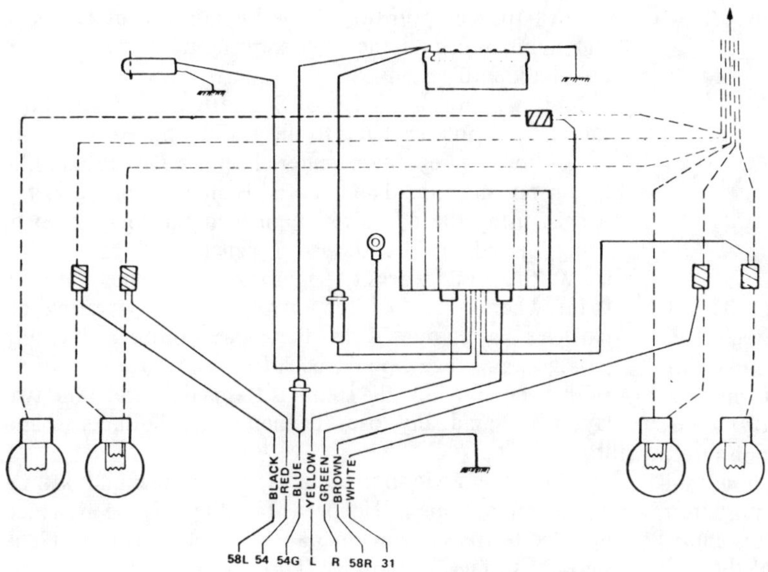

Fig. O:23 Flashmaster wiring diagram

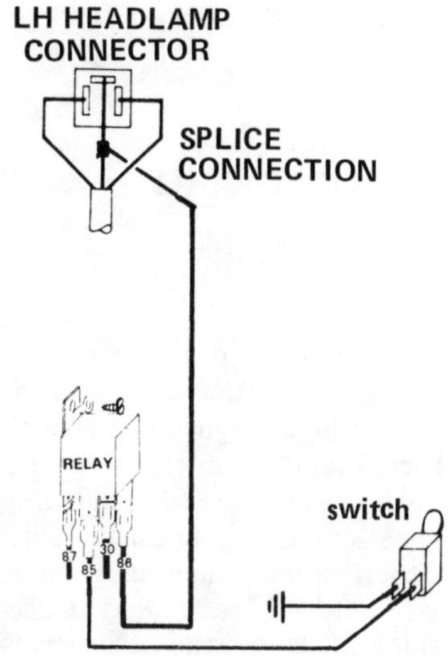

Fig. O:24 Headlamp/foglamp relay connections

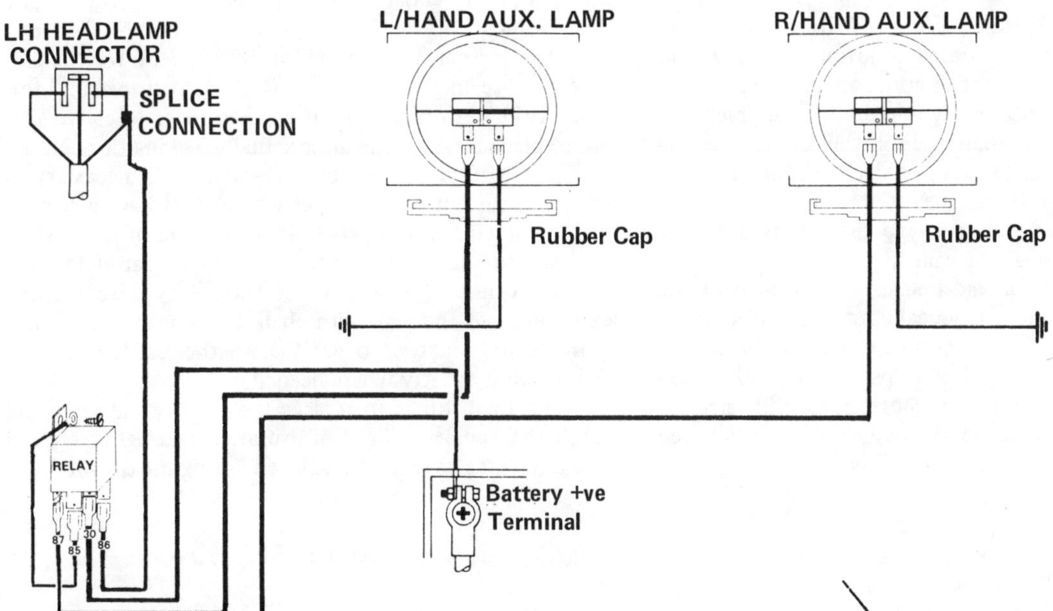

Fig. O:25 Hella driving lamp circuit

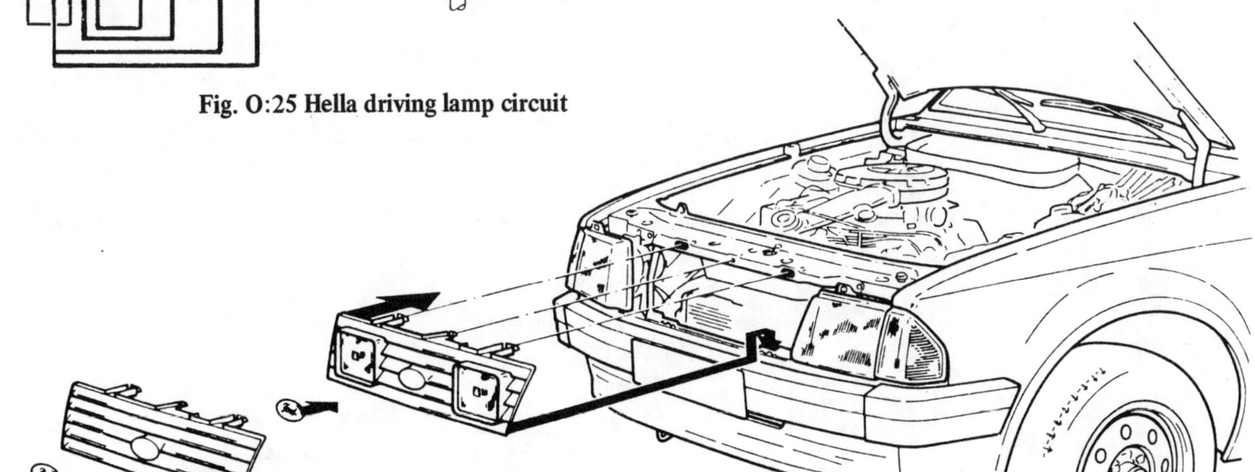

Fig. O:26 Hella auxilliary light grille fitting

point. Do not make the earth connection from terminal 85 on the relay as shown in Fig. O:25.

SOUNDPROOFING KIT.................[10]

Fitting

In any car, a certain amount of noise is transmitted to the passenger compartment and, if of a high enough level, can not only be annoying, but also tiring on long journeys. Therefore, the elimination or reduction of this noise is desirable for more enjoyable and safer motoring.

The passengers in a car can be subjected to noise from various sources: the wind rushing round the body and blowing round badly sealed doors; mechanical noise from the engine and transmission; the exhaust; tyres drumming on the road; and noise cause by the vibration of the car's sheet metal panels. By insulating the body of the car, it is possible to eliminate or considerably reduce a large proportion of this noise.

Sound Service (Oxford) Ltd., of Witney, Oxon, are leading manufacturers of sound proofing materials, and their Autosound kit for the Escort contains a variety of these, each designed specifically to reduce or eliminate the various types of noise experienced in the car. The following is based on information supplied by the company.

Before doing anything, the front seats should be removed as well as all the carpets and floor mats. Then the floor should be thoroughly cleaned.

When the floor/bulkhead are clear, check for any holes and seal them with the mastic sealing strip supplied in the Autosound kit. This should also be applied to all rubber grommets where cables and pipes pass through the bulkhead and to the area at the bottom of the steering column. The object is to obtain an air-tight seal between the engine and passenger compartments.

All the components of the kit are pre-cut to shape and numbered, and the next step should be to lay them out and identify them.

Fit the pre-cut pieces of Sound Barrier Mat to both sides of the bulkhead and to the gearbox hump. These are made from a grey foam material with a stiff rubber-like facing on one side. The material is glued in place on the panels, using the adhesive supplied in the kit, the foam being placed against the steel panel. In practice, this allows the sound waves to pass through the panel, into the foam where some of the energy of the vibrating air particles is dissipated as they pass through its tiny passages, and then bounce back from the stiff outer layer. In this way much of the sound is trapped between the two layers and gradually loses its energy as it bounces to and fro.

For the under-bonnet area, the Autosound kit includes some items in Neoprene faced felt. These are glued in place in the same manner as the Sound Barrier Mat, the smooth surface outermost. Like the foam material, the felt has tiny passages between its fibres that help dissipate energy of the vibrating air particles. This material cuts down multiple reflections of sounds within the engine compartment, preventing the build up of noise.

One further item provided is a roll of Weatherseal tape for sealing doors, etc. Rather then placing this around the entire opening, it is often more effective simply to place it where the air is actually leaking through the seal, and to find this is a simple matter. It is necessary to drive the car with a passenger who should have a section of normal garden hose. By holding one end of the hose to his ear and the other to the door seal, he can determine exactly where the seal is deficient. By marking the extremities of the leak with chalk or some similar means, it is a simple matter to cut the Weatherseal tape to size and install it exactly where needed.

The final job is to replace the carpets and seats and check the operation of all instruments and lights, etc., in case any wires have been displaced during the operation.

Technical Data

GENERAL DIMENSIONS - in (mm) . . . . [1]

Length
- Saloon 156.6 (3972)
- Estate 158.8 (4035)
- Van 164.0 (4166)

Width 64.6 (1641)

Height
- Saloon 54.4 (1382)
- Estate 54.5 (1384)
- Van 62.4 (1585)

Wheelbase - (except van) 94.4 (2398)
- Van 98.4 (2499)

Track -
- Front 54.5 (1385)
- Rear - (except van) 56.3 (1430)
- Van 54.5 (1385)

Turning circle - (except van) . . . . . . 33.0 feet (10,05 m)
- Van . . . . 34.1 feet (10,39 m)

Gross vehicle weight
- 1100 saloon . . . . 2750 lb (1250 kg)
- 1100 estate . . . . 3025 lb (1375 kg)
- 1300/1600 saloon/estate . . . . 3025 lb (1375 kg)

Maximum towing limit (2-up)
- 1100 saloon/estate . . . . 660 lb (300 kg)
- 1300/1600 saloon/estate . . . . 1980 lb (900 kg)

Maximum roof rack load . . . . 165 lb (75 kg)

ROUTINE MAINTENANCE DATA . . . . [2]

Specified Lubricants

Engine oil . . . . Multigrade SAE 10W-30 or 20W-50
Manual gearbox/final drive . . . . EP gear oil SAE 80
Steering rack . . . . Hypoid gear oil SAE 90
Anti-freeze (Pink colour) . . . Ford spec SSM-97B-9103A
Brake fluid . . . . Universal type, Ford spec. ESEA-M6C-1001A/SAM-6C9101A
Fuel octane 97 (4-star)

Capacities - pints (litres)

Engine oil (inc filter)
- 1100 OHV engine 5.7 (3,25)
- 1100/1300/1600 CVH engines 6.6 (3,75)

Cooling system (inc heater)
- 1100 engines* 11.8 (6,7)
- 1300 engine 13.4 (7,6)
- 1600 engine 13.7 (7,8)
- *1100 engines with 'towing pack' 13.9 (7,9)

Manual gearbox
- 4 - speed 4.9 (2,8)
- 5 - speed 5.5 (3,1)

Fuel tank . . . . 8.8 gal (40)

Tyre Pressures - lb/in^2 (bar)

Saloon/Estate:	Front	Rear
Loaded with up to 3 persons and normal usage	26 (1,8)	26 (1,8)
Fully laden - normal usage	28 (2,0)	33 (2,3)
Van:		
Fully laden - normal usage	26 (1,8)	37 (2,6)

TUNE-UP . . . . [3]

Compression Pressures - lb/in^2 (bar)

1100 OHV engine . . . . 156 - 180 (10,75 - 12,40)
All CVH engines . . . . 156 - 205 (10,75 - 14,13)

Spark Plugs

Make . . . . Motorcraft

Type:
- 1100 OHV engine . . . . AGRF 22
- 1100 CVH engine . . . . AGP 12C or Super AGPR 12C
- 1300 CVH engine . . . . AGP 22C
- 1600 CVH engine . . . . Super AGPR 12C
- 1600 CVH engine - XR3 . . . . Super AGP 12C or Super AGPR 12C

Electrode gap:
- 1100 OHV engine (AGRF 22) . . . 0.030 in (0,75 mm)
- All other engines . . . . 0.025 in (0,6 mm)

Valve Clearances - in (mm) Cold

OHV engine only
- Inlet valve 0.009 (0,22)
- Exhaust valve 0.023 (0,59)

Distributor

Contact breaker gap
- 1100 OHV engine . . . . 0.016 - 0.020 in (0,4 - 0,5 mm)
- 1100 CVH engine . . . . 0.016 - 0.020 in (0,4 - 0,5 mm)
- 1300/1600 engines . . . . N/A breakerless ignition

Dwell angle . . . . 48° - 52°
Rotor rotation . . . . Anti-clockwise

Firing order:
- 1100 OHV engine . . . . 1 - 2 - 4 - 3
- 1100/1300/1600 CVH engines . . . . 1 - 3 - 4 - 2

Ignition Timing

Static - all engines . . . . 12° BTDC

Idle Settings

Idle speed
- 1100 OHV engine . . . . 750 - 850 rpm
- All CVH engines . . . . 775 - 825 rpm

OHV ENGINE . . . . [4]

General Specifications - in (mm)

Engine identification code . . . . GLB
Capacity . . . . 1117 cc
Bore diameter 2.912 (73,96)
Stroke 2.56 (64,98)
Compression ratio 9.5:1

Cylinder Head - in (mm)

Identification marking 77BM-6090-DF
Valve seat angle. . . . . 45°
Valve spring free length 1.6535 (42)
Valve stem to guide clearance:
 Inlet valve 0.0008-0.0027 (0,21-0,070)
 Exhaust valve 0.0016-0.0036 (0,043-0,092)
Valve stem diameter: (Standard)
 Inlet valve 0.3097-0.3096 (7,868-7,866)
 Exhaust valve 0.3088-0.3096 (7,846-7,864)
Valve rocker clearance (Cold)
 Inlet valves. 0.009 (0,22)
 Exhaust valves 0.023 (0,59)

Oil Pump - in (mm)

Relief valve opening pressure . . . . 34 - 39 lb/in^2 (2,34 - 2,69 bar)
Warning light switch pressure. . . . . .4.5 - 7.5 lb/in^2 (0,31 - 0,52 bar)
System pressure (min Hot)-idling . . . .8.5 lb/in^2 (0,6 bar)
 2000 rpm. . . . . 21.3 lb/in^2 (1,5 bar)
Clearances
 Rotor to housing clearance . . . . .0.0055-0.0102 (0,14 - 0,26)
 Inner to outer rotor gap . . . . .0.0020-0.0050 (0.051 - 0,127)
 Rotors to end cover endfloat . . . . .0.0009-0.0023 (0,025 - 0,06)

Camshaft - in (mm)

Endfloat 0.0024-0.0075 (0,062-0,193)
Thrust plate thickness 0.1754-0.1774 (4,457-4,508)
Camshaft bearing diameter . . . . .1.5596 - 1.5604 (39,615 - 39,636)
Bearing internal diameter-standard . . . . . .1.5614-1.5635 (39,662 - 39,713)
Cam lift - inlet 0.2356 (5,985)
 exhaust 0.2320 (5,894)

Crankshaft - in (mm)

Endfloat . . . . 0.0035 - 0.0109 (0,079-0,279)
Main bearing journal diameter
 std 2.2437-2.2440 (56,990-57,000)
 Yellow spot 2.2433-2.2437 (56,980-56,990)
Undersizes . . . 0.010, 0.020, 0.030 (0,254, 0,508, 0,762)
Clearance. . . . . 0.0003-0.0018 (0,009 - 0,046)
Big-end bearing journal diameter
 std 1.6925-1.6933 (42,99-43,01)
Undersizes . . . 0.010, 0.020, 0.030 (0,254, 0,508, 0,762)
Thrust washer thickness 0.1102-0.1122 (2,80-2,85)
 oversize. 0.1177-0.1196 (2,99-3,04)

Cylinder Block - in (mm)

Identification marking 77BM-6015-CC
Number of main bearings 3
Bore diameter:
 std, size 1. . . . . 2.9110-2.9114 (73,940-73,950)
 size 2 . . . . 2.9114-2.9118 (73,950-73,960)
 size 3 . . . . 2.9118-2.9122 (73,960-73,970)
 size 4 and service. . . . 2.9122-2.9125 (73,970-73,980)
Permissible oversizes 0.0196-0.0393 (0,5, 1,0)

Pistons, Rings and Con-Rods - in (mm)

Piston to bore clearance . . . 0.0005-0.0019 (0,015-0,050)
Ring gap (fitted in bore)
 Top and centre rings 0.0098-0.0177 (0,25-0,45)
 Bottom ring 0.0078-0.0157 (0,20-0,40)
Ring gap position on piston
 Top . . . . 180° to oil ring gap
 Centre. . . . . 90° to oil ring gap
 Bottom . . . . .In-line with piston pin
Piston pin diameter:
 White 0.8118-0.8120 (20,622-20,625)
 Red 0.8120-0.8121 (20,625-20,628)
 Blue 0.8121-0.8122 (20,628-20,631)
 Yellow 0.8122-0.8123 (20,631-20,634)
Interference in piston at 21^oC 0.0005-0.0017 (0,013-0,045)
Pin to piston clearance at 21^oC . . . . .0.0001-0.0004 (0,005 - 0,011)
Big-end bore diameter . . 1.8379-1.8387 (46,685-46,705)
Small-end bore diameter 0.8105-0.8113 (20,589-20,609)

Tightening Torques - lb ft (kg m)

Main bearing cap . . . . 64.9 - 75.3 (8,8 - 10,2)
Connecting rod bolts . . . . 19.2 - 24.4 (2,6 - 3,2)
Rear oil seal carrier . . . . 11.8 - 14.8 (1,6 - 2,0)
Flywheel . . . . 47.2 - 51.7 (6,4 - 7,0)
Clutch pressure plate . . . . .5.2 - 7.4 (0,6 - 1,0)
Chain tensioner. . . . . .5.2 - 6.6 (0,7 - 0,9)
Camshaft thrust plate. . . . . .3.0 - 3.7 (0,4 - 0,5)
Camshaft sprocket. . . . . 11.8 - 14.8 (1,6 - 2,0)
Timing cover . . . . .5.2 - 7.4 (0,7 - 1,0)
Water pump . . . . .5.2 - 7.4 (0,7 - 1,0)
Crankshaft pulley . . . . 39.9 - 43.5 (5,4 - 5,9)
Water pump pulley . . . . .5.2 - 7.4 (0,7 - 1,0)
Starter motor . . . . 25.8 - 33.2 (3,5 - 4,5)
Fuel pump . . . . 11.8 - 14.8 (1,6 - 2,0)
Distributor. . . . . .5.2 - 7.4 (0,7 - 1,0)
Distributor clamp bolt . . . . .2.2 - 3.0 (0,3 - 0,4)
Oil pump . . . . 11.8 - 14.8 (1,6 - 2,0)
Oil pump cover . . . . .5.9 - 8.9 (0,8 - 1,2)
Sump - Stage 1 . . . . .4.4 - 5.9 (0,6 - 0,8)
 Stage 2 . . . . .5.9 - 8.1 (0,8 - 1,1)
 Stage 3 . . . . .5.9 - 8.1 (0,8 - 1,1)
Sump plug . . . . 15.5 - 20.7 (2,1 - 2,8)
Oil pressure switch. . . . . .9.6 - 11.1 (1,3 - 1,5)
Temperature sender . . . . .8.9 - 11.8 (1,2 - 1,6)
Rocker shaft supports . . . . 25.1 - 29.5 (3,4 - 4,0)
Cylinder head bolts - Stage 1 . . . . . . 7.4 - 11.1 (1,0 - 1,5)
 Stage 2 . . . . 29.5 - 36.9 (4,0 - 5,0)
 Stage 3 . . . . 59.0 - 66.4 (8,0 - 9,0)
 Stage 4 (after 10 to 20 mins) . . . . 73.8 - 81.2 (10,0 - 11,0)
Rocker cover . . . . .3.0 - 3.7 (0,4 - 0,5)
Exhaust manifold . . . . 10.3 - 12.6 (1,4 - 1,7)
Inlet manifold. . . . . 11.8 - 14.8 (1,6 - 2,0)
Carburettor . . . . 12.6 - 15.5 (1,7 - 2,1)
Thermostat housing . . . . 12.6 - 15.5 (1,7 - 2,1)
Spark plugs . . . . 11.1 - 14.8 (1,5 - 2,0)

CVH ENGINE........................[5]

General Specifications - in (mm)

Engine identification code: ('P' Engine)
- 1100 cc (HC) .... GPA
- 1300 cc .... JPA
- 1600 cc .... LPA
- 1600 cc XR3 .... LUA

Capacity - 1100 .... 1117 cc
- 1300 .... 1296 cc
- 1600 .... 1597 cc

Bore diameter
- 1100 cc 2.9118 (73,96)
- 1300/1600 cc 3.148 (79,96)

Stroke
- 1100 cc 2.56 (64,98)
- 1300 cc 2.54 (64,52)
- 1600 cc 3.13 (79,52)

Compression ratio 9.5:1

Cylinder Head - in (mm)

Valve seat angle .... 45°

Valve spring free length 1.8582 (47,2)

Valve stem-to-guide clearance
- Inlet valve .... 0.0007 - 0.0024 (0,020 - 0,063)
- Exhaust valve .... 0.0018 - 0.0035 (0,046 - 0,089)

Valve stem diameter (Standard)
- Inlet valve .... 0.3159 - 0.3166 (8,025 - 8,043)
- Exhaust valve .... 0.3149 - 0.3156 (7,999 - 8,017)

Oversizes .... 0.0078 - 0.0157 (0,2 - 0,4)

Oil pump - in (mm)

Minimum oil pressure - idling .... 14.22 lb/in^2 (1,0 bar)
- 2000 rpm .... 41.0 lb/in^2 (2,8 bar)

Camshaft - in (mm)

Endfloat .... 0.0019 - 0.0059 (0,05 - 0,15)

Thrust plate thickness .... 0.1964-0.1972 (4,99 - 5,01)

Camshaft bearing diameter:
- No 1 bearing 1.7618 (44,75)
- No 2 bearing 1.7716 (45,00)
- No 3 bearing 1.7814 (45,25)
- No 4 bearing 1.7913 (45,50)
- No 5 bearing 1.8011 (45,75)

Cam lift-inlet/exhaust - All except XR3 .... 0.2279 (5,79)
- XR3 model 0.2397 (6,09)

Crankshaft - in (mm)

Endfloat .... 0.0035 - 0.0118 (0,09 - 0,30)

Main bearing journal diameter
- std 2.2826-2.2834 (57,98-58,00)

Undersizes 0.009, 0.019, 0.029 (0,25, 0,50, 0,75)

Clearance .... 0.0004 - 0.0022 (0,011 - 0,058)

Big-end journal diameter
- std 1100 cc .... 1.6925 - 1.6933 (42,99 - 43,01)
- std - All .... 1.8854 - 1.8862 (47,89 - 47,91)

Undersizes ... 0.009; 0.019; 0.029 (0,25; 0,50; 0,75; 1,00)

Clearance .... 0.0002 - 0.0023 (0,006 - 0,060)

Thrust washer thickness
- std .... 0.0905 - 0.0988 (2,301 - 2,351)
- oversize .... 0.0980 - 0.1000 (2,491 - 2,541)

Cylinder Block - in (mm)

Identification marking
- 1100 cc 81SM-6015-CKA
- 1300 cc 81SM-6015-CLA
- 1600 cc 81SM-6015-CMA

Number of main bearings 5

Bore diameter:
- 1100 cc std 1 .... 2.9110 - 2.9 114 (73,94-73,95)
- 1100 cc std 2 .... 2.9114 - 2.9118 (73,95-73,96)
- 1100 cc std 3 .... 2.9118 - 2.9122 (73,96-73,97)
- 1100 cc std 4 .... 2.9122 - 2.9125 (73,97-73,98)
- 1300/1600 cc std 1 .... 3.1472-3.1476 (79,94-79,95)
- 1300/1600 cc std 2 .... 3.1476-3.1480 (79,95-79,96)
- 1300/1600 cc std 3 .... 3.1480-3.1484 (79,96-79,97)
- 1300/1600 cc std 4 .... 3.1484-3.1488 (79,97-79,98)

Pistons, Rings and Con-Rods - in (mm)

Piston to bore clearance 0.0003-0.0017 (0,010-0,045)

Ring gap (fitted in bore)
- 1100 cc rings 1 and 2 .... 0.0098-0.0177 (0,25-0,45)
- 1100 cc oil control ring .... 0.0078-0.0157 (0,2 - 0,4)
- 1300/1600 cc rings 1 and 2 .... 0.0118-0.0196 (0,30-0,50)
- 1300/1600 cc oil control ring .... 0.0157-0.0551 (0,4 - 1,4)

Piston pin diameter:
- White 0.8118-0.8120 (20,622-20,625)
- Red 0.8120-0.8121 (20,625-20,628)
- Blue 0.8121-0.8122 (20,628-20,631)
- Yellow 0.8122-0.8123 (20,631-20,634)

Interference in piston at 21°C .... 0.0005 - 0.0017 (0,013 - 0,045)

Pin to piston clearance at 21°C .... 0.0001 - 0.0004 (0,005 - 0,011)

Big-end bore diameter
- 1100 cc 1.8379-1.8387 (46,685-46,705)
- 1300/1600 cc 2.0035-2.0043 (50,890-50,910)

Small-end bore diamter .... :0.8105-0.8113 (20,589-20,609)

Tightening Torques - lb ft (kg m)

Main bearing caps .... 66.4 - 73.8 (9,0 - 10,0)

Big-end bearing caps .... 19.2 - 24.3 (2,6 - 3,3)

Oil pump .... 5.9 - 8.1 (0,8 - 1,1)

Oil pump suction head to engine block .... 11.8 - 14.8 (1,6 - 2,0)

Oil pump suction head to oil pump .... 5.9-8.9 (0,8 - 1,2)

Rear oil seal .... 5.9 - 8.1 (0,8 - 1,1)

Sump 1 st stage .... 5.9 - 8.1 (0,8 - 1,1)
- 2 nd stage .... 5.9 - 8.1 (0,8 - 1,1)

Flywheel .... 61.2 - 66.4 (8,3 - 9,0)

Cylinder head - tighten in 4 stages in specified order.
- 1st stage 18 (2,5)
- 2nd stage 41 (5,5)
- 3rd stage .... + 90°
- 4th stage .... + 90°

(mark bolt heads with chalk in same direction as check)

Camshaft thrust plate .... 7.4 - 11.1 (1,0 - 1,5)

Camshaft toothed belt pulley. . . . . 39.8 - 43.5 (5,4 - 5,9)
Toothed belt tensioner . . . . 11.8 - 14.8 (1,6 - 2,0)
Water pump . . . . .5.2 - 7.4 (0,7 - 1,0)
Rocker studs in cylinder head . . . . . 7.4 - 11.1 (1,0 - 1,5)
Rocker arms. . . . . .18 - 21 (2,5 - 2,9)
Rocker cover . . . . .4.4 - 5.9 (0,6 - 0,8)
Timing cover . . . . .5.2 - 6.6 (0,7 - 0,9)
Petrol pump. . . . . .3.7 - 5.2 (0,5 - 0,7)
Distributor. . . . . .3.7 - 5.2 (0,5 - 0,7)
Exhaust manifold . . . . 10.3 - 12.5 (1,4 - 1,7)
Inlet manifold. . . . . 11.8 - 14.8 (1,6 - 2,0)
Carburettor . . . . .8.9 - 15.5 (1,2 - 2,1)
Thermostat housing . . . . .5.2 - 7.4 (0,7 - 1,0)

ENGINE ELECTRICS. . . . . [6]

Battery

Type. . . . . .12 volt, 35, 43 or 52 A.h

Alternator - Bosch (Ford Type A)

Type. G1-28A, K1-35A, K1-45A, K1-55A
Rated output:
G1-28A. . . . . 28 amps
K1-35A. . . . . 35 amps
K1-45A. . . . . 45 amps
K1-55A. . . . . 55 amps
Minimum brush length 0.197 (5)
Regulated voltage . . . . .13.7 - 14.6 volts

Alternator - Lucas (Ford Type B)

Type. A115/28, A115/36, A133/45, A133/55
Rated output:
A115/28 . . . . 28 amps
A115/36 . . . . 35 amps
A133/45 . . . . 45 amps
A133/55 . . . . 55 amps
Minimum brush length 0.197 (5)
Regulated voltage . . . . .13.7 - 14.5 volts

Alternator - Motorola (Ford Type C)

Type. 9AR 2650D, 9AR 2652F, 9AR 2627G
Rated output:
9AR 2650D . . . . 28 amps
9AR 2652F . . . . 35 amps
9AR 2627G . . . . 45 amps
Minimum brush length 0.16 (4)
Regulated voltage . . . . .13.7 - 14.6 volts

Starter Motor - Lucas

Type. 8M90, 9M90
Minimum brush length 0.32 (8,0)
Armature endfloat. 0.01 (0,25)

Starter Motor - Bosch

Type. . . . . 0.8 Kw, 0.85 Kw, 0.9 Kw
Minimum brush length 0.39 (10,0)
Armature endfloat. 0.01 (10,3)

Starter Motor - Nippondenso

Type. . . . . .0.6 Kw, 0.9 Kw
Minimum brush length:
0.6 Kw 0.39 (10.0)
0.9 Kw 0.35 (9,0)
Armature endfloat. .0.02 (0,6)

Distributor - OHV Engine

Type. Bosch 81BF-12100-AA
Condenser capacity . . . . 0.235-0.315 mfd

Distributor - CVH 1100 cc Engine

Conventional Ignition

Type. . . . . Lucas 81SF - 12100 - AAA
Bosch 81SF - 12100 - ABA
Condenser capacity . . . . Lucas 0.21 - 0.29 mfd
Bosch 0.235 - 0.315 mfd

Distributor - CVH 1300/1600 Engine

Breakerless Ignition

Type-1300 cc . . . . Lucas 81SF-12100-AGA
1600 cc . . . . .Lucas 81SF-12100-AJA
1600 cc XR3 . . . . Lucas 81SF-12100-ALA
Type-1300 cc . . . . Bosch 81SF-12100-AHA
1600 cc. . . . . Bosch 81SF-12100-AKA
1600 cc XR3. . . . . Bosch 81SF-12100-AMA

Ignition Coil - 1100 OHV Engine

Make. . . . . Bosch, Lucas, Cajavec, Polmot, Femsa
Type. . . . . Low voltage, use with 1.5 ohm ballast resistor
Output . . . . 23.0 kilovolt (minimum)
Primary resistance . . . . 1.20 - 1.40 ohm
Secondary resistance . . . . 5000 - 9000 ohm

Ignition Coil - 1100 cc CVH Engine

Make. . . . . Bosch, Lucas, Cajavec, Polmot, Femsa
Type. . . . . Low voltage, use 1.5 ohm ballast resistor
Output . . . . 25.0 kilovolt (minimum)
Primary resistance . . . . 1.20 - 1.40 ohm
Secondary resistance . . . . 5000 - 9000 ohm

Ignition Coil - 1300/1600 cc CVH Engine

Type. . . . . High output 'breakerless' type ignition coil.
Output . . . . 30.0 kilovolt (minimum)
Primary resistance . . . . 0.72 - 0.88 ohm
Secondary resistance . . . . 4500 - 7000 ohm

Tightening Torques - lb ft (kg m)

Distributor to head bolts. . . . . 4.0 - 5.0 (0,51 - 0,71)
Distributor cap screws - Bosch . . . . 2.0 - 3.0 (0,26 - 0,41)
Lucas . . . . 1.0 - 1.5 (0,18 - 0,23)
Module fixing screws . . . . 0.75 - 1.0 (0,1 - 0,14)

COOLING SYSTEM . . . . [7]

Type - 1100 cc OHV engine. . . . . . Semi-pressurised with crossflow type radiator
Type - 1100 cc CVH engine. . . . . .Semi-pressurised with overflow tank
Type - 1300/1600 cc CVH engine. . . . . Fully pressurised with degas tank

Capacity (inc. heater and overflow tank)

1100 cc OHV engine . . . . .9.27 pints (5,27 litres)
1100 cc CVH engine . . . . 10.9 pints (6,2 litres)
large radiator . . . . 12.7 pints (7,2 litres)
1300 cc CVH engine . . . . 12.5 pints (7,1 litres)
1600 cc CVH engine . . . . 12.1 pints (6,9 litres)

Thermostat

Type. . . . . Wax element
Opening temperature . . . . 185°-192°F (85-89°C)
Fully open temperature . . . . . . 217°-223°F (99-102°C)
For used thermostats, add 5°F (3°C)

Pressure cap

1100 cc OHV engine . . . . 13 lb/in^2 (0,9 bar)
CVH engines - releasing pressure. . . . .12.0-15.7 lb/in^2 (0,83 - 1,08 bar)

Water Pump Drive Belt Tension

1100 cc OHV Engine Only

load for deflecting a new belt 0.4 (10). . . . .79-101 lbs (36 - 46 kg)
load for deflecting a used belt 0.4 (10) . . . . . . .56 lbs (25,5 kg)

Tightening Torques lb ft (kg m)

OHV engine:
Water pump . . . . 5.2 - 7.4 (0,71 - 1,02)
Thermostat housing. . . . . 12.5 - 15.5 (1,73 - 2,14)
Water pump pulley . . . . 6.3 - 7.8 (8,5 - 10,6)
CVH engine:
Radiator to front panel. . . . . .5.0 - 7.0 (6,8 - 9,5)
Thermostat housing retaining bolts . . .5.0 - 7.0 (6,8 - 9,5)
Water pump retaining bolts . . . . .5.0 - 7.0 (6,8 - 9,5)
Fan shroud to radiator bolts . . . . . . .5.0 - 7.0 (6,8 - 9,5)
Fan motor to shroud . . . . 5.9 - 7.4 (8,0 - 10,0)

FUEL SYSTEM. . . . . [8]

Fuel Pump

Delivery pressure (idle speed). . . . . .3.2 - 5.2 lb/in^2 (0,22 - 0,35 bar)

Carburettor Type:

1100 cc OHV engine . Ford VV (1V) 79BF-9510-KCB
1100 cc CVH engine . Ford VV (1V) 81SF-9510-KAA
1300 cc CVH engine . Ford VV (1V) 81SF-9510-KCA
1600 cc CVH engine . Ford VV (1V) 81SF-9510-KFA
1600 cc XR3 engine. . .Weber (2V) 81SF-9510-AA/AB
Idle speed - all models (fan on) . . . . 800 rpm ± 25 rpm
Idle CO content - all except XR3 . . . . . . .1.5 ± 0.5% CO
XR3 model . . . . 1.25 ± 0.25% CO

General Specifications (Ford VV)

1100 cc OHV . . . . FCH
1100 cc CVH . . . . FDA
1300 cc CVH . . . . FDK
1600 cc CVH . . . . FCX
Choke system:
Choke gauging (twist drill)-1100 cc OHV 0.120 (3,0)
All CVH 0.135 (3,4)
Choke fast-idle/pull down (twist drill):
1100 cc OHV . . . . 0.130 (3,3) clockwise
1100/1300 CVH . . . . 0.170 (4,4) anti-clockwise
1600 cc CVH . . . . 0.130 (3,3) clockwise
Choke housing alignment - all Ford VV . . . . . . On index

General Specifications (Weber 2V)

Choke tube diameter:
Primary throttle . . . . 24 mm
Secondary throttle. 25 mm
Choke plate pull-down . . . . .5,5 mm ± 0,3 mm
Choke phasing (twist drill) . . . . .2,0 mm ± 0,5 mm
Main jet (- 10.81) 112/125
(10.81 -) 115/125
Compensating jet. 160/150
Slow running jet 50/60
Emulsion tube F30/F30
Float level setting . . . . 35 mm ± 0,5 mm

CLUTCH & GEARBOX . . . . [9]

Clutch

Type. . . . . .Diaphragm/single dry plate
Operation .Self-adjusting cable
Clutch lining thickness (nominal) 0.12 (3,20)
Clutch plate diameters - in (mm)
1100 cc OHV/CVH saloon 6.5 (165)
1100 cc Estate and Van/all 1300 cc CVH. . . 7.5 (190)
1600 cc CVH and XR3. . . . . 7.9 (200)
Clutch pedal stroke (nominal) 6.1 (155)

Gearbox - 4 - Speed ('J' gearbox)

Ratios-all models:
Fourth 0.95:1
Third 1.35:1
Second 2.05:1
First 3.58:1
Reverse 3.77:1

Speedometer Drive

Drive worm - number of teeth 21
Driven gear pinion teeth - OHV engine:
with 145 SR - 13 tyres 20
with 155 SR - 13 tyres 19
Driven gear pinion teeth: - all CVH engines:
with 145 SR - 13 tyres 19
with 155 SR - 13, 175/70 HR - 13, 185/60 HR - 14, 165 RR - 13 tyres. 20

Tightening Torques - 4 Speed Gearbox lb ft (kg m)

Pressure plate to flywheel - 165 mm dia. . 7 - 8 (0,9 - 1,1)
190 and 200 mm dia . . . . .13 - 15 (1,6 - 2,0)
Gearbox to engine flange bolts. . . . . . .26 - 33 (3,5 - 4,5)
Cover - clutch housing . . . . .26 - 33 (3,5 - 4,5)

Starter motor bolts . . . . .26 - 33 (3,5 - 4,5)
Front gearbox mounting to gearbox (4 bolts) . . .30 - 37.5 (4,1 - 5,1)
Gearbox mounting bolts (front and rear) . . . . 38 - 47 (5,2 - 6,4)
Spindle carrier to ball joint . . . . .52 - 66 (7,0 - 9,0)
Gearshift stabiliser to gearbox . . . . . . .37 - 44 (5,0 - 6,0)
Clamp bolts - shift rod to gearbox. . . 10 - 12.5 (1,4 - 1,7)
Axle drive bevel wheel to
differential carrier . . . . .72 - 94 (9,8 - 12,8)
Small housing section to large
housing section . . . . .15 - 20 (2,1 - 2,7)
Cover to housing . . . . .9 - 11 (1,2 - 1,45)
Cap nut (selector shaft locking mechanism) 22 (3,0)
Gearshift housing to floor pan . . . . . . .10 - 12 (1,3 - 1,7)
Oil filler plug . . . . .17 - 22 (2,3 - 3,0)

Gearbox - 5 - Speed ('K' gearbox)

Ratios:	1300	1600
Fifth	0.76:1	0.76:1
Fourth	0.95:1	0.95:1
Third	1.35:1	1.28:1
Second	2.04:1	1.91:1
First	3.58:1	3.15:1
Reverse	3.62:1	3.62:1

Speedometer Drive

Drive worm, number of teeth 21
Driven gear pinion teeth . . . . 19 (20 with 145 SR - 13 tyres only)

Tightening Torques - 5 Speed Gearbox lb ft (kg m)

Flange bolts - gearbox to engine . . . 25.8 - 33.2 (3,5 - 4,5)
Cover - clutch housing . . . . 25.8 - 33.2 (3,5 - 4,5)
Support bracket to clutch housing . 14.0 - 18.4 (1,9 - 2,5)
Starter motor bolts . . . . 25.8 - 33.2 (3,5 - 4,5)
Front gearbox mounting to gearbox (4 bolts) . . 30.2-37.6 (4,1 - 5,1)
Gearbox mounting bolts (front and rear) . . . . 38.3 - 47.2 (5,2 - 6,4)
Spindle carrier to ball joint . . . . . . 51.7 - 66.4 (7,0 - 9,0)
Gearshift stabiliser to gearbox . . . . 36.9 - 44.3 (5,0 - 6,0)
Clamp bolts-shift rod to gearbox 10.3-12.5 (1,4-1,7)
Selector block to main selector shaft. . .8.8-11.1 (1,2-1,5)
Axle - drive bevel wheel to
differential carrier . . . . 72.3 - 94.5 (9,8 - 12,8)
Small housing section to large
housing section . . . . 15.5 - 19.9 (2,1 - 2,7)
5th gear housing to gearbox. . . . . 8.5 - 10.7 (1,15 - 1,45)
5th gear selector pin. . . . . 10.3 - 14.8 (1,4 - 2,0)
5th gear selector plate to housing . . 19.9 - 25.1 (2,7 - 3,4)
Cover to 5th gear housing . . . . .5.9 - 8.1 (0,8 - 1,1)
Cap nut (selector shaft locking mechanism) . . .22.1 (3,0)
Gearshift housing to floor pan . . . . 11.1 - 16.2 (1,5 - 2,2)
Oil filler plug . . . . 17.0 - 22.1 (2,3 - 3,0)
Switch - reversing light . . . . 17.0 - 22.1 (2,3 - 3,0)

STEERING . . . . [10]

Type. . . . . .Rack and pinion

Steering Rack Adjustments - in (mm)

Rack slipper shim sizes:
71BB-3N597-AA. . . . . .Steel 0.005 (0,127)
71BB-3N597-BA . . . . Steel 0.007 (0,19)
71BB-3N597-CA . . . . Steel 0.010 (0,25)
71BB-3N597-DA. . . . . Steel 0.015 (0,38)
71BB-3N597-EA . . . . Steel 0.020 (0,50)
77FB-3581-AA . . . . .Paper gasket 0.0055 (0,14)
Track rod articulation - pull. . . . . 5.1 lbs (2,3 kg)
Track rod turning torque . . 3 - 4.3 lb ft (0,36 - 0,6 kg m)
Pinion turning torque. . . . . . . 5 - 12 lb in (6 - 13 kg cm)

Tightening Torques - lb ft (kg m)

Steering gear to bulkhead . . . . .33 - 37 (4,6 - 5,1)
Track-rod end to steering arm . . . . . . .18 - 22 (2,5 - 3,0)
Coupling to pinion spline . . . . .33 - 41 (4,6 - 5,7)
Steering wheel to steering shaft . . . . . .20 - 25 (2,8 - 3,4)
Track - rod end locknut . . . . .42 - 50 (5,8 - 6,9)
Pinion bearing cover. . . . . 12.5 - 17.6 (1,7 - 2,4)
Rack slipper cover . . . . .4.5 - 6.7 (0,6 - 0,9)

FRONT SUSPENSION & DRIVESHAFTS . . [11]

Front Spring Specifications

1100 cc engine: Colour Code
All saloons except estate car
and heavy duty . . . . Blue/violet
Estate car and heavy duty . . . . .Orange
1300/1600 cc engine:
All saloons except estate car
and heavy duty . . . . Blue/green
Estate car and heavy duty . . . . Green/red/brown
All vans . . . . .Green/red/violet

Tightening Torques - lb ft (kg m)

Hub retaining nut (With greased driveshaft
threads) . . . . 151 - 173 (20,5 - 23,5)
Lower arm pivot bolt . . . . .38 - 47 (5,1 - 6,4)
Lower arm ball joint clinch bolt . . . . . .35 - 44 (4,8 - 6,0)
Lower arm ball joint to lower arm . . . .55 - 66 (7,5 - 9,0)
Suspension strut to spindle carrier . . . .59 - 66 (8,0 - 9,0)
Tie bar to mounting bracket . . . . .32 - 40 (4,5 - 5,6)
Tie bar mounting bracket to body . . . .33 - 41 (4,5 - 5,6)
Piston rod nut. . . . . .30 - 38 (4,0 - 5,2)
Top mount retaining bolts. . . . . .15 - 18 (2,0 - 2,4)
Stabiliser bar clamp to bracket. . . . . . .33 - 41 (4,5 - 5,6)
Stabiliser bar to lower arm . . . . .66 - 81 (9,0 - 11,0)

Lubricants

Front wheel bearings . . . . . S-MIC-4515-A (High melting point lithium grease)
Constant velocity joints . . . S-MIC-75-A/SQM-1C-9004-A (High quality lithium based grease containing molybdenum disulphide) (40 grms (1,5 oz) per joint)

REAR SUSPENSION . . . . [12]

Rear Springs Specifications

	Free length	Colour Code
1100 cc Base saloon, except heavy duty	9.6 (244,1)	Red/red/blue
1100 cc L, GL, Ghia except heavy duty	9.7 (246,5)	Red/red/violet
1300/1600 cc saloon, except heavy duty	9.2 (233,5)	Red/red/yellow
1300/1600 cc 'S' pack and XR3	9.7 (245,9)	Violet/orange
All heavy duty saloons	9.7 (246,9)	Blue/orange
Estate car	9.3 (238,1)	Blue/green
Heavy duty estate car	9.9 (251,1)	Blue/pink

NOTE: Free lengths are nominal and may vary by up to 0.39 (10,0) either way.

Rear Spring Specifications - Van

	Centre distance between eyes	Colour Code
Type 35 van	47.0 (1194)	Red/white
Type 55 van	47.0 (1194)	Blue

Tightening Torques - Saloon/Estate - lb ft (kg m)

Lower arm to body bolts . . . . .52 - 66 (7,0 - 9,0)
Lower arm to stub axle bolts . . . . .44 - 52 (6,0 - 7,0)
Shock absorber top mounting nuts . . . .33 - 38 (4,2 - 5,2)
Shock absorber to stub axle bolts . . . . .52 - 66 (7,0 - 9,0)
Back plate bolts . . . . .33 - 40 (4,5 - 5,5)
Tie bar to body nuts . . . . .52 - 66 (7,0 - 9,0)
Tie bar to stub axle nuts . . . . .52 - 66 (7,0 - 9,0)

Tightening Torques - Van - lb ft (kg m)

'U' bolts .27 - 33 (3,7 - 4,6)
Spring eye bolts (front or rear) . . . . . . .52 - 66 (7,1 - 9,2)
Shackle plate to body/spring . . . . .30 - 37 (4,1 - 5,1)
Shock absorber top mounting . . . . . . .30 - 37 (4,1 - 5,1)
Shock absorber lower mounting . . . . . .30 - 37 (4,1 - 5,1)
Shock absorber plate to body . . . . . . .15 - 18 (2,0 - 2,6)
Carrier plate bolts . . . . .33 - 41 (4,6 - 5,6)

BRAKES . . . . [13]

Master Cylinder

Master cylinder bore diameter 0.813 (20,64)
Brake servo diaphragm diameter 7.87 (200)

Load Sensitive Valve (Van Only)

Checking dimension . . . . 0.65 - 0.85 (16,5 - 21,5)
Setting dimension (existing linkage) . . . . 0.76 - 0.84 (18,5 - 20,5)

Front Brakes

Disc diameter 9.43 (239,45)
Minimum disc thickness
solid 0.34 (8,7)
ventilated 0.89 (22,7)
Maximum disc run-out (installed) 0.006 (0,15)
Caliper cylinder bore diameter 2.21 (54,0)
Brake pad minimum thickness 0.60 (1,5)

Rear Brakes

Drum diameter:
saloon/estate 7.2 (180,0)
van 8,0 (203,2)
Wheel cylinder bore diameter:
saloon 0.75 (19,05)
van 0.875 (22,2)
Shoe lining thickness (new) leading shoe:
saloon/estate (Mintex. M79/1) 0.17 (4,35)
van (DON 242) 0.17 (4,35)
Shoe lining thickness (new) trailing shoe:
saloon/estate (Mintex M79/1) 0.11 (2,80)
van (DON 242) 0.11 (2,80)
Minimum lining thickness 0.4 (1,0)

Tightening Torques - lb ft (kg m)

Caliper to front suspension unit . . . . . .38 - 45 (5,2 - 6,2)
Brake carrier plate to rear wheel spindle:
saloon/estate . . . . .15.18 (2,1 - 2,5)
van . . . . .33 - 40 (4,5 - 5,5)
Brake bleed valves . . . . Sufficient to seal
Brake pressure control valve to bracket . . . . 15 - 18 (2,0 - 2,5)
Piston housing mounting bolts . . . . . . .15 - 18 (2,0 - 2,5)

GENERAL ELECTRICS . . . . [14]

Replacement Bulbs - Exterior Lights

Headlamps	Wattage	Fixing
Halogen H4	60/55W	spring clip
Tunsten	50/45W	clip ring
Parking Lamp	4W	bayonet
Front indicator Lamp	21W	bayonet
Side Repeater Flasher (where fitted)	5W	glass socket
Stop/Tail Lamp	21/5W	bayonet
Reversing Lamp	21W	bayonet
Rear Fog Lamp	21W	bayonet
Rear Indicator Lamp	21W	bayonet
Licence Plate Lamp	5W	glass socket
Auxiliary Lamp Halogen H3	55W	clip
Fog Lamp Halogen H3	55W	clip

Replacement Bulbs - Interior Lights

Lamp designation	Wattage	Fitting
Warning lamps (instrument cluster)	1,3W	glass socket
Warning lamps (switches) hazard flashers	1,3W	glass socket
Panel illumination lights	2,6W	glass socket
Cigarette lighter illumination light	1,4W	bayonet
Glove compartment light	2W	bayonet
Luggage compartment light	10W	bayonet
Interior light	10W	bayonet

Index